AF787575

Electrochemical Capacitors

Online at: https://doi.org/10.1088/978-0-7503-5042-6

Electrochemical Capacitors

Edited by
Jianmin Ma
School of Chemistry, Tiangong University, Tianjin, People's Republic of China

Junliang Yang
School of Physics and Electronics, Central South University, Changsha, People's Republic of China

IOP Publishing, Bristol, UK

© IOP Publishing Ltd 2023

All rights reserved. No part of this publication may be reproduced, stored in a retrieval system or transmitted in any form or by any means, electronic, mechanical, photocopying, recording or otherwise, without the prior permission of the publisher, or as expressly permitted by law or under terms agreed with the appropriate rights organization. Multiple copying is permitted in accordance with the terms of licences issued by the Copyright Licensing Agency, the Copyright Clearance Centre and other reproduction rights organizations.

Permission to make use of IOP Publishing content other than as set out above may be sought at permissions@ioppublishing.org.

Jianmin Ma and Junliang Yang have asserted their right to be identified as the authors of this work in accordance with sections 77 and 78 of the Copyright, Designs and Patents Act 1988.

ISBN 978-0-7503-5042-6 (ebook)
ISBN 978-0-7503-5040-2 (print)
ISBN 978-0-7503-5043-3 (myPrint)
ISBN 978-0-7503-5041-9 (mobi)

DOI 10.1088/978-0-7503-5042-6

Version: 20230901

IOP ebooks

British Library Cataloguing-in-Publication Data: A catalogue record for this book is available from the British Library.

Published by IOP Publishing, wholly owned by The Institute of Physics, London

IOP Publishing, No.2 The Distillery, Glassfields, Avon Street, Bristol, BS2 0GR, UK

US Office: IOP Publishing, Inc., 190 North Independence Mall West, Suite 601, Philadelphia, PA 19106, USA

Contents

Preface

We already use clean energy, and its use will continue to increase. However, it needs to be stored before use. Thus, energy storage systems are developed to meet this demand. There is a variety of potential technology choices, such as lithium-ion batteries, sodium-ion batteries, aqueous batteries, or electrochemical capacitors. Compared with batteries, electrochemical capacitors have some excellent characteristics, such as long cyclic life, low cost, and high power density. In addition, many new materials are being used for electrochemical capacitors. Thus, there are many opportunities to develop high-performance electrochemical capacitors.

Against this background, I invited some experts to write this book on electrochemical capacitors, and divided it according to the different types of electrochemical capacitor. Prof. Jiabiao Lian from Jiangsu University wrote chapter 1 on electric double-layer capacitors. Dr Wei Ni from the Research Institute of Vanadium & Titanium (Iron & Steel) mainly wrote chapter 2 on aqueous electrochemical capacitors based on porous carbon anodes and battery cathodes. Prof. Qinghong Wang from Jiangsu Normal University prepared chapter 3 on aqueous electrochemical capacitors based on battery anodes and porous carbon cathodes. Fangyuan Su and Prof. Chengmeng Chen from the Institute of Coal Chemistry (CAS) wrote chapter 4 on 'rocking-chair' hybrid capacitors. Dr Zhanheng Yan mainly wrote chapter 5 on non-aqueous electrochemical capacitors based on battery anodes and porous carbon cathodes. Prof. Linfen Lai from Nanjing Tech University completed chapter 6 on low-temperature electrochemical capacitors. Chapter 7 on flexible electrochemical capacitors was completed by Prof. Xinwen Peng (South China University of Technology). As we know this book has sound content, we commend it to the reader. Finally, we would especially like to thank the excellent contributors for this book.

Editor biographies

Dr Jianmin Ma

Dr Jianmin Ma is a professor at Tiangong University, Changsha, China. He serves as the Associate Editor of *Chinese Chemical Letters*, Academic Editor for *Rare Metals* and is an editorial board member for the *Journal of Energy Chemistry, Chinese Chemical Letters, Nano-Micro Letters, Journal of Physics: Condensed Matter, Journal of Physics: Energy, Chemistry-An Asian Journal*, and others. His research interests focus on energy storage devices and components including metal anodes and electrolytes, and theoretical calculations from density functional theory and molecular dynamics to finite element analysis.

Dr Junliang Yang

Dr Junliang Yang is full professor in the School of Physics and Electronics at Central South University. His research interests are focused on flexible and printed electronics, as well as energy material and devices. He has published over 260 papers in international journals, such as *Chemical Reviews, Advanced Materials, Advanced Energy Materials, ACS Nano, Nano Energy*, etc, (over 160 first-author or corresponding-author papers), been granted 35 patents, and has co-founded a company. His published papers are cited over 10 000 times and his H-index is 53. He has received over 20 research fundings from governments and companies. He is a Fellow of the Royal Society of Chemistry, and was selected as one of the 'Top 100000 Scientists in the World' and 'World's Top 2% Scientist'. He has been included in the National Top Young Talents and the Excellent Talents of the Ministry of Education.

List of contributors

Yongfeng Bu
Institute for Energy Research, Jiangsu University, 301 Xuefu Road, Zhenjiang 212013, People's Republic of China

Chengmeng Chen
CAS Key Laboratory of Carbon Materials, Institute of Coal Chemistry, Chinese Academy of Sciences, Taiyuan 030001, Shanxi, People's Republic of China

Yingxue Cui
Institute for Energy Research, Jiangsu University, 301 Xuefu Road, Zhenjiang 212013, People's Republic of China

Sherif A El-Khodary
Institute for Energy Research, Jiangsu University, 301 Xuefu Road, Zhenjiang 212013, People's Republic of China
and
The Building Physics and Environment Institute, Housing and Building National Research Center (HBRC), Dokki, Cairo 12311, Egypt

Linfei Lai
Key Laboratory of Flexible Electronics and Institute of Advanced Materials (IAM), Jiangsu National Synergistic Innovation Center for Advanced Materials (SICAM), Nanjing Tech University, 5 XinMofan Road, Nanjing 210009, People's Republic of China
and
CNRS-International-NTU-Thales Research Alliance, 50 Nanyang Drive Singapore 637553, Singapore

Jiabiao Lian
Institute for Energy Research, Jiangsu University, 301 Xuefu Road, Zhenjiang 212013, People's Republic of China

Xinhua Liu
School of Transportation Science and Engineering, Beihang University, Beijing, People's Republic of China

Jianmin Ma
School of Chemistry, Tiangong University, Tianjin, 300387, People's Republic of China

Wei Ni
State Key Laboratory of Vanadium and Titanium Resources Comprehensive Utilization, ANSTEEL Research Institute of Vanadium & Titanium (Iron & Steel), Chengdu 610031, People's Republic of China

Xinwen Peng
State Key Laboratory of Pulp and Paper Engineering, South China University of Technology, Guangzhou 510640, People's Republic of China

Ling-Ying Shi
College of Polymer Science and Engineering, State Key Laboratory of Polymer Materials Engineering, Sichuan University, Chengdu 610065, People's Republic of China

Fangyuan Su
CAS Key Laboratory of Carbon Materials, Institute of Coal Chemistry, Chinese Academy of Sciences, Taiyuan 030001, Shanxi, People's Republic of China

Qinghong Wang
School of Chemistry and Chemical Engineering, Jiangsu Key Laboratory of Green Synthetic Chemistry for Functional Materials, Jiangsu Normal University, Xuzhou 221116, People's Republic of China

Zhanheng Yan
School of Science, Henan University of Engineering, Zhengzhou 451191, People's Republic of China

Junliang Yang
School of Physics and Electronics, Central South University, Changsha, People's Republic of China

Wang Yang
State Key Laboratory of Pulp and Paper Engineering, South China University of Technology, Guangzhou 510640, People's Republic of China

Wu Yang
State Key Laboratory of Pulp and Paper Engineering, South China University of Technology, Guangzhou 510640, People's Republic of China

Jiaming Zeng
State Key Laboratory of Pulp and Paper Engineering, South China University of Technology, Guangzhou 510640, People's Republic of China

IOP Publishing

Electrochemical Capacitors

Jianmin Ma and Junliang Yang

Chapter 1

Electric double-layer capacitors

Sherif A El-Khodary, Yingxue Cui, Yongfeng Bu and Jiabiao Lian

In this chapter, electric double-layer capacitors (EDLCs) based on carbon materials are discussed in depth, and brief information is given about their storage mechanisms and structural configurations. This chapter also highlights all the kinds of electrode (both aqueous and non-aqueous) currently used for EDLCs, showing their advantages and disadvantages. Recent research directions have focused on synthesizing nanosized carbons with active surface functional groups via different physical and chemical approaches. The utilization of nanosized carbon materials with different dimensionality and plenty of surface redox groups leads to shorter diffusion pathways and enhanced ion transport kinetics due to the improved surface area and structural/surface modification. In addition, recent studies and significant research developments are fully covered and summarized, which will help to guide the development of advanced carbon-based EDLCs.

1.1 Introduction

The rapid growth of portable electronics in the last two decades required the fast development of efficient electrochemical energy storage (EES) systems, such as lithium-ion batteries [1, 2], supercapacitors [3, 4], hybrid metal–ion capacitors [5, 6], fuel cells [7], and more recently, metal–air batteries [8–10], in order to meet the massive demands of portable devices in the electronics market. Among these EES systems, supercapacitors have attracted considerable attention in recent years because of their low cost, high safety, and high specific power density (>10 kW g^{-1}) [11, 12], as shown in figure 1.1. Supercapacitors (SCs), also called electrochemical capacitors or ultracapacitors, are basically classified into two classic categories: (i) EDLCs and (ii) pseudocapacitors, according to the charge storage mechanism and the active materials applied [13–15]. In effect, EDLCs store charges physically via the electrostatic adsorption of accumulated charges at the electrode/electrolyte interface (based on the Helmholtz principle) [3], while pseudocapacitors store charges chemically through the fast and reversible surface/near-surface faradic reaction, as exemplified

doi:10.1088/978-0-7503-5042-6ch1

© IOP Publishing Ltd 2023

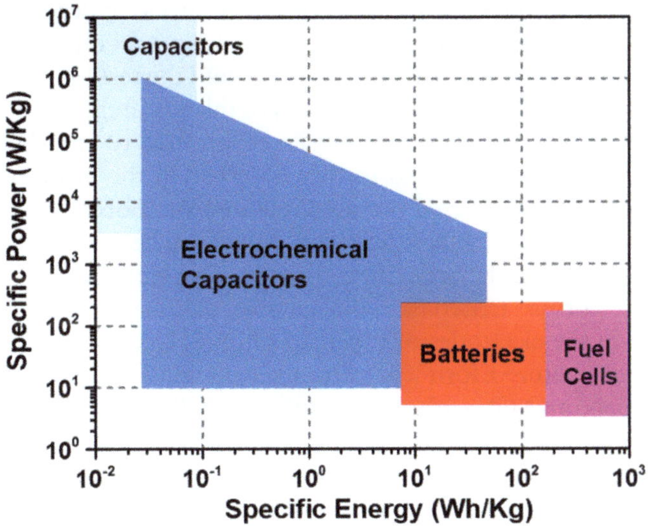

Figure 1.1. A simplified Ragone plot of various electrical energy storage devices.

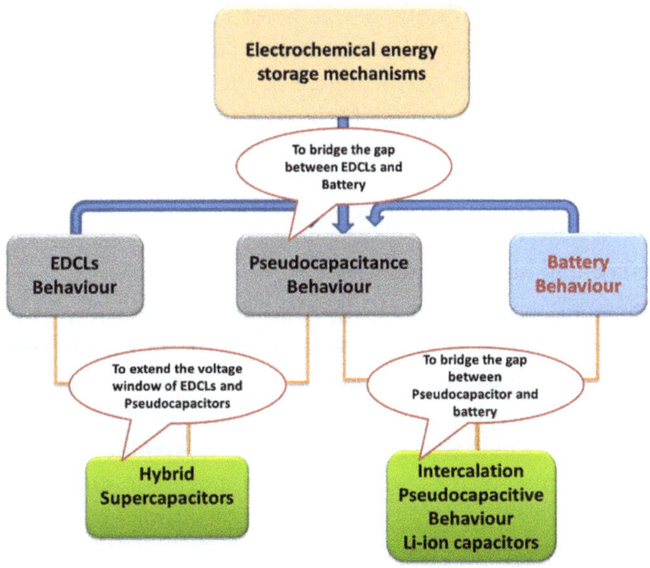

Figure 1.2. The classifications and relationships of various energy storage systems and their mechanisms.

by MnO_2 and RuO_2, known as a surface redox reaction [16–18]. Recently, a new pseudocapacitor storage mechanism, known as intercalative pseudocapacitance, has emerged (figure 1.2) [19–21]. Intercalative pseudocapacitors store charge via the successful insertion of electrolyte cations into the tunnels or layers of porous electrode materials [22, 23]. Unlike batteries, the reaction kinetics in intercalative pseudocapacitance is not controlled by the diffusion process of electrolyte cations (e.g. Li^+ or Na^+)

within the crystalline framework structure of the electrode material [24]. This means that the storage mechanism of intercalated cations is not accompanied by a change in the structural framework of the electrode material, and the resultant capacity is largely independent of the scan rate. In the early stages of this research, this phenomenon was found in nanosized TiO_2 that had a particle size of less than 10 nm when measured in a Li-based organic electrolyte [25] and further observed in the orthorhombic phase of Nb_2O_5, which has a highly ordered porous structure [26]. Indeed, this new mechanism offers a new concept that combines the pseudocapacitive behaviors of supercapacitor-type electrodes and the faradic behaviors of battery-type electrodes [16, 22]. Therefore, high-surface-area porous materials with tunable electronic conductivity are very desirable for use in the electrodes of supercapacitors [11, 27].

Different nanostructured forms of carbonaceous material, such as high-surface-area activated carbon (AC), carbon nanotubes (CNTs), carbon nano-onions (also called onion-like carbon, OLC), and graphene, are widely utilized in EDLCs [28]. Due to its high surface area and microporous structure, AC has been widely used in commercial EDLCs [29]. However, its low output energy densitycannot satisfy the ever-increasing demands of consumer electronics and other future technological systems. Alternative strategies, including modifying the surface (e.g. by the creation of defects or dangling bonds), modulating the porosity, increasing the material conductivity through the use of nonmetallic heteroatomic dopants, and/or extending the potential window by changing the electrolyte substance, are of great importance in boosting the electrochemical performance of the abovementioned carbon materials and thus obtaining the required high energy density. More details of these processes are described in the following sections.

1.2 The basic principles of EDLCs

1.2.1 The structure and assembly of EDLCs

EDLCs generally comprise two electrodes (negative and positive) and a separator filled with an electrolyte. The separator is made of an electrically insulating material such as glass fiber or cellulose filter papers in order to prevent internal short circuits, which are caused by direct physical contact between the electrodes inside the assembled cell. EDLC devices may have either a symmetric or an asymmetric configuration, according to the types of both negative and positive electrodes [12, 30]. Symmetric devices have two identical electrodes, which means that the negative and positive electrodes are made of the same material and have analogous storage mechanisms [31–33]. In contrast, asymmetric devices are fabricated using two different types of electrode material with different storage mechanisms in order to enlarge the operating voltage window [12, 34]. Indeed, the main difference between symmetric and asymmetric systems is the operating voltage window. Symmetric devices have limited voltage windows, mostly less than 1.0 V, and they are only used with aqueous electrolytes [14]. In contrast, asymmetric devices have voltage windows of up to 2.0 V [24, 35, 36]. The most common structural configuration for asymmetric devices is the combination of a carbonaceous material in the anodic part with a pseudocapacitive material in the cathodic part [16, 25, 37]. Asymmetric devices are a well-recognized design for next-generation supercapacitors due to their wider voltage window and higher energy density.

1.2.2 Electrolytes

The electrolyte, which is a salt dissolved in a solvent, is one of the key elements of SCs; it creates ionic conductivity and therefore facilitates cation and anion transfer inside the cell. The kind and concentration of the electrolyte substance are the key factors used to enhance the electrochemical activity, operational cell voltage, and capacitance of supercapacitor devices. More recently, it has been found that the nature of the electrolyte makes a major contribution to the charge storage mechanism of the electrode material [38, 39]. Three kinds of liquid electrolyte, including aqueous electrolytes, organic electrolytes, and ionic liquids (ILs), are widely used (figure 1.3) [40]. Furthermore, solid/quasi-solid-state electrolytes are now attracting much attention to the currently available SCs, as they provide improved safety and lighter cells compared to their liquid counterparts [40, 41], but they are not discussed in this chapter.

Aqueous electrolytes can be classified into three types: (i) alkaline electrolytes (such as KOH dissolved in water), (ii) acidic electrolytes (such as H_2SO_4 dissolved in water), and (iii) neutral electrolytes (such as Na_2SO_4 or KCl aqueous solutions) [17, 42–44]. Indeed, aqueous electrolytes can offer high ionic conductivity and low equivalent series resistance (ESR), achieving high specific capacitance and power density. It is interesting to note that the ionic conductivity of 1 M H_2SO_4 at room temperature can reach a value of 0.73 S cm^{-1}, while the ionic conductivity of the most popular alkaline electrolyte, KOH, has a value of 0.54 S cm^{-1}, and this value is strongly dependent on the concentration of KOH, which is 6.0 M, and the ambient conditions (the value given was measured at 25 °C) [45]. In addition to its high ionic conductivity, the use of H_2SO_4 as an aqueous electrolyte has a major advantage for carbon-based electrode materials due to the high reactivity of protonic ions (H^+)

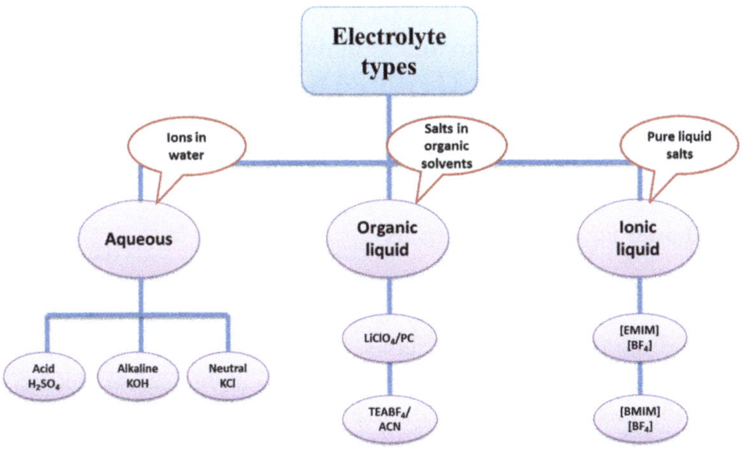

Figure 1.3. The classification of the electrolytes used for EDLCs.

with highly negative oxygen functional groups and consequently makes an extra pseudocapacitive contribution to the total capacitance value [46]. Quinone-type functional groups, for example, have a pronounced pseudocapacitive effect in acid-based electrolytes, and this interplay is influenced by the pH value of the electrolyte [47]. However, aqueous electrolytes suffer from narrow voltage windows due to the limited bandgap energy of water (E_g = 1.23 eV), thus leading to the lower energy density of supercapacitors based on aqueous electrolytes. Organic electrolytes, on the other hand, can offer wider voltage windows (i.e. more than 2.0 V).

Due to the above factors, symmetric and asymmetric SCs with organic electrolytes have prompted extensive investigations in the last few years with the aim of increasing the energy density of SC devices by extending the operational voltage window. Propylene carbonate (PC) and acetonitrile (AN) are the solvents most used for organic electrolytes; they are mainly used in conjunction with $LiClO_4$ [21, 40]; other inorganic salts, such as $LiPF_6$, LITFSI, and $TEABF_4$, have also been utilized [48–51]. Notably, the AN-based electrolytes have higher ionic conductivity than the PC-based electrolytes due to the high dielectric constant of the AN solvent [40]. However, the high toxicity of AN solvents is a major obstacle to the widespread application of AN-based electrolytes [52, 53]. In addition, the low output energy densities associated with the limited voltage windows of both AN- and PC-based electrolytes (~2.5–2.8 V) have led to the introduction of ionic liquid (IL) electrolytes to overcome these issues and further extend the operational cell voltage beyond this limit.

Ionic electrolytes, also known as room-temperature molten salts, are desirable electrolyte materials for next-generation supercapacitors due to their low flammability, nontoxicity, high thermal and chemical stabilities, and more importantly, their large operating voltage window (more than 3 V) [54]. Thus, solvent-free ILs could have major advantages for application in SCs. Notably, ILs can be classified into three types: (i) aprotic, (ii) protic, and (iii) zwitterionic [45, 54]. In addition, the chemical and physical properties of ILs are highly tunable due to the many combinations of variant cations and anions in the same electrolyte medium [54]. Nevertheless, the relatively low ionic conductivity of ionic liquid electrolytes, which arises from their high viscosity at room temperature, is still a major drawback for their practical use in SCs. The most popular ILs, [EMIM][BF_4], have an ionic conductivity of 14 ms cm^{-1} (at 25 °C), which is far from the value achieved by $TEABF_4$/AN (59.9 ms cm^{-1} at 25 °C), an organic electrolyte [55]. Therefore, more work is still needed to improve the ionic conductivity of ILs to reach the same level as organic electrolytes or even beyond that.

1.2.3 EDLCs based on carbon materials

As stated above, carbon materials are the main active component of EDLCs. Different forms of carbon (e.g. zero-dimensional (0D) carbon nano-onions or fullerenes, one-dimensional (1D) CNTs, two-dimensional (2D) graphene, and three-dimensional (3D) carbon aerogel) have been widely used as electrode materials for EDLCs in the last two decades (figure 1.4). It has been found that the shape, morphology, porosity, and hybridization type play major roles in the electrochemical activity of electrodes, including their capacitance value, cyclic stability,

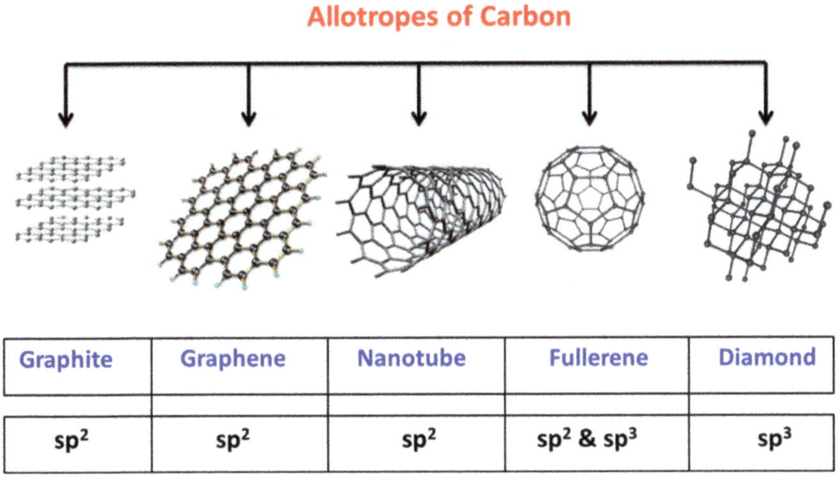

Figure 1.4. The structures and hybridization types of some selected allotropes of carbon.

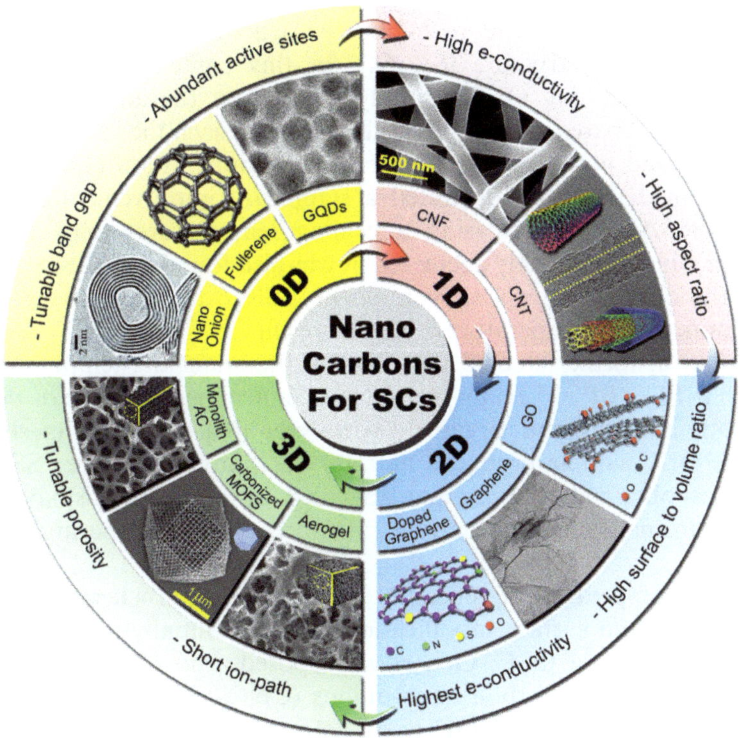

Figure 1.5. An overview of various carbon materials of different dimensionalities used in EDLCs.

and rate performance. A change in the crystal structure of the arranged carbon atoms is followed by a change in the hybridization type of the resultant carbon material, known as a carbon allotrope [56]. Graphite, graphene, fullerene, CNTs, and diamond are all allotropes of carbon (figure 1.5) [57]. The following sections discuss the abovementioned carbon materials in more detail in terms of their structural/electrochemical relationships.

1.3 The electrochemical performances of carbon materials of different dimensionalities

1.3.1 Zero-dimensional carbon nanomaterials

Zero-dimensional nanostructured carbon materials are defined as round particles (with almost perfect sphere-like structures) with particle sizes less than 10 nm, such as fullerenes, OLC, and carbon-based quantum dots (C-QDs). Among the different 0D carbon materials, OLC, also known as multishelled fullerene material, has generated tremendous interest in its application in EDLCs [58–61]. The symmetric structure and small particle size (on the scale of a few nanometers) of OLC may offer better electrolyte infiltration, thus boosting ion transport kinetics and enhancing the device rate performance [59, 60, 62]. The most common and efficient method for the fabrication of OLC is the thermal treatment of artificial nanodiamonds (NDs) either in a vacuum or in an argon atmosphere at an elevated temperature (~1700 °C) to ensure the transformation of the sp^3-hybridized structure of the NDs into OLC with a fully sp^2-hybridized structure [63, 64]. In this context, the physicochemical properties of the resulting OLC strongly depend on the annealing temperature used. Portet *et al* prepared carbon onion-like structures via the vacuum annealing of 5 nm nanodiamond powders at a wide range of temperatures (1200 °C–2000 °C) [65]. The obtained OLC demonstrated a significant improvement in the specific surface area (SSA) with increasing annealing temperatures until 1800 °C was reached, after which it suddenly decreased; it was also found that the electrochemical activity of the obtained samples, such as the ESR, was impressively reduced by increased annealing temperatures as a result of structural modifications. McDonough *et al* further studied the effect of the annealing temperature of NDs in a vacuum (~10^{-5}–10^{-6} Torr) on the electrochemical performance of the obtained OLC in both aqueous and organic electrolytes [66]. The results showed that although the OLC sample produced at 1500 °C exhibited a higher SSA than the other samples obtained, the sample annealed at 1800 °C displayed the best capacitive performance at high current rates because it had the highest electrical conductivity.

The reaction temperature significantly impacts the electrochemical performance of OLC material, and surface modification can substantially improve the electrochemical performance of OLC. Goa *et al* proposed a novel synthesis strategy for adjusting the porosity of carbon nano-onions (CNOs) [68]. CNOs with a porous structure were obtained using KOH as an activator to introduce porosity into the outer surface of CNOs. After activation, the results showed that the ACNO-6M sample had a high SSA of 804 $m^2\ g^{-1}$ with a large pore volume (1.27 $cm^3\ g^{-1}$), resulting in a maximum specific capacitance of 122 F g^{-1} for the activated sample. This work opened the way for the successful construction of porous CNOs. Anjos *et al* proposed an effective method for increasing the specific capacitance of OLC in acidic aqueous electrolytes by decorating its surface with a quinone substance [69].

The results showed that the 9,10-phenanthrenequinone-decorated carbon onion (PQ-OLC) electrode had additional pseudocapacitance compared with the capacitance of pure OLC and that the improvement amounted to nearly an order of magnitude. In addition, attaching extra functional groups, such as oxygen groups, to the surface of CNOs is another way of increasing their capacitance in acidic media. Moussa *et al* used a special thermal treatment strategy in oxygen and hydrogen atmospheres in order to modify the surface of CNOs to investigate the effect of oxygen functional groups on the electrochemical performance of obtained samples in both aqueous and organic electrolytes (figure 1.6(a)) [67]. The resulting CNOs (denoted by C1700-Tair), which were annealed at 580 °C for 1 h in an oxygen atmosphere, demonstrated improved electrochemical characteristics in terms of their capacitance value and rate performance due to the presence of oxygen functional groups; indeed, the effect was more highly pronounced in the aqueous electrolyte than in the organic one (figures 1.6(b) and (c)) due to the high reactivity of oxygen functional groups in acidic media, as mentioned above. Van Aken *et al* provided another important strategy for controlling the surface functional groups on the outer surface of OLC [60]. Surface-modified OLC was successfully synthesized using the attrition milling process in the presence of NaCl salt, followed by acid treatment and then sonication (the result was called Sonicated-OLC). The Sonicated-OLC sample possessed a uniform aggregate size (around 100 nm), and the surface decorated with carboxyl groups was conducive to exposing more active sites and accelerating ion transport kinetics. Furthermore, a Sonicated-OLC electrode displayed excellent

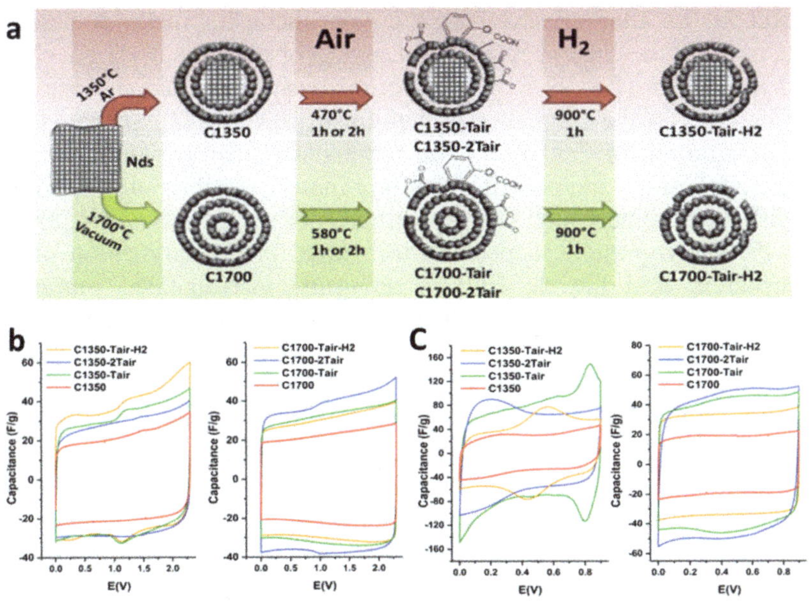

Figure 1.6. (a) A schematic illustration of the synthesis of CNOs under different thermal treatment conditions; (b) cyclic voltammetry curves of CNO materials modified by different treatments at a scan rate of 20 mV s^{-1} in an organic electrolyte (1.5 M Net_4BF_4/AN) and (c) in an aqueous electrolyte (0.1 M H_2SO_4) [67] reproduced with permission © Elsevier.

electrochemical performance, including high specific capacitance (52 F g^{-1} at a scan rate of 10 mV s^{-1}) and superior rate performance due to the synergistic effect of oxygen functional groups. Other methods and strategies have been further used for the synthesis of OLC, such as laser irradiation [70, 71], vapor deposition [72], stainless steel autoclave reactions [73], *in situ* high-resolution transmission electron microscopy [74], and flash pyrolysis [75]; nevertheless, the limited production and purity of the output material is still a major drawback for the large-scale application of OLC in different energy storage applications, specifically EDLCs.

On the other hand, C-QDs have attracted great interest in the last few years due to their unique features, such as high electron mobility, tunable bandgap energy, abundant active sites, and ultrasmall particle sizes [76]. In general, C-QDs are classified into three categories: (i) carbon dots (CDs), (ii) carbon quantum dots (C-QDs), and (iii) graphene quantum dots (GQDs) [77]. These classifications are based on the arrangement of carbon atoms, their dimensionality, and their crystalline structure. Among the various types of C-QD, GQDs are the most promising materials for EDLCs due to their high surface area, good electrical conductivity, and impressive mechanical properties [78–80]. The synthesis strategies for GQDs can be divided into two categories: (i) top-down methods, which usually use any source of solid carbon (e.g. CNTs, graphene nanosheets, carbon nanofibers) as a precursor, which is then cut down into smaller sizes through the use of hydrothermal, solvothermal, or electrochemical techniques, or (ii) bottom-up methods, which generally employ any small organic molecule (such as glucose, citric acid, pyrene, and fullerenes) as a precursor via dehydration, polymerization, and then carbonization (as one process) or alternatively using solution chemistry methods based on acid treatments (such as the cage opening of fullerenes) or even the pyrolysis of polycyclic aromatic hydrocarbons [78, 79, 81]. The electrodeposition technique is one of the most attractive methods for the fabrication of high-quality GQDs [82]. Lui *et al* reported the successful preparation of GQDs as electrode materials for symmetric EDLCs via a simple electrodeposition method; the resulting materials achieved outstanding cyclic stability with an extremely high rate performance of up to 1000 V s^{-1} [83]. Lee *et al* employed the electrophoretic deposition technique for the fabrication of GQDs to build transparent and flexible EDLCs alongside interdigitated graphene sheets (figures 1.7(a)–(c)) [84]. The EDLCs had a high transparency of 92.97% at 550 nm and exhibited a capacitance of 9.09 μF cm^{-2} with extremely high cyclic stability (almost 100% after 10 000 cycles), as shown in figures 1.7(d) and (e). This research opens the way for the efficient creation of transparent microsupercapacitors. Chen *et al* successfully synthesized GQDs for EDLCs on a 3D graphene electrode via the electrochemical deposition method [85]. The results showed that the 3D assembled structure (GQDs/3DG-10h) had a high SSA of 292 m^2 g^{-1} with a tunable microporous structure. The electrochemical results further revealed that the symmetric GQD/3DG-10h device produced possessed high specific capacitance (268 F g^{-1}) and retained almost 90% of its initial capacitance after 5000 cycles.

Despite the above features of GQDs, their current electrochemical performance is still far from the practical level of currently available EDLCs due to the limited storage mechanism of the undoped GDQ surface. Therefore, heteroatom-doped GQDs are

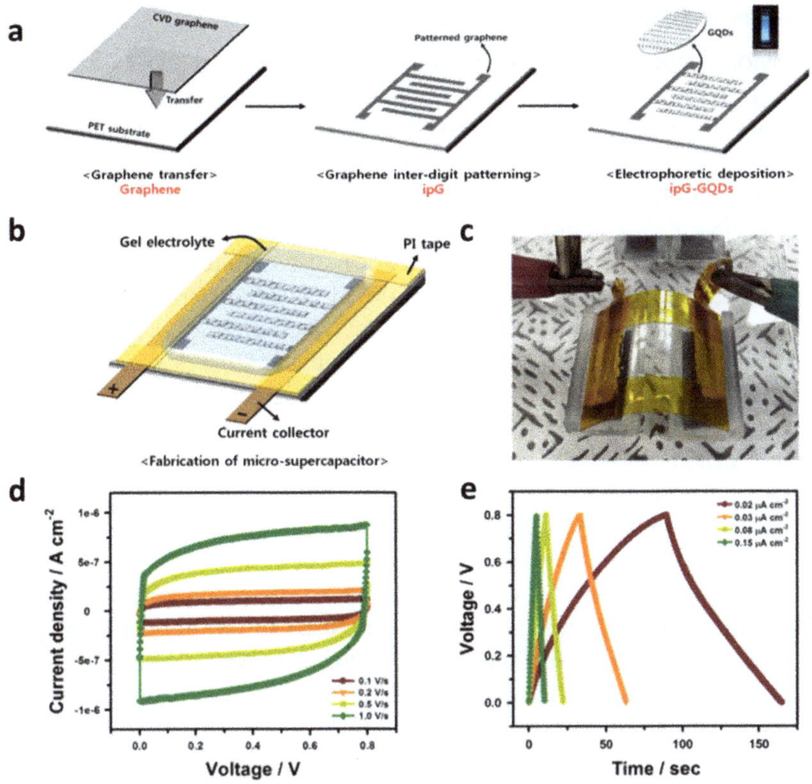

Figure 1.7. (a) A schematic illustration of the fabrication of an interdigitated pattern of graphene (ipG)-GQD microsupercapacitor (MSC); (b) a schematic illustration of a fabricated ipG-GQD MSC; (c) an optical image of an ipG-GQD MSC in the bent state; (d) cyclic voltammetry (CV) curves of the device collected at various scan rates; and (e) galvanostatic charge–discharge curves collected at various current densities [84], reproduced with permission © Elsevier.

promising materials for supercapacitors due to their modified surfaces, which can profoundly enhance their electrochemical properties. Li *et al* reported on the synthesis of nitrogen-doped GQDs (N-GQDs) via the alkali-catalyzed molecular fusion method, which is classified as one of the bottom-up methods, using pyrene and nitric acid as carbon and nitrogen sources, respectively [86]. Following synthesis, the prepared N-GQDs were loaded onto a carbonized metal–organic framework (cMOF-5) material using the electrodeposition technique and applied as electrode materials for SCs. The composite sample (N-GQDs/cMOF-5) had a cubic porous framework structure with a high SSA of 704 $m^2\ g^{-1}$ and good electrical conductivity. Furthermore, the composite electrode produced showed superior electrochemical performance, namely a specific capacitance of 780 F g^{-1} in an acidic aqueous electrolyte; indeed, the excellent electrochemical performance of N-GQDs/cMOF-5 was attributed to the presence of O and N heteroatoms in the framework structure of the composite sample, which provided extra pseudocapacitance to the electrode material via the surface redox reactions of pyrrolic-N, pyridone-N, and quinone functional groups.

1.3.2 One-dimensional carbon nanomaterials

One-dimensional nanostructured carbon materials, such as CNTs and nanofibers, have had a tremendous impact on the field of freestanding and binder-free energy storage devices due to their unique physical and chemical attributes, such as high SSA, tunable electronic structure, high electrical conductivity, high chemical stability, and finally, the ability to be integrated into a freestanding membrane in different ways [87, 88]. The hollow structure and high graphitization degree of CNTs give them advantages for electrolyte penetration and help to create a large electrode/electrolyte interface, leading to a high rate of storage performance [44, 89]. In addition, the highly covalent sp^2 C–C bonds between individual carbon atoms offer high electronic conductivity, which minimizes the total resistance of the system [90, 91]. CNTs are classified into three categories according to the number of walls they have: single-walled CNTs (SWCNTs, only a single tube), double-walled CNTs (DWCNTs, which have a double tube structure), and multiwalled CNTs (MWCNTs, which have more than three rolled graphene layers) [92]. Different strategies for CNT synthesis have been used to obtain high-quality CNTs, such as arc discharge [93], chemical vapor deposition CVD [94], and the laser ablation technique [95]. Among those, CVD is the most effective and widely used technique for synthesizing MWCNTs and SWCNTs due to its suitability for mass production with high purity and low surface defects [94, 96]. Remarkably, CNTs can be assembled and integrated into a wide range of shapes with various dimensionalities, including 1D fibers and yarns, 2D freestanding films, and 3D sponges and aerogels, allowing flexible devices to be produced for advanced wearable electronics [97].

As we stated above, catalytic CVD is the most effective technique for the large-scale fabrication of CNTs due to its high productivity [98, 99]. However, removing the metal catalyst and opening the end tips are the main obstacles to acquiring high-quality CNTs [91, 100]. In this regard, Zhang *et al* successfully synthesized MWCNTs on the surface of carbon cloth (CC) via the CVD technique followed by an acid treatment to remove the metal catalyst and to open the end tips, as shown in figures 1.8(a)–(d) [101]. The results showed that the obtained sample functionalized and tip-open CNTs (FTO-CNTs) had high specific capacitance in lithium-based aqueous electrolytes as well as in other kinds of electrolyte due to the opened tips and the increased amount of active ionic species (figures 1.8(e)–(g)). Therefore, this research proves that opening the end tips and adding surface functional groups play a crucial role in enhancing the electrochemical activity of CNTs. In the same way, Ragavan *et al* further increased the specific capacitance of MWCNTs in an acidic aqueous electrolyte after opening the end tips via acid treatment [102]. The improved performance was further attributed to the higher nitrogen content (15.8 at.%) of the obtained CNTs. In addition to using an acid treatment to open the end tips of CNTs, alkaline media, e.g. KOH, have also been applied to open them. Yang *et al* prepared a hierarchical composite of CNTs grown over the surface of electrospun carbon nanofibers via catalytic CVD followed by chemical treatment in an alkaline solution to open their end tips and further increase the SSA [103]. Electrochemical characterization was then performed in a highly concentrated acidic electrolyte (2 M H_2SO_4) at room temperature. The treated sample

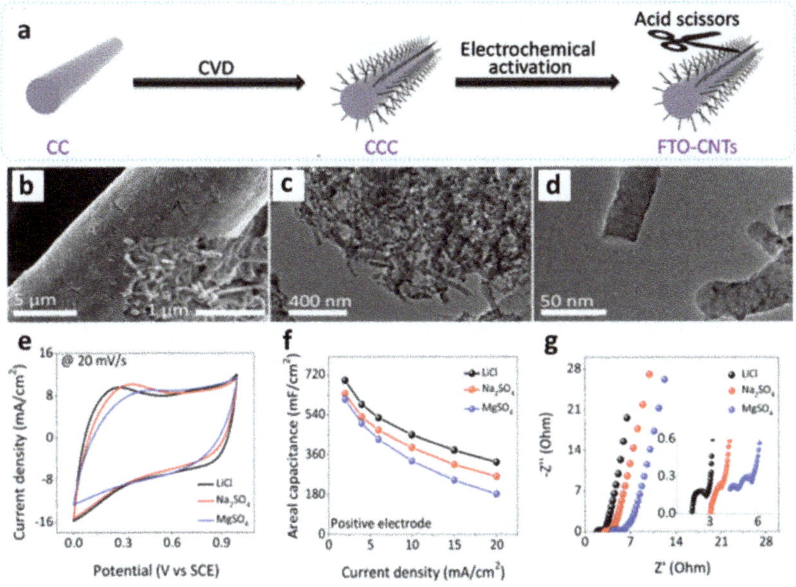

Figure 1.8. (a) A schematic illustration of the fabrication process used to produce the FTO-CNTs. The morphology of FTO-CNTs: (b) a scanning electron microscope (SEM) image of FTO-CNTs (the inset is a high-magnification SEM image); (c) a transmission electron microscopy (TEM) image of FTO-CNTs; (d) a high-resolution (HR) TEM micrograph of FTO-CNTs. (e) The performance of FTO-CNTs as a positive electrode in different electrolytes. (f) The areal capacitance of FTO-CNTs at different current densities in different electrolytes. (g) Nyquist plots (the insets show the high-frequency parts) of FTO-CNTs in different electrolytes [101], reproduced with permission © Royal Society of Chemistry.

exhibited a higher double-layer capacitance value than that of an untreated sample due to the opened tips. It was also found that opening the end tips of CNTs could facilitate the penetration of the electrolyte and accelerate the diffusion of cations.

On the other hand, CNTs have been considered to be an excellent platform for rapid developments in 1D flexible EDLCs in the last decade, owing to their outstanding mechanical properties [97]. Carbon nanotube yarns spun from CNT forests possess remarkable textile properties for the construction of highly efficient flexible energy storage devices, specifically for linear EDLC applications [104]. Indeed, linear EDLCs always demonstrate outstanding electrochemical activity for wearable EDLCs; however, it has been found that their electrochemical performance strongly depends on the yarn length. Therefore, the key challenge is to extend their lengths to those used in the textile manufacturing process while simultaneously maintaining their original electrochemical performance. In this context, Zhang *et al* suggested a new design to scale up the yarn length and considerably enhance the electrochemical performance of CNT yarn-based liner EDLCs [105]. They based their strategy on embedding a very thin metal wire into the center of the carbon nanotube yarn in order to improve electrolyte accessibility and shorten the charge transport kinetics. Research shows that if the new strategy is used, extending the yarn length does not affect the capacitance value. To further improve the electrochemical performance of CNT yarn-based linear supercapacitors, Miao and co-workers prepared a flexible device using a wide range of metal filaments, as shown in figure 1.9(a) [106]. The results showed that two EDLC devices constructed from Cu

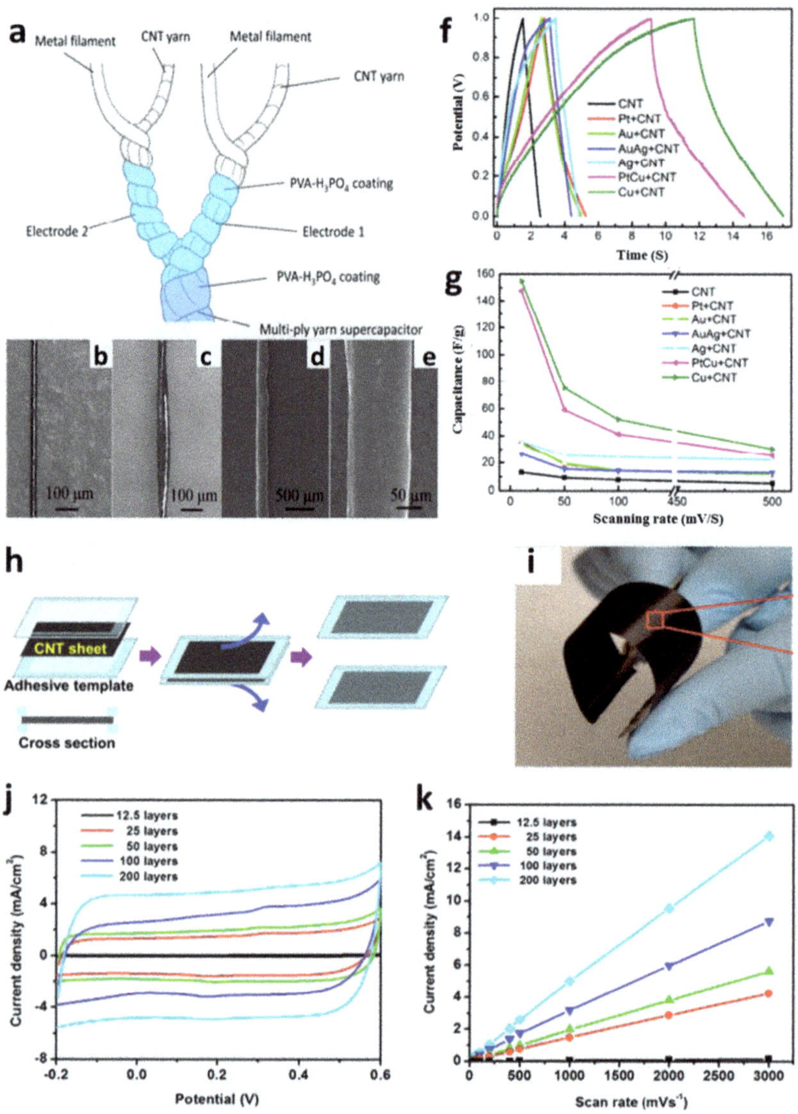

Figure 1.9. (a) Architecture of the CNT yarn/metal filament linear supercapacitor; (b) SEM image of bare metal filament Cu; (c) SEM image of Cu^+ CNT electrode; (d, e) SEM images of final Cu^+ CNT linear supercapacitor; (f) galvanostatic charge/discharge curves of supercapacitors at constant current density of 3.57 A/g; (g) specific capacitances of all supercapacitors at different scanning rates; [106] reproduced with permission © American Chemical Society; (h) schematic illustration of the inverse-ordered fabrication of CNT sheets; (i) a 200-layered CNT sheet which bent without a specific binder; (j) CVs of various-layered CNT sheets at constant scan rate, 1000 mV s^{-1}; (k) current densities of CNT sheets at 0.2 V potential [113], reproduced with permission, © American Chemical Society.

and PtCu metal filaments exhibited better electrochemical performances than devices made using other metal filaments, such as Pt, Ag, Au, and AuAg, as shown in figure 1.9(b). In addition, further structural/electrochemical investigations showed that the formation of CuO species over the surface of the Cu filament during the fabrication procedure created an additional active material (CuO), which provided an extra pseudocapacitance contribution via the surface redox reaction of CuO.

Two-dimensional freestanding CNT films, on the other hand, are currently deemed to be promising electrode materials for highly advanced, binder-free, and flexible EDLCs due to their ease of fabrication, low weight, and high mechanical robustness [107, 108]. The most common and available methods for the fabrication of freestanding CNTs films are: solution processing techniques, spray deposition, dip coating, and vacuum filtration of a powdered CNT material [98, 109–111]. However, film weakness due to the structural distortion of CNTs is the most common obstacle facing the abovementioned techniques, which limits their widespread utilization. As stated above, the CVD approach can be used to prepare high-quality freestanding CNT films with high electrical conductivity and mechanical robustness. For this reason, it has been extensively applied in the preparation of freestanding CNT films [111, 112]. Using this framework, Yun and his co-workers successfully synthesized freestanding CNT films with different thicknesses for use as EDLC electrodes. First, a MWCNT array was prepared via a highly controlled CVD technique. Freestanding films with different numbers of CNT layers were smoothly attached to a polytetrafluoroethylene (PTFE) substrate by an inverse-order process [113]. Electrochemical results showed that films with smaller numbers of layers (50 layers) had high electrochemical activity with a specific capacitance of 51.37 F g^{-1}; furthermore, they exhibited prolonged stability. Moreira *et al* used catalytic CVD to fabricate a vertically aligned freestanding MWCNT electrode using camphor oil as the carbon source [114]. Electrochemical measurements were then performed in a two-electrode system using PVA/H_3PO_4 as a gel polymer electrolyte. The results showed that the freestanding aligned CNT electrode exhibited higher gravimetric power density and energy density than those of a randomly packed electrode due to the high electronic conductivity of the binder-free film. The use of polymer additives in the randomly packed electrode increased the overall electrode resistance and thus reduced the contact strength between the active materials. Therefore, the use of the binder-free strategy is highly effective and provides better electron and ion transport dynamics during the charge/discharge process in a two-electrode device. Again, as mentioned above, solution-processing techniques such as spray deposition and vacuum filtration always face numerous issues in generating high-quality freestanding electrodes based on CNTs. Nevertheless, Wang and his co-workers designed a new strategy for the preparation of freestanding composite electrodes composed mainly of MWCNTs and supported by hollow carbon nanocages (HCNCs) to avoid and tackle the stacking phenomenon of freestanding CNT films [115]. A hybrid structure was obtained via a spray deposition method in which the HCNC was *in situ* added into the CNT film during the deposition process (the resulting sample was called purified carbon nanotube (PCNT)-HCNC). Supercapacitor measurements were performed in a three-electrode system in an acidic aqueous electrolyte (0.5 M H_2SO_4). The specific capacitance of the PCNT-HCNC-1.0 freestanding electrode (which had a moderate loading of HCNC) was 183.7 F g^{-1} at a scan rate of 10 mV s^{-1}, which was approximately twice that of an HCNC free electrode.

In addition to the use of CNTs as a raw material for the effective fabrication of freestanding and binder-free electrodes, electrospun carbon nanofibers (CNFs) are further considered to be a promising candidate at the practical level. In addition to its simplicity and ease of utilization, the electrospinning technique is regarded as the leading approach for the fabrication of high-quality CNFs (figure 1.10) [116]. Polymer precursors, such as polyacrylonitrile (PAN), pitch, lignin, poly(vinylidene fluoride) (PVDF), poly(vinyl alcohol) (PVA), poly(vinyl-butyral) (PVB), and cellulose, are the main sources of carbon used in the electrospinning technique [116–121], whereas PAN-based nanofiber is the polymer source most often used for the manufacture of high-quality CNFs due to its high carbon yield, commercial viability, and the superior mechanical attributes of the resulting CNFs [122, 123]. Normally, following the spinning step, the spun PAN nanofiber mats are generally subjected to a two-step process: (i) stabilization, which is usually performed at a temperature between 200 °C and 300 °C in air to transform the electrospun PAN into a ladder compound; and (ii) carbonization to generate the final freestanding carbon mats [124]. The carbonization conditions are utterly different from the stabilization conditions. Elevated temperatures, usually starting from 800 °C and reaching as high as 2800 °C, and an inert atmosphere (e.g. nitrogen and/or argon) are the main conditions for the carbonization process used to transform the ladder structure into a graphite-like one [125]. Notably, the most crucial feature of the above technique is that the resulting carbon mats are easy to handle and use directly in different ways due to their interconnected fiber network structure. This means that no further treatment steps are needed to produce freestanding electrodes for EDLCs. Unfortunately, the low SSA of the resulting fibrous mats limits their use as electrode materials for EDLCs. Barranco *et al*

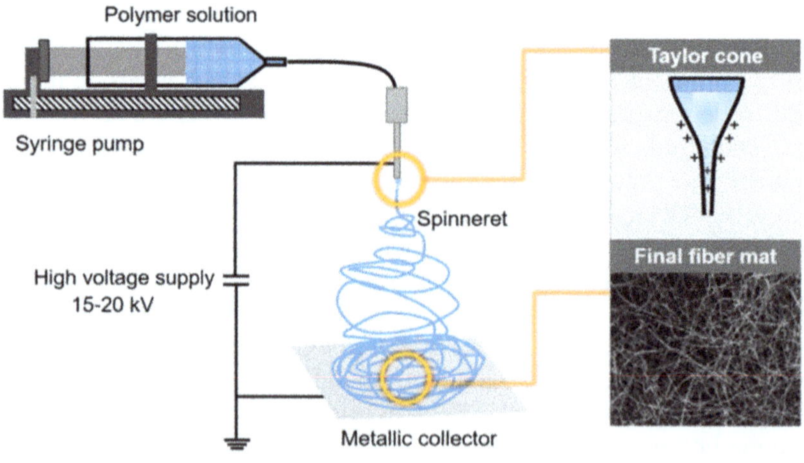

Figure 1.10. A schematic diagram of the electrospinning step based on the use of a Taylor cone [116], reproduced with permission © Elsevier.

proposed an effective method to increase the SSA of PAN-based CNFs by treating their surface using KOH as an activator at a temperature of 750 °C [126]. The results revealed that the effective surface area was increased by increasing the amount of KOH activator, reaching 1520 $m^2\ g^{-1}$ at a relative ratio of 3:1 of KOH to CNFs. As a result, the specific capacitance was enhanced. The activated sample showed superior electrochemical performance in different electrolyte media (such as KOH and H_2SO_4), reflecting the importance of increasing the SSA of carbon nanofibers. In another report, Yun *et al* used KOH and melamine as an activator and a nitrogen source, respectively, to modulate the SSA of PAN-based carbon nanofibers by changing the activation temperature [127]. An N-HPCNF1200 sample activated at a temperature of 1200 °C possessed a high SSA (2527.7 $m^2\ g^{-1}$) and a low specific electrical resistance ($1.8 \times 10^2\ \Omega$ m) compared to those of the other activated samples. Consequently, the N-HPCNF1200 demonstrated high electrochemical performance with superior energy density (113 Wh kg^{-1}) and power density (105 kW kg^{-1}). Wu and his co-workers designed an *in situ* activation strategy to produce highly activated porous carbon nanofibers via the electrospinning of an H_3PO_4-containing PAN precursor [128]. The results demonstrated that adding a small amount of H_3PO_4 (1% of the total weight of the PAN precursor) was conducive to enhancing the SSA of CNFs. Electrochemical measurements were performed using a highly concentrated alkaline electrolyte (6.0 M KOH) in a two-electrode system, and the activated sample (P-CNF-10) exhibited a high specific capacitance value (156 F g^{-1}) at 0.5 A g^{-1}. Indeed, the improved performance was attributed to the high SSA (709 $m^2\ g^{-1}$) and large pore volume (0.356 $cm^3\ g^{-1}$) of the activated sample (P-CNF-10) as compared to those of other *in situ* activated samples.

On the other hand, supporting the nanofibers with other kinds of carbon nanomaterial, such as CNTs, graphene, or GQDs, is a highly effective approach and has shown great potential to improve the electrochemical performance of freestanding carbon nanofiber electrodes by increasing their electrical conductivity. As mentioned above, GQDs possess numerous physical and chemical attributes, such as high electron mobility, high SSA, and abundant active sites as a result of their ultrasmall particle sizes and surface functional groups. In this context, embedding GQDs into the framework structure of CNF fabrics can greatly enhance the capacitive energy storage of the resulting electrode. Zhao *et al* designed a synthesis strategy for a highly porous and mechanically robust CNF electrode using GQDs as a reinforcement filler [129]. The results showed that the embedded GQDs significantly impacted the construction of a conductive network and synchronously reinforced the framework structure of the CNFs. Electrochemical measurements were performed in a three-electrode system using 6.0 M KOH as an aqueous electrolyte and the obtained CNF mats as binder-free electrodes. Due to its unique building block structure and interconnected nanopores, the optimized AGRCNF-3 showed an ultrahigh specific capacitance (358 F g^{-1} at 1.0 A g^{-1}) and a remarkable rate performance of 76% at 100 A g^{-1}. In another study, Cuo and co-workers reported that the electrochemical performance of a freestanding CNF electrode could be enhanced by regulating its internal network structure and decorating the outer surface of CNF mats with GCDs via the electrospinning technique, as shown

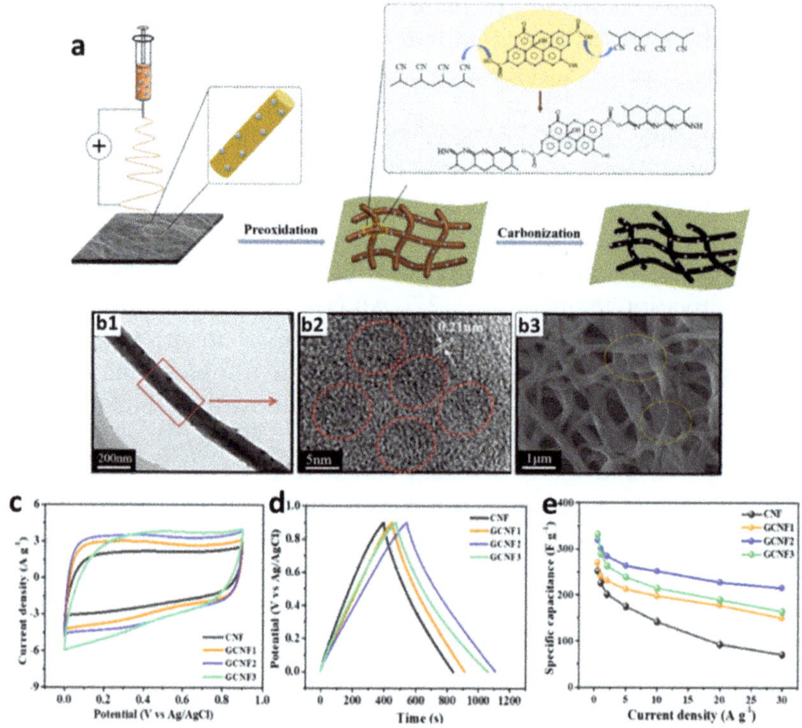

Figure 1.11. (a) A schematic illustration of the procedure used to prepare GCNF mats. (b) TEM, HR TEM, and SEM images of the GCNF2 sample. Electrochemical measurements: (c) CV curves measured at a scan rate of 20 mV s^{-1}, (d) charge/discharge curves acquired at a current density of 0.5 A g^{-1}, and (e) specific capacitances at various current densities [130], reproduced with permission © American Chemical Society.

in figure 1.11(a) [130]. Systematic structural and morphological investigations showed that the resulting CNF mats were highly cross-linked and welded to each other as a result of using GQDs as conductive fillers (figure 1.11(b)). At the same time, the highly GQD-decorated surface provided high electrical conductivity and facilitated fast ion transport kinetics. The electrochemical results showed that the GCNF2 composite sample displayed an excellent specific capacitance of 319 F g^{-1} at 0.5 A g^{-1} and a remarkable rate performance (214 F g^{-1} even at 30 A g^{-1}), as portrayed in figures 1.11(c)–(e).

Another strategy is to use CNTs as a conductive filler to simultaneously improve the electrical and mechanical properties of CNFs. Indeed, the addition of other electrochemically active components that have the same 1D nanostructure form (such as CNTs) into the framework structure of CNFs not only increases the electronic conductivity of CNF fabrics but also affects their architecture. Guo *et al* first reported the fabrication of hybrid carbon nanofibers containing CNTs as a composite freestanding EDLC electrode in 2009 [131]. In this study, CNTs were first functionalized by highly concentrated acid (6 M HNO_3) in order to ensure good dispersion of the PAN precursor. The spun CNT-embedded PAN nanofiber mats were then carbonized at 700 °C for 30 min in an Ar atmosphere, and the activation process was performed at 650 °C for 20 min in a specific atmosphere made of hydroperoxide and a water–steam mixture. The sample thus obtained (CNT/PAN-CENM) exhibited high specific capacitance (310 F g^{-1} at 0.1 A g^{-1}) with good cyclic stability. Its excellent electrochemical properties were attributed to the high electrical conductivity and large SSA caused by the synergistic effect of two different carbon materials. In addition to the use of a PAN polymer as a carbon source in the electrospinning process, cellulose (CA) was further utilized in the presence of MWCNTs to produce a highly porous CNF/MWCNT composite electrode [132]. It was found that the addition of 6 wt.% of MWCNTs led to a massive increase in the SSA from ~870 (for the filler-free sample) to ~1120 m^2 g^{-1}. As a result, a composite electrode made from this material (CNF/MWCNT) exhibited a high specific capacitance of 145 F g^{-1}, which was higher than that of a CA-derived CNF electrode (105 F g^{-1}). It is notable that superior electrochemical performance was further achieved using SWCNTs instead of MWCNTs in electrospun CNFs. By electrospinning a PAN solution containing SWCNTs and zinc acetate, followed by a calcination process, a highly porous and conductive CNF/SWCNT composite electrode was produced [133]. The freestanding electrode thus obtained displayed a specific capacitance of 417 F g^{-1} at 0.5 A g^{-1} in a 6.0 M KOH aqueous electrolyte, which was markedly higher than that of a pristine CNF electrode (100 F g^{-1}) measured at the same current density.

As in the case of CNTs, the addition of graphene (a single layer of graphite and the most popular 2D carbon nanostructure) has a further influence on the physical properties and architecture of electrospun CNFs. Zhou *et al* reported porous graphene/CNF (C/CNF) composite electrodes prepared via the electrospinning of a PAN solution containing oxidized graphene sheets, followed by carbonization at 800 °C in an Ar atmosphere [134]. Electrochemical measurements showed that a freestanding composite electrode made of this material had a specific capacitance that reached up to 263.7 F g^{-1} at 0.1 A g^{-1}. Furthermore, the capacitance retention declined by only 13% after 2000 cycles, demonstrating good stability. The improved performance was attributed to the unique structure of the composite sample 'plan-to-line' due to the introduction of graphene sheets, resulting in high electrical conductivity and a large SSA. In another study, Pan and his co-workers studied the effect of nitrogen doping on the electrochemical performance of graphene/CNF-based freestanding composite electrodes [135]. A composite sample nitrogen-doped electrospun reduced graphene oxide (NG/CNFs) was fabricated by electrospinning a PAN precursor in the presence of graphene oxide (GO) followed by thermal treatment under an ammonia atmosphere (800 °C for 2 h). The sample thus obtained shows a distinctive 'plan-to-line' hierarchical structure with high electrical conductivity and a large SSA. Consequently, the supercapacitor electrode (NG/CNFs) displayed an initial specific capacitance of 338 F g^{-1} in 1 M NaCl, demonstrating the effect of nitrogen doping in improving the electrochemical performance of graphene/CNF-based freestanding electrodes. Due to their high electrochemical activity, CNTs were used to further enhance the electrochemical performance of CNFs in the presence of GO [136]. In this study, CNTs and GO were embedded in PAN-based carbon nanofibers to form a flexible electrode (GO-CNT/CNF) for use with 0.5 M Na_2SO_4 aqueous electrolyte in a supercapacitor. The electrochemical results for a composite sample made of this material (GO-CNT/CNF) showed a high specific capacitance of 120.5 F g^{-1}, which resulted from the combination of GO and CNT in the framework structure of the freestanding electrode. In effect, the electrochemically active surface functional groups of GO and the high electrical conductivity of the CNTs significantly improved the total electrochemical performance of the composite electrode.

1.3.3 Two-dimensional (2D) carbon nanomaterials

Two-dimensional nanomaterials are materials that have thicknesses on the nanometer scale and a large size in two dimensions (length and width). Compared with bulk materials, the most significant feature of 2D nanomaterials is their extremely high surface-to-volume ratio, which is beneficial for mass transfer, heat transfer, and ion diffusion. Therefore, they have many applications in areas such as adsorption, catalysis, and energy storage. Among numerous carbon materials, 2D carbon nanomaterials have advantages such as good conductivity, high SSAs, porous layered structures, and good chemical stability. The increased area–volume ratio of 2D carbon nanomaterials can promote contact between the electrode and the electrolyte, shorten the ion diffusion distance, and provide higher energy and power density. Therefore, 2D carbon nanomaterials are eminently suitable for use in EDLCs [137, 138].

1.3.3.1 Pristine graphene

Graphene is the 2D carbon nanomaterial most used for EDLCs because of its unique 2D structure, high mechanical strength, large SSA, good electrical conductivity, and excellent chemical stability [139]. Theoretically, the electronic double-layer (EDL) capacitance of graphene could reach about 550 F g^{-1} if its entire surface area could be fully utilized [140]. To date, many methods have been developed to prepare graphene electrodes for EDLCs, such as CVD growth [141–146], oxidation–exfoliation–thermal reduction [147–150], liquid-phase exfoliation [151], electrochemical exfoliation [152, 153], and laser reduction [154–157].

Since 2009, the preparation of graphene electrodes for EDLCs by the CVD method has set off a research boom due to its advantages of good controllability and large-scale synthesis [141–146]. Cai *et al* reported a radio-frequency plasma-enhanced chemical vapor deposition (RF-PECVD) method that was used to grow vertically oriented graphene nanosheets (VOGNs) on nickel substrates using C_2H_2 as the feedstock; they then applied VOGNs as electrodes for EDLCs (figure 1.12(a)) [142]. The specific capacitance of the VOGNs was 265 μF cm^{-2} at a film height of 2 μm and a growth temperature of 850 °C at 120 Hz for 10 min, which provided the prospect of a specific capacitance of more than 1 mF cm^{-2} for graphene-based EDLCs with sufficient AC filtering ability (figure 1.12(b)). Yang *et al* employed different plasma sources, i.e. microwave and radio-frequency plasma, to prepare VOGNs with different edge densities [143]. The edge density of VOGNs obtained from a microwave plasma source was greater than that of those obtained from a radio-frequency plasma.

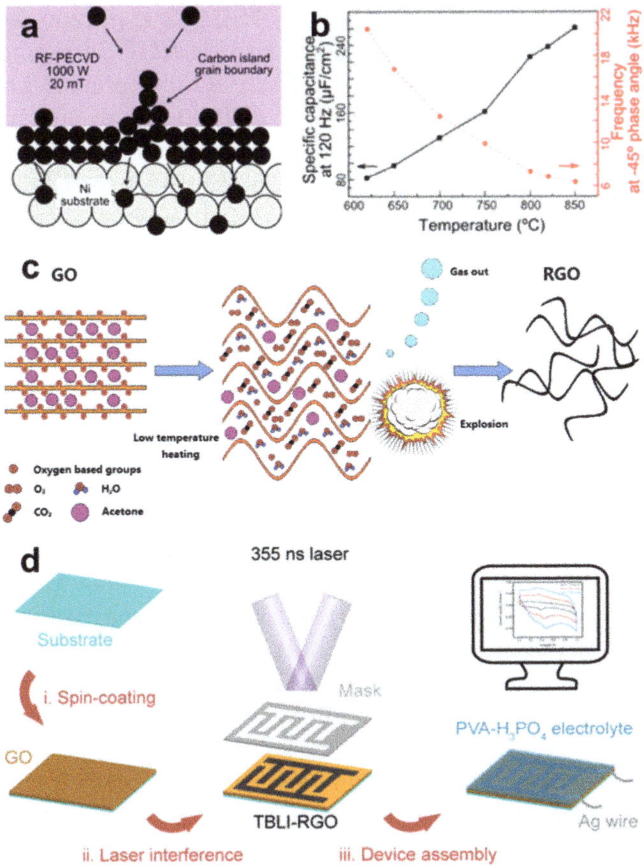

Figure 1.12. (a) A schematic diagram of graphite island impingement and upward growth; (b) the specific capacitance of VOGNs at a film height of 2 μm at 120 Hz for 10 min [142], reproduced with permission © American Chemical Society. (c) The formation process of reduced graphene oxide (rGO) nanosheets [147], reproduced with permission © John Wiley and Sons. (d) A schematic diagram of photoreduced GO-based planar EDLCs prepared by dual-beam laser interference with masking techniques [155], reproduced with permission © AIP Publishing.

The increased edge density contributed to a larger mass and greater area specific capacitance. Faraezi *et al* fabricated single-layer graphene-based EDLCs on gold substrates using the CVD method but with CH_4 as the raw material [144]. The authors compared the capacitive performances of graphene in different aqueous electrolytes. They found that the KCl electrolyte system had the largest area specific capacitance at different scan rates and the fastest ion diffusion kinetics.

The CVD method can meet the requirements of the large-scale preparation of high-quality graphene, but its cost is high and its procedure complex. The oxidation–exfoliation–thermal reduction method for the preparation of graphene is inexpensive and easy to use. A stable graphene suspension can be obtained, which solves the problem that graphene is difficult to disperse; this approach is conducive to the synthesis of graphene-based composites [147–150]. Jia *et al* reported a method for chemically reducing GO to reduced graphene oxide (rGO) using acetone at a low temperature of 80 °C (figure 1.12(c)) [147]. The rGO thus prepared had a large SSA of 318 m^2 g^{-1} and a mass specific capacitance of 100 F g^{-1} even at a current density of 50 A g^{-1}. Kil *et al* proposed liquid-phase oxidation accompanied by a low-temperature reduction method to produce rGO electrodes for EDLCs [148]. A GO coating could be completely reduced to rGO without any chemical reduction under air at the relatively low temperature of 300 °C. EDLCs consisting of rGO electrodes and poly(vinyl alcohol)/phosphoric acid (PVA/H_3PO_4) electrolyte gels had a high mass specific capacitance of about 240 F g^{-1}, an energy density of 33.3 Wh kg^{-1}, and an excellent power density of approximately 833.3 W kg^{-1}.

The disadvantage of the oxidation–exfoliation–thermal reduction method is that large-scale preparation can easily cause liquid waste pollution problems. The graphene thus prepared has certain defects, which leads to the loss of some of the electrochemical properties of graphene, so the applications of graphene-based EDLCs are limited. The liquid-phase exfoliation method can be applied to prepare high-quality graphene. No defects on the surface of graphene are introduced during the whole liquid-phase exfoliation process, which provides favorable prospects for graphene-based EDLCs [151]. Li *et al* prepared high-concentration inks of 1 mg ml^{-1} with liquid-phase exfoliated graphene nanosheets to be used for printing micro EDLCs [151]. However, this process needed continuous stirring for 40 h, resulting in a low exfoliation yield of 1% and a small nanosheet size of less than 1 μm.

Recently, electrochemical exfoliation has become an attractive way to produce graphene with a large nanosheet size, good electrical conductivity, and a high carbon-to-oxygen ratio [152, 153]. Tamgadge *et al* proposed an electrochemical exfoliation method to fabricate VOGNs with low electrolyte resistance [152]. The authors found that a low-pH solution allowed graphite to expand better, leading to a greater VOGN density, while a high-pH solution could lead to the formation of loose graphene nanosheets. The optimally exfoliated VOGN electrode presented an area specific capacitance of 752 mF cm^{-2} at a current density of 1.0 mA cm^{-2}. Furthermore, a symmetric device consisting of two VOGN electrodes provided a volumetric specific capacitance of 1.56 F cm^{-3} at a current density of 2.0 mA cm^{-2}.

In recent years, the laser manufacture of graphene-based EDLCs has shown significant advantages due to its simple steps and its ability to integrate with future electronics [154–157]. Fu *et al* reported a method for the preparation of photo-reduced GO-based planar EDLCs based on a combination of dual-beam laser interference and masking techniques (figure 1.12(d)) [155]. The hierarchical structures of photoreduced GO helped to increase the contact surface area between the electrolyte and the electrode, which was conducive to enhancing the EDL capacitance. In addition, a planar device with photoreduced GO-based interdigital current collectors was conducive to fast ion diffusion kinetics. As a result, photoreduced GO-based planar EDLCs achieved an area specific capacitance of 3.97 mF cm^{-2} at a scan rate of 10 mV s^{-1}.

1.3.3.2 Doped graphene

The specific capacitances of pristine graphene-based EDLCs in aqueous and organic electrolytes are 135.0 F g^{-1} and 99.0 F g^{-1}, respectively, due to the restacking of graphene nanosheets caused by π–π interactions and van der Waals forces, resulting in the incomplete entry of electrolyte ions into the graphene surface [140]. Heteroatom-doped graphene is an effective way to improve the electrochemical performance of SCs for the following reasons: (i) the defects produced in graphene by doping make it more porous, thereby enhancing the access of electrolyte ions to the graphene surface during electrochemical processes; (ii) doping produces more active sites, increasing the reactivity of graphene; (iii) the electrical conductivity of graphene can be improved by doping; (iv) heteroatomic doping can generate pseudocapacitance in addition to the EDL capacitance of graphene; (v) heteroatomic doping also enhances the wettability of graphene, which is conducive to electrochemical reactions. So far, many nonmetallic elements have been doped into graphene to improve the electrochemical performance of SCs, such as nitrogen [158–161], sulfur [162], boron [163, 164], and phosphorus [165–167].

Li *et al* reported a simple and efficient method for the synthesis of nitrogen-doped rGO (N-rGO) that combined immediate thermal exfoliation and covalent bond conversion of melamine–GO mixtures [158]. The authors compared the capacitive performances of N-rGO, thermal-rGO (T-rGO), and chemical-rGO (C-rGO) in different electrolytes (5 M H_2SO_4 and 1 M tetraethyl ammonium tetrafluoroborate ($TEABF_4$) in propylene carbonate (PC)). The relatively abundant oxygen-containing functional groups on C-rGO formed obvious ion diffusion barriers, resulting in poor electrochemical performance of rGO in the $TEABF_4$/PC electrolyte. N-rGO presented a large SSA and a relatively low density of oxygen-containing groups, resulting in the excellent electrochemical performance of organic EDLCs made using N-rGO. The capacitance retention of N-rGO in the $TEABF_4$/PC electrolyte reached 71.1% when the scan rate was increased to 200 mV s^{-1}, indicating that N-rGO could address the problem of the relatively low power of EDLCs based on organic electrolytes. The energy and power density of N-rGO-based symmetric devices in the $TEABF_4$/PC electrolyte reached 25 Wh kg^{-1} and 10 kW kg^{-1}, respectively. Youn *et al* prepared N-rGO with a large SSA (1007 m^2 g^{-1}), high electrical conductivity (1532 S m^{-1}), and low oxygen content (1.5 wt.%) using solid-state microwave irradiation accompanied by NH_3 heat treatment and applied N-rGO as electrodes for EDLCs [159]. The N-rGO showed a specific capacitance of 291 F g^{-1} at a current density of 1 A g^{-1} and maintained a specific capacitance of 261 F g^{-1} at 50 A g^{-1}, indicating its excellent rate capability. N-rGO also exhibited good cyclic stability and retained 96% of its initial mass capacitance after 100 000 cycles. The outstanding electrochemical performance of N-rGO was attributed to its large SSA, low oxygen content, and high electrical conductivity.

Some researchers have reported that sulfur-doped graphene electrodes can also improve the electrochemical performance of SCs. Zhang *et al* proposed a method for tailoring the doping of sulfur species in graphene [162]. In this doping strategy, sulfur doping is first achieved by the heat treatment of H_2SO_4-graphite intercalation compound (GIC) compounds, followed by the post-customization of the chemical state of the doped sulfur by Hummers' oxidation. The unique oxidized sulfur species (–C–(O)$_x$–C–) transformed from the thiophene-S species by Hummers' oxidation made outstanding contributions to the electrochemical performance of SCs. Sulfur-doped graphene with a high content of –C–(O)$_x$–C– presented a specific capacitance retention of 100.8% after 10 000 cycles, showing excellent long-term cyclic stability.

Boron doping in graphene can also enhance the electrochemical performance of SCs. Thirumal *et al* prepared boron-doped graphene nanosheets by a hydrothermal method using boric acid (H_3BO_3) and GO [163]. Compared to GO, boron-doped graphene showed a larger specific capacitance and a smaller charge transfer resistance (R_{ct}). Balaji *et al* used H_3BO_3 as a boron precursor to prepare boron-doped graphene by supercritical fluid (SCF) processing and hydrothermal methods [164]. The boron-doped graphene synthesized via SCF processing sample fabricated by supercritical method using boric acid with graphene oxide (SCBAGO) showed a specific capacitance of 286 F g^{-1} at 1.0 A g^{-1} in a 20% KOH aqueous electrolyte, which was higher than that of B-doped graphene prepared by a hydrothermal method sample fabricated by hydrothermal method using boric acid with graphene oxide (HTBAGO, 142 F g^{-1}), and presented a high specific capacitance retention of 96% after 10 000 cycles at 20.0 A g^{-1}. The energy density of a symmetric SC (with SCBAGO as the electrode) in a 1 M $NaClO_4$ aqueous electrolyte was 21.2Wh kg^{-1}, which was four times higher than that in a 20% KOH aqueous electrolyte (5.3Wh kg^{-1}). The specific capacitance and energy density of the symmetric device (with SCBAGO as the electrode) in an ionic liquid electrolyte (1-ethyl-3-methyl imidazolium tetrafluoroborate, $EMIMBF_4$) were 138 F g^{-1} and 43.1 Wh kg^{-1}, respectively, which were higher than those in aqueous electrolytes (1 M $NaClO_4$ and 20% KOH).

Co-doping/dual-atom doping is also a common strategy for improving the electrochemical performance of graphene. Xia *et al* used $(NH_4)_3PO_4$ as the nitrogen and phosphorus source to synthesize N, P co-doped graphene by a two-step method [165]. Adjusting the surface oxygen content (C/O ~ 5.7) of the precursor functionalized graphene nanosheets could effectively promote the activation of $(NH_4)_3PO_4$, thereby achieving a high SSA of 320 m^2 g^{-1} and efficient N, P co-doping. The N, P co-doped graphene showed specific capacitances of 219 F g^{-1} (123 F cm^{-3}) and 175 F g^{-1} (98 F cm^{-3}) at 0.25 A g^{-1} and 10 A g^{-1}, respectively. A N, P co-doped graphene-based symmetric supercapacitor device was able to achieve an energy density of 8.2 Wh kg^{-1} (4.6 Wh l^{-1}) at a power density of 162 W kg^{-1} (91 W l^{-1}) and retained 6.7 Wh kg^{-1} at 6.5 kW kg^{-1} at a voltage window of 1.3 V. Yu *et al* prepared sulfur and phosphorus dual-doped graphene (SPG) films using an ice template and heat activation method and applied the SPG films to flexible electrochemical capacitors [167]. The SPG-based flexible devices achieved a high specific capacitance (169 F g^{-1}), an excellent rate capability (91.7%), and long cyclic stability (92.5%). Even at a bending angle of 120°, no noticeable change in specific capacitance was observed. The excellent electrochemical performance of SPG was attributed to the porous structures formed during freeze-drying and steam activation, the enhanced surface roughness due to the heteroatom doping, and the formation of additional active sites by functional groups containing S and P. Cheng *et al* successfully demonstrated a strategy for the preparation of nitrogen–sulfur co-doped rGO (NSG) by electron beam irradiation using L-cysteine and GO as precursors (figure 1.13(a)) [168]. As shown in figure 1.13(b), an NSG-210 (produced using an irradiation dose of 210 kGy) -based symmetric supercapacitor provided an energy density of 8.3 Wh kg^{-1} at a power density of 510 W kg^{-1} and 5.8 Wh kg^{-1} at a higher power density of 10 440 W kg^{-1}. Series devices could illuminate a red light-emitting diode (LED) indicator for a few minutes (figure 1.13(c)). The capacitance retention of the NSG-210-based symmetric supercapacitor was approximately 94% after 10 000 cycles at a current density of 5.0 A g^{-1} (figure 1.13(d)). This work has inspired the design and large-scale production of heteroatom-doped and co-doped graphene-based materials.

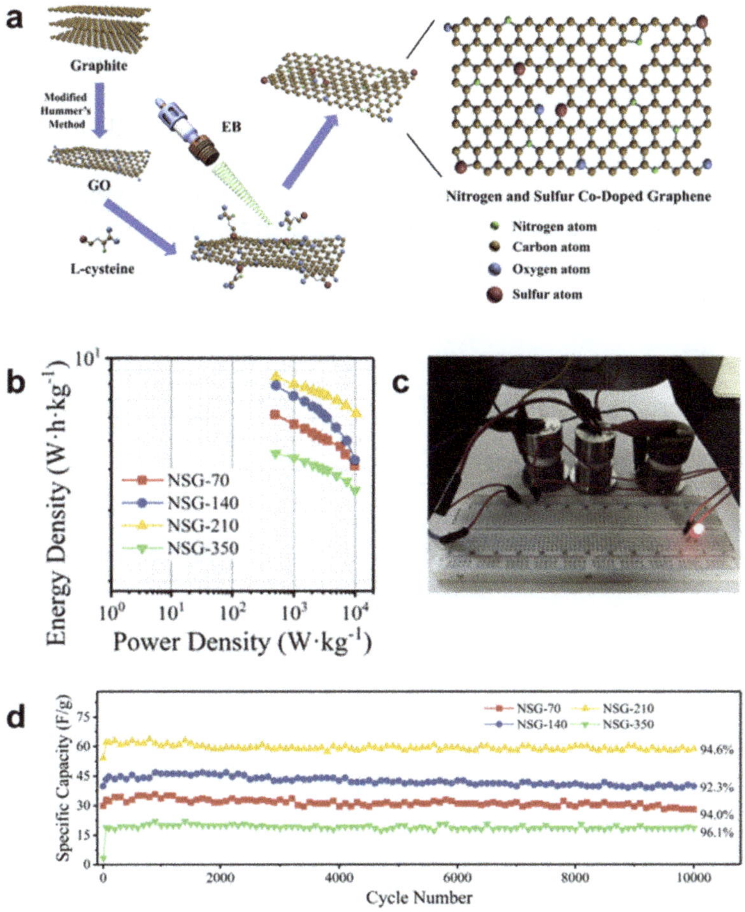

Figure 1.13. (a) A schematic diagram of the preparation of N, S co-doped graphene, (b) Ragone plots of the relationship between the power and energy density of NSG-based symmetric supercapacitors, (c) an image of a red LED lit by three NSG-210-based symmetric supercapacitors connected in series, and (d) the cyclic life of NSC-based symmetric supercapacitors at a current density of 5.0 A g^{-1} [168], reproduced with permission © Elsevier.

1.3.3.3 Carbon nanosheets

Although graphene has excellent properties as an energy storage material, its complex preparation method is not conducive to large-scale production, which limits its applications. Therefore, researchers are trying to find other 2D carbon nanomaterials to simplify the preparation process and reduce the cost of material preparation without reducing the performance. Carbon nanosheets (CNSs) have become a substitute for graphene due to their simple preparation and low cost. The synthesis methods of CNSs include the templating method, the hydrothermal method, direct carbonization, the CVD method, RF sputtering, the molten salt method, and the sol–gel method [169, 170]. For example, Wang *et al* used montmorillonite as a hard template to synthesize mesoporous CNSs [171]. Montmorillonite has a layered structure with a large interlayer distance of 1.47 nm. The small-molecule F127 micelles and Resol can easily enter the wide interlayers and assemble in an orderly way, as shown in figure 1.14(a). After calcination and template removal, ordered mesoporous CNSs were obtained, which displayed a high capacitance of 130 F g^{-1} at 0.5 A g^{-1}.

The carbon sources used in the preparation of CNSs are mainly fossil fuel sources, such as methane, ethylene, phenol, pitch, and polymers [172]. Recently, many studies have reported the use of biomass as a resource in the preparation of CNSs, which greatly reduces production costs and energy consumption. Biomass can

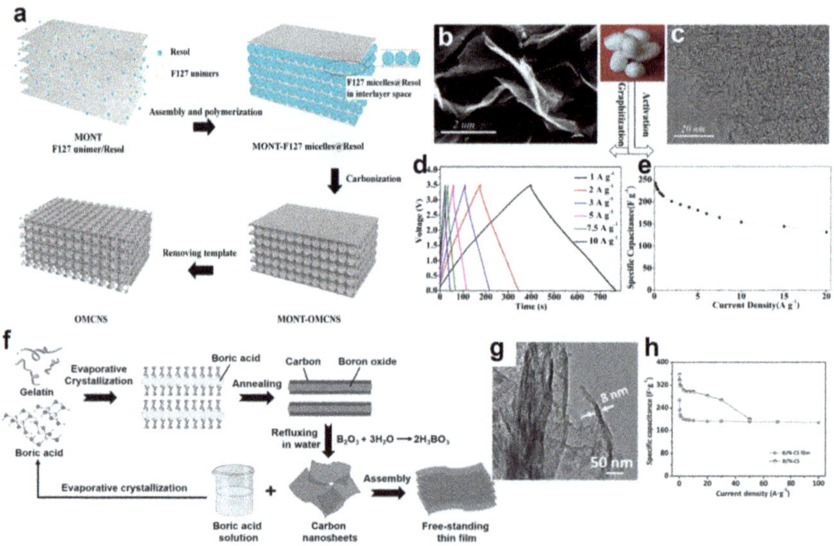

Figure 1.14. (a) A schematic representation of the preparation of mesoporous CNSs using the hard template method [171], reproduced with permission © John Wiley and Sons. (b) An SEM image of mesoporous CNSs, (c) a TEM image of mesoporous CNSs, (d) charge/discharge curves, and (e) specific capacitances at different current densities of porous CNSs synthesized from natural silk [173], reproduced with permission © American Chemical Society. (f) The synthesis process of N/B co-doped CNSs, (g) a TEM image of N/B co-doped CNSs, and (h) the rate performances of N/B co-doped CNS and N/B co-doped CNS films [179], reproduced with permission © John Wiley and Sons.

provide a green and sustainable source of carbon materials through photosynthesis, and the diversity of biological tissues can evolve into carbon materials with rich structures through special pathways. Lignin, sugars, cellulose, chitin, proteins, and other biomasses have been widely used to prepare CNSs. Hou *et al* successfully prepared porous CNSs by the direct carbonization of natural silk after treatment in $ZnCl_2/FeCl_3$ solution (see figures 1.14(b)–(e)) [173]. The CNSs thus prepared showed a high SSA of up to 2494 $m^2\ g^{-1}$ and a high pore volume of 2.28 $m^3\ g^{-1}$ with various pore structures. The capacitance of this material in the $EMIMBF_4$ electrolyte reached 242 F g^{-1} at 0.5 A g^{-1} for EDLCs due to the highly exposed surface. Yan's group employed human hair as the carbon source and synthesized 2D carbon flakes with outstanding EDLC performance [174]. Moreover, nut shells, leaves, peanut skins, corn cobs, etc. can be used as biomass-derived carbon raw materials.

For 2D nanomaterials with a high area/volume ratio, there is a significant issue that cannot be ignored. Although they exhibit fast ion and electron transfer in the direction of the 2D layer plane, there are obstacles to ion or electron transfer in the vertical direction of the 2D layers. As a result, the reaction does not easily penetrate the interior of the electrode, and quick storage is difficult to achieve. This is very detrimental to improving the power density and energy density of supercapacitors and the rapid realization of energy storage. Meanwhile, the assembly of CNSs leads to significantly reduced electrode active surfaces due to their dense structure, which lengthens the ion transport pathway and increases ion transport resistance. A porous structure can alleviate the problem of the restacking of 2D nanomaterials, enabling ions or electrons to be transported faster to the inside of the electrode, greatly shortening the ion or electron transport pathway [175]. In addition, a porous structure can effectively promote the penetration of electrolytes into the interior of the electrode, relieving the volume changes of electrode materials and releasing structural strain during the charge–discharge process.

In the last decade, researchers have achieved an increase in the active surface area, a reduction in ion transport pathways, and a reduction of ion transport kinetics by a hole-punching treatment of CNSs. A series of porous-structure CNSs have demonstrated the feasibility of applying CNSs in energy storage devices to achieve high performance. In general, the preparation of porous CNSs takes place via three methods: the hard template method, the soft template method, and the template-free method. The hard template method is versatile in terms of material and size, and can easily be scaled up. Commonly used templates include montmorillonite, porous silica, NaCl, GO/rGO sheets, layered metal oxides, etc [176]. Under suitable conditions, small amphiphilic molecules or block copolymers can self-assemble to produce lamellar structures, which are very suitable for use as soft templates. These structures comprise a planar bilayer consisting of hydrophobic alkyl chains or blocks sandwiched between two hydrophilic shells composed of the hydrophilic segments of the amphiphiles. The template-free method mainly uses the activation of carbon precursors, which is the most convenient way to increase the SSA of CNSs and adjust the meso/micropore ratio and pore size.

Depending on the activation mechanism used, the template-free methods can be divided into three types: (i) chemical activation; (ii) physical activation;

(iii) self-activation. KOH is a commonly used chemical activation reagent [177]. The carbon lattice is etched to form a rich porous structure by a reaction between carbon and KOH. The intermediate metal K produced during activation can be inserted into the carbon framework to expand the carbon lattice. The SSA of the AC material is greatly increased. In addition, the pore structure can be adjusted by controlling the activation conditions (such as the ratio of the precursor to the activating agent, activation temperature, activation time, etc). In addition to KOH activation, $ZnCl_2$ activation is also an effective strategy for increasing the SSA of carbon materials [178]. Because $ZnCl_2$ is a Lewis acid, it can promote the dehydration of biomass precursors at low temperatures and react at high temperatures.

Carbon dioxide and water vapor activation are the main physical activation methods. During the activation process, CO_2/H_2O can react with carbon to remove carbon atoms and form a porous structure. Compared to chemical activation, porous CNSs prepared by physical activation have higher yields and bulk densities. However, the etching of carbon atoms is not very efficient, so the SSA of the material after physical activation is relatively low. Self-activation does not require additional activation reagents, simplifying the activation process and reducing costs. The self-activation process includes two activation mechanisms: (i) activation using the gases produced during pyrolysis, also known as physical self-activation; (ii) *in situ* activation using inorganic materials (e.g. K^+) already present in the precursors, also known as chemical self-activation.

In addition to the SSA and pore structure, the surface properties of CNSs also play an essential role in their electrochemical performance. Recent studies have shown that heteroatom doping (using, for example, N, O, S, B, and P) or surface functional groups can significantly modify the surface properties of CNSs. Heteroatoms can provide a lone pair of electrons, effectively improving the electron donor and acceptor characteristics of CNSs, thereby introducing pseudocapacitive reactions and correspondingly improving the material's capacitance. In addition, heteroatoms and surface functional groups can improve the surface hydrophilicity of CNSs, making more of the SSA an effective SSA, forming a larger double layer. As such, heteroatom doping has been successfully applied in the electrode materials of supercapacitors. It can not only improve the specific capacitance values of the materials in aqueous electrolytes but also affect the capacitance of non-aqueous systems.

Recently, the co-doping of various heteroatoms has been adopted as a relatively new strategy for improving the EDLC performance of CNSs. The effects of various dopant atoms can be comprehensively utilized by co-doping. For instance, a N atom can be easily and stably combined with a C atom, which promotes pseudocapacitive charge transfer and increases the basicity of porous CNSs, simultaneously increasing conductivity and wettability. The B atom is another common heteroatom that can be incorporated into the carbon framework by substituting carbon at the trigonal site. B atoms have three kinds of valence electron, which can be used as electron acceptors and induce changes in the Fermi energy level, modify the electronic structure of doped carbon materials, and promote charge transfer, thereby improving the electrochemical performance of CNSs. However, when 2D carbon is doped with single N atoms, the N content often decreases as the pyrolytic temperature of the N-containing precursors increases. The activation methods used to improve the SSA and pore structure can also lead to a decrease in the N content. When 2D carbon is doped with single B atoms, the relatively low reactivity of H_3BO_3 (the B source) results in a low B doping level. In addition, B atoms are difficult to introduce into the carbon structure at temperatures below 1500 °C. Compared to single atom doping, the co-doping of N/B heteroatoms can effectively solve these issues (figures 1.14(f)–(h)) [179]. The co-doping of N and B heteroatoms into carbon materials can generate N–B bonds in the carbon framework, improving the stability and doping level of the heteroatoms. In addition, synergistic effects between N/S, P/N, and O/N have also been confirmed to enhance the overall performance of CNSs [180–182].

1.3.4 Three-dimensional (3D) porous carbon

1.3.4.1 Activated carbons

AC has an abundantly microporous structure and a high SSA; it can also be regarded as a 3D porous carbon material. AC can be prepared by the thermal decomposition and partial oxidation of organic compounds, including a wide range of natural and synthetic precursors. As one of the oldest and most common types of porous carbon, it was used in Egypt as early as 1550. The United States realized the industrial production of AC in 1913. It carried out extensive research to improve the activation process, but the precise adjustment of the pore structure is still challenging. Most of the pores in AC are concentrated in the range of 0.4–4 nm. The pore size distribution is usually relatively wide, which may limit the capacitive performance in EDLCs. Common natural precursors of AC synthesis include coal, petroleum coke, nut shell, etc. and the more unusual natural precursors include starch, sucrose, straw, etc [183–192]. The materials produced can inherit the 3D pore structures of the precursors, such as eggshells (figure 1.15) [191]. Carbonized eggshell membrane (CESM) was a three-dimensional macroporous carbon film composed of interwoven connected carbon fibers containing around 10 wt.% oxygen and 8 wt.% nitrogen. Despite its relatively low surface area of 221 $m^2\ g^{-1}$, exceptional specific capacitances of 297 F g^{-1} and 284 F g^{-1} were achieved in basic and acidic electrolytes, respectively, in a three-electrode system. Furthermore, the electrodes demonstrated excellent cyclic stability: only 3% of capacitance fading was observed after 10 000 cycles at 4 A g^{-1}. These very attractive electrochemical properties were discussed in the context of the unique structure and chemistry of the material.

AC has the advantages of a large SSA, rich micropores, stable chemical properties (against acids, alkalis, and heat), high mechanical strength, insolubility in water and organic solvents, and good adsorption of inorganic or organic substances and colloidal particles in the gas and liquid phases. In addition to its 80%–90% content of elemental carbon, AC also contains two types of admixture: the first consists of chemically combined elements, e.g. oxygen and hydrogen, remaining in the carbon due to incomplete carbonization, while the second consists of external non-carbon

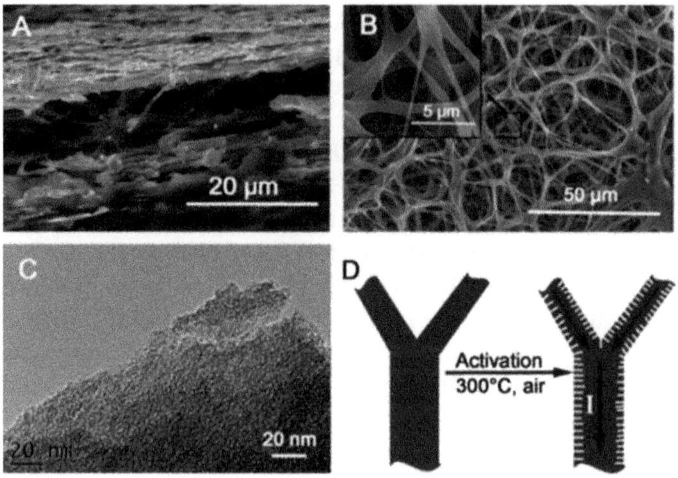

Figure 1.15. SEM images of activated CESM in cross-sectional view (A) and plan view (B). The inset of (B) is a high-resolution image of the selected area. (C) A TEM image of activated CESM. (D) An illustration of the carbon–carbon core–shell structure of activated CESM [191], reproduced with permission © John Wiley and Sons.

elements that are chemically combined with the surface of the AC during activation. For example, when water vapor is used as an activator, the surface of the AC is oxidized; other inorganic components that cannot be carbonized are called ash. AC materials with high SSA, high conductivity, and low ash content are ideal materials for EDLC electrodes. At present, commercial EDLC electrode materials are still mainly high-quality coconut shell AC, which shows that AC plays an important role in the development of EDLC electrode materials. Activation processes are usually divided into two categories: physical activation and chemical activation. The preparation of AC by physical activation includes precursor carbonization (the removal of non-carbon substances by thermal decomposition in an inert atmosphere) and gasification (the formation of porosity by the partial corrosion of carbon by an oxidant during annealing) (figure 1.16). The oxidant used for gasification may be CO_2, H_2O, or a mixture of both [193]. The preparation of AC by chemical activation usually involves the reaction of precursors with chemical reagents (KOH, H_3PO_4, $ZnCl_2$, H_2SO_4, etc.) at high temperatures [194–198], resulting in the formation of small and uniform pores and high specific capacitance in aqueous and organic electrolytes.

Due to the low cost, large-scale production, and high cyclic stability of AC, AC powder is used in almost all commercial EDLCs. In the process of preparing electrodes, AC powder is mixed with polytetrafluoroethylene (PTFE) to form a slurry, then coated on the current collector, dried and assembled into a capacitor, which is finally filled with an organic or ionic liquid electrolyte. The SSA of commercial AC is usually 700–2200 $m^2\ g^{-1}$, and its specific capacitance is 70–200 $F\ g^{-1}$ in aqueous electrolytes; its capacity in organic electrolytes is slightly less than that in water, specifically 50–120 $F\ g^{-1}$ [14, 199, 200]. In addition, recent studies of AC have shown that its specific capacitance in water and an organic electrolyte can

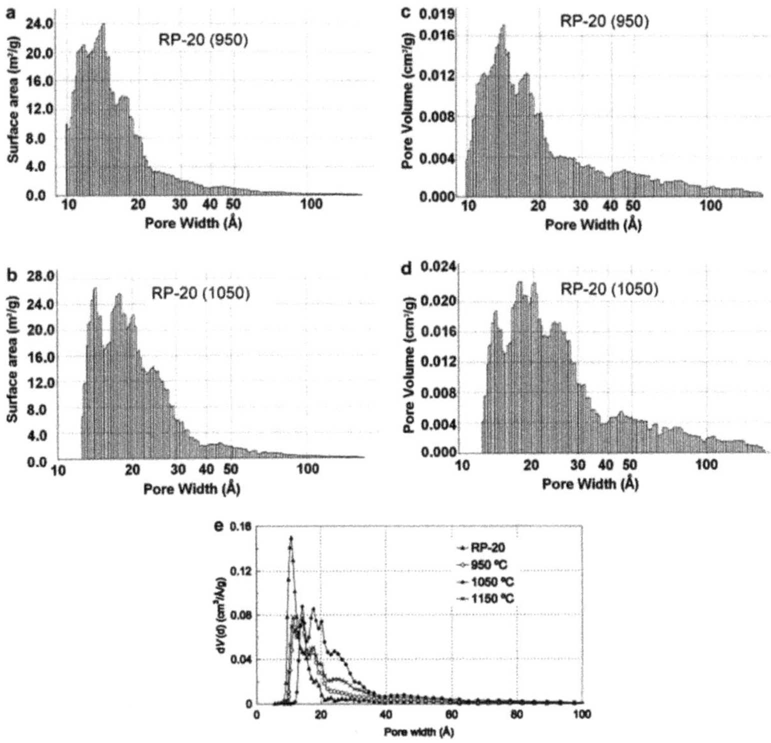

Figure 1.16. (a, b) Pore size distributions, (c, d) pore volumes, and (e) differential pore size distribution (the pore size distribution plot was obtained using the non-local density function theory (NLDFT) method) [193], reproduced with permission © Elsevier.

reach 250–300 F g^{-1} [186, 201–206]. Overall, AC is still the mainstream electrode material of choice for EDLCs.

Low mass energy density is the main disadvantage of AC-based EDLCs; it is usually less than 5 Wh kg^{-1}. A large SSA is conducive to increasing the capacitance. However, a study by Shi and Béguin showed that the capacitance, which usually varies between 94 and 413 F g^{-1}, does not always increase as the SSA is improved (figure 1.17). When the SSA reaches a certain value and is then further improved, an increase in the specific capacitance is not evident [200, 207, 208]. There are two possible reasons for this: the first is that the SSA measured by the nitrogen adsorption–desorption isotherm method is not the area that can be soaked by an electrolyte; the second reason is that the specific electrosorption of ions is closely related to the pore size. Therefore, Shi proposed a model that could be used to measure the change of specific capacitance and pore size using the C/S value, which could obtain a better linear relationship; it also showed that micropores play an important role in the capacity and power performance of double-layer capacitors [207]. In addition, the conductivity of AC, which is mainly determined by the activation temperature and time, is another important factor that affects the performance of EDLCs. AC that can meet the requirements can usually be obtained

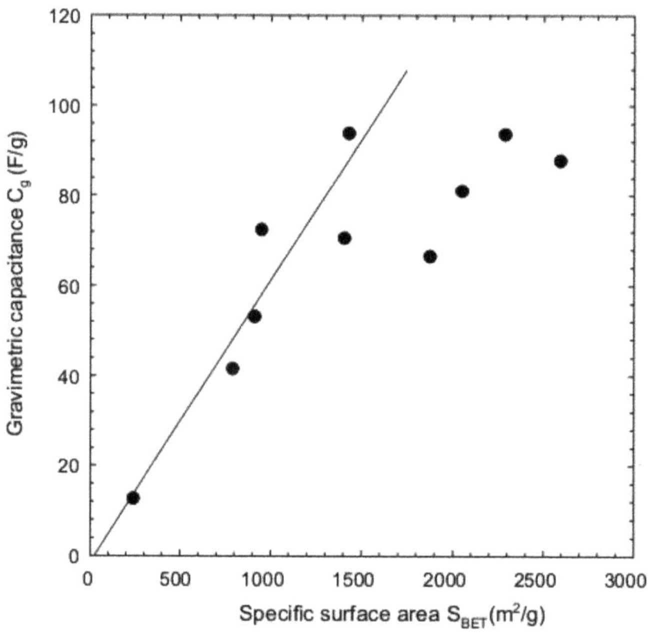

Figure 1.17. The relationship between the mass specific capacitance and the SSA of AC materials [207], reproduced with permission © Elsevier.

by increasing the annealing time. It can be seen that regulating and understanding pores is still the research focus for improving the performance of EDLCs.

1.3.4.2 Carbon-nanotube-based 3D architectures

CNTs are a 1D linear allotrope of carbon; they can be regarded as nanoscale seamless tubes made of curled graphite layers. Generally, the macro preparation of CNTs has been achieved through the chemical vapor deposition of CO and hydrocarbon gas precursors (such as methane and acetylene) with the aid of catalysts. Compared with advanced AC, carbide-derived carbon (CDC), and template carbon, CNTs have relatively low SSA and density, and it is difficult to make them into thicker electrodes using traditional electrode manufacturing methods. However, a new technique for CNT growth is able to prepare CNT arrays perpendicular to the current collector; such arrays can also be regarded as 3D supercapacitive electrode materials (figure 1.18) [212]. In fact, the low SSA and bulk density of CNTs limit their volumetric capacitance and energy density, but their high electrical conductivity and open two-terminal feature facilitate the construction of high-power EDLCs [87, 209–226].

The earliest study showed the potential of thinner multiwalled CNT electrodes (25 μm) to provide high power and high rate performance. The specific capacitance reached 102 F g^{-1}, and the operating frequency reached 100 Hz, indicating that multiwalled CNTs have high power characteristics [87]. Acid oxidation is usually used to obtain hydrophilic CNTs. For example, treating multiwalled CNTs with 69% nitric acid at 80 °C for 1 h can add a large number of oxygen-containing functional groups

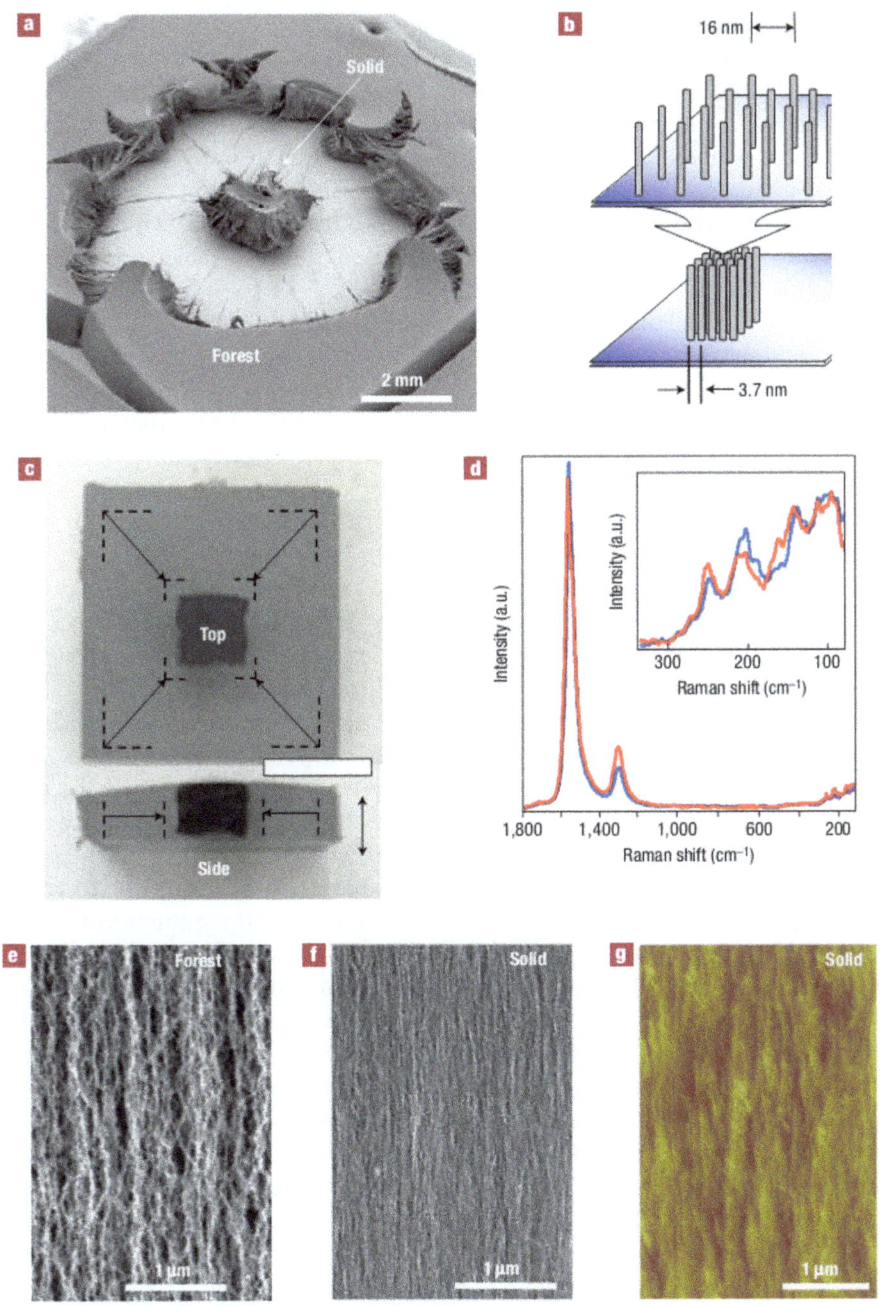

Figure 1.18. The process through which a CNT forest is shrunk by a liquid capillary force. (a) An SEM image of a single droplet after evaporation; (b) a schematic diagram of the density increase caused by the shrinkage of the CNT forest; (c) top and side views after shrinkage; (d) a Raman image of the CNT forest before and after shrinkage; (e) an SEM enlarged image of the side view before shrinkage; (f and g) SEM and atomic force microscopy (AFM) images of the CNT forest after shrinkage, respectively [212], reproduced with permission © Springer Nature.

to their surfaces, and their specific capacitance can be increased to 137 F g^{-1} [219]. However, it should be noted that the introduction of oxygen-containing functional groups may reduce the electrical conductivity of CNTs and the self-discharge performance of assembled capacitors. Therefore, the use of the acid oxidation method to treat CNTs used as EDLC electrode materials still has limitations.

Single-walled CNTs have higher theoretical SSAs, but they easily form bundles during the preparation process, resulting in lower actual SSAs, which limit electrolyte immersion and affect ion transport when used as EDLC electrodes. The specific capacitances of single-walled CNTs in 1 M NaCl, 7 M H_2SO_4, and 5 M KOH are around 19–41 F g^{-1}. High-purity single-wall CNTs have high SSAs (about 462–782 m^2 g^{-1}), a high rate capability, and stability in 1 M $LiClO_4$, but their specific capacitance (45–75 F g^{-1}) is lower than that of multiwalled CNTs. Since 2002, chemical activation methods have been introduced to increase the specific capacitance of CNTs while maintaining high electrical conductivity [227]. Their SSA was increased to 1050 m^2 g^{-1}, and their specific capacitances in 6 M KOH and 1.4 M tetraethyl ammonium tetrafluoroborate ($TEABF_4$) electrolytes were 90 and 65 F g^{-1}, respectively [228].

Growing CNTs on conductive substrates is beneficial to improving the electrochemical performance of CNTs by reducing the contact resistance between the active material and the current collector. For example, the specific capacitance of multiwalled CNTs grown on graphite substrates reached 116–153 F g^{-1} in 1 M H_2SO_4 [215]. Similarly, CNTs grown on metal substrates have also been developed, but their EDLC performance was mediocre, with a specific capacitance of only 18 F g^{-1} in 6 M KOH [223]. In contrast, the specific capacitance of CNTs grown in Inconel reached 83 F g^{-1} [213]. Regarding the volume specific capacitance of CNTs, various multiwalled CNTs, including CNTs grown on aluminum foil with SSAs of 47–1064 m^2 g^{-1} (whose corresponding densities were 0.1–0.9 g cm^{-3}) had volumetric capacitances of only 8–16 F cm^{-3}; this value is more than three times lower than that of commercial AC, so it is difficult for CNTs to become a mainstream EDLC electrode material [211]. However, in 2006, it became possible to increase the density of single-walled CNT forests by a factor of 20 using the capillary force of water evaporation, and the Brunauer–Emmett–Teller SSA reached 1000 m^2 g^{-1}; the resulting material exhibited a capacity of 80 F g^{-1} in $TEABF_4$ electrolyte [212]. In addition, plasma-treated CNTs have also been used as EDLC electrode materials (figure 1.19) and have exhibited specific capacitances of up to 207 F g^{-1} [220]. In conclusion, various methods have been applied to develop CNTs as electrode materials for EDLCs, but the progress in this field has also revealed the limitations of CNTs. With the rise of new 2D graphene materials, research interest in the use of CNTs for electrochemical storage has reduced further.

1.3.4.3 3D graphene and graphene-like composites

Ideally, graphene is a layer of graphite that has a thickness of one atom; its two-dimensional plane corresponds to the (002) crystal plane of graphite. Graphene fragments can be considered to be the basic structural units of many carbon materials with regular structures. For example, graphene crystals can be formed

Figure 1.19. Electron microscope images of CNTs directly grown on Ni substrates: (a)–(c) SEM images of CNTs grown on Ni substrates etched by an NH_3 plasma for (a) zero (bare Ni foil), (b) five, and (c) ten minutes, respectively; (d) an HR TEM image of the CNT that appears in (b) [220], reproduced with permission © Elsevier.

by stacking alternate sides. Single-walled CNTs with openings at both ends can be formed when the sides are connected. Distortion, folding, and connection on the nanoscale and the microscale can form porous types of carbon such as AC and CDC. Graphene has a lot of synthesis methods, such as tape stripping, CVD, graphite oxide reduction, and so on. From the perspective of mass production, the graphite redox method is a promising and mainstream way of preparing graphene electrode materials for EDLCs. The reduction method has an important effect on the morphology, pore structure, SSA, and electrical conductivity of graphene electrodes. With the development of graphene research, the preparation of three-dimensional porous graphene electrodes has become one of the most important development directions. Such materials include graphene hydrogel/aerogel and graphene foam/sponge. Three-dimensional self-assembled graphene and graphene hierarchical porous materials can be obtained because of innovations in the preparation of porous carbon materials. In addition to the new preparation methods used for electrode materials, EDLCs have also developed from traditional capacitors into microcapacitors, all-solid-state wear-resistant capacitors, and 3D printed capacitors, so this section introduces the latest developments and applications of carbon materials in EDLCs.

Studies of the electrochemical performance of graphene in different electrolytes began in 2008. Taking advantage of the fact that the oxygen-containing functional groups of GO decompose to produce gas at a high temperature (1050 °C), graphene electrode materials with high SSA (925 $m^2\ g^{-1}$) were obtained by the thermal stripping of GO. At 1 M H_2SO_4, specific capacitances of 100–117 F g^{-1} could be generated at sweep speeds of 1–0.01 V s^{-1}, which were comparable to those of OLC and CNT electrode materials. A high specific capacitance (75 F g^{-1}) was obtained

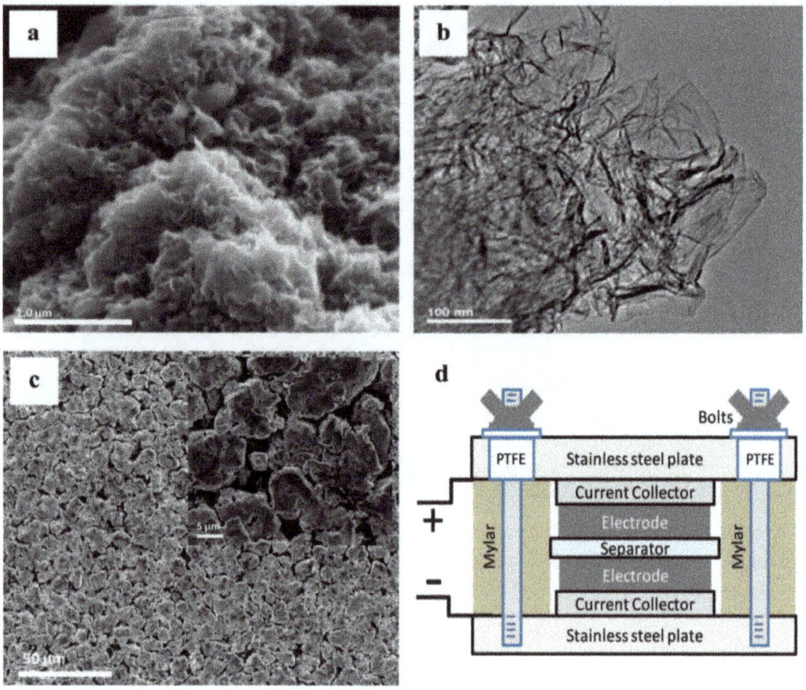

Figure 1.20. (a) An SEM image of a chemically modified graphene (CMG) particle surface, (b) a TEM image showing individual graphene sheets extending from a CMG particle surface, (c) low- and high- (inset) magnification SEM images of a CMG particle electrode surface, and (d) a schematic of a test cell assembly [230], reproduced with permission © American Chemical Society.

even in ionic liquids [229]. Second, 3D porous graphene electrode materials with high SSA (705 m^2 g^{-1}) could also be obtained by reducing a GO aqueous dispersion using hydrazine hydrate (figure 1.20), and a relatively high specific capacitance could also be obtained in aqueous (100–135 F g^{-1}) and organic system electrolytes (90 F g^{-1}) [230]. However, due to their low density, these graphene electrode materials have a lower volume ratio capacitance, which at its lowest is less than the volume performance of AC. A possible reason for this is that the GO water dispersion reunites easily during reduction, forming larger particles. The formation of a porous structure depends on the intergranular pores formed during reunion, which is why the graphene material proportion is higher, the pores are larger, and the microporous and mesoporous slants are reduced; the pores are too big to be reduced, which would restore the bulk density of graphene oxide materials.

Thus, it can be seen that the key to realizing the application of GO is to reduce aggregation in the restoration process. For example, GO was reduced using gas-phase zinc, and it was claimed that a specific capacitance of 205 F g^{-1} could be obtained in 30% KOH [231]. Its SSA was only 320 m^2 g^{-1}, its cyclic stability was poor, but only 170 F g^{-1} of capacitance was lost after 1200 charges and discharges. Analysis showed that the higher specific capacitance than those of other methods might be caused by incomplete GO reduction, resulting in a significant contribution of pseudocapacitance. Another method used to mitigate the agglomeration phenomenon is to change the charge state of the GO surface using the principle of electrostatic repulsion. For example, GO adsorbs p-phenylenediamine with the result that its surface becomes positively charged. Using electrophoretic deposition, GO was deposited on a nickel foam to directly prepare EDLC electrodes, and the electrode thickness could be controlled according to the conditions. This electrode exhibited a high specific capacitance (164 F g^{-1}) in 6 M KOH [232]. Using the principle of π–π interaction, 1-pyrene formic acid was used to nondestructively functionalize GO to stabilize it in an aqueous solution, and the specific capacitance of graphene porous materials generated by reduction reached 120 F g^{-1} [233]. Some novel reduction methods can also efficiently reduce GO, yielding graphene with high electrical conductivity. For example, graphene materials with low aggregation and high specific capacitance can be obtained using *in situ* electrochemistry [234], hydrobromic acid [235], and dimethyl formamide (DMF) solvothermal reduction [236].

In addition to the use of functionalization to improve aggregates, a unique new method for the preparation of porous graphene materials is graphene sheet bending. For example, the bent 3D graphene electrical material designed by Jang *et al* (figure 1.21) had a specific capacitance of 100–250 F g^{-1} [237]. In an organic electrolyte, the operating voltage reached 4.0 V, and its energy density was as much as 87–136 Wh kg^{-1}, which matches the level achieved by nickel—metal hydride batteries. The nitrogen atom doping of graphene is considered to be an effective way of further improving graphene-based electrode materials for supercapacitors. Using the ion-beam treatment method, Kang *et al* were able to add 1.86–2.51 at.% N atoms to the graphene surface and used the contribution of pseudocapacitance to increase the specific capacitance of a graphene electrode material to 250 F g^{-1} with 6 M KOH. Even in TEATFB-based organic electrolytes, the specific capacitance was able to reach 220 F g^{-1}. At the same time, it showed high cyclic stability, with a capacitance retention rate as high as 95% after 10 000 charges and discharges [238]. Analysis has shown that among various nitrogen species, pyridine nitrogen contributes the most to the improvement of capacitance contrast. The specific capacitance of nitrogen-doped graphene-like materials increased to 313 F g^{-1} while maintaining high rate performance and cyclic life [239].

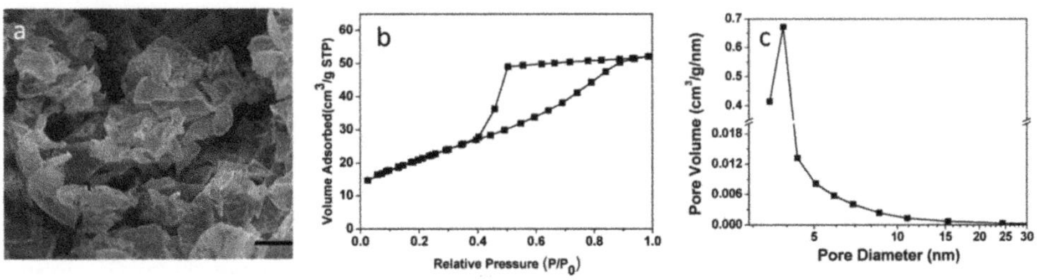

Figure 1.21. (a) An SEM image of 3D curved graphene sheets (scale bar 10 μm); (b) nitrogen adsorption isotherm; and (c) pore size distribution of a mesoporous graphene material [237], reproduced with permission © American Chemical Society.

KOH chemical activation energy was used to further increase the SSA of graphene to 3100 m^2 g^{-1}; the graphene pores mainly composed of 0.6 nm micropores and 4 nm mesopores. This material achieved a specific capacitance of 160 F g^{-1} (60 F cm^{-3}) and a corresponding energy density of 70 Wh kg^{-1} [240]. Kaner *et al* used a cold casting method to convert trace moisture in the GO film into ice, which was used as the template of the porous structure to prepare a three-dimensional porous graphene electrode with ultrahigh power (figure 1.22) [241]. The results of CV characterization showed that its sweep speed could reach 20 V s^{-1}, confirming its ultrahigh power density. Its charge–discharge current density reached 500 A g^{-1}, i.e. its average power density was 282 kW kg^{-1}. The high power density was attributed to the large pore structure of the material, which provided an electrolyte reserve at large currents and reduced the diffusion distance for a large number of electrolyte ions. At the same time, the material exhibited high specific capacitance (284 F g^{-1} at 1.0 A g^{-1}) and good cyclic stability.

A high-rate capacitor electrode was produced through the use of three-dimensional porous nickel foam as a skeleton in the preparation of graphene gel (figure 1.23) [242]. The unique advantage of the electrode was that the three-dimensional porous active material was synthesized directly on the fluid collector, which not only helped to reduce the interfacial contact resistance of the material but also facilitated the diffusion of high-concentration ions at large currents. The specific capacitance of the graphene electrode was greater than 40 mF cm^{-2}, and it maintained a good rectangle at a sweep speed of 2 V s^{-1}. The maximum current density of the discharge reached 281.7 A g^{-1}.

The large-scale synthesis of porous carbon materials on graphene surfaces through a simple air oxidation method not only increased the diffusion rate of electrolyte ions along the Z-axis but also increased the bulk densities of graphene materials (1.2 g cm^{-3}) [243] by 700% compared with those of untreated carbon

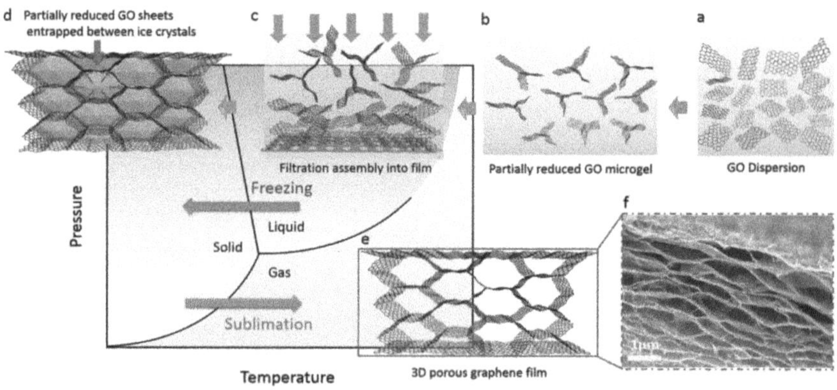

Figure 1.22. Schematic diagrams of the preparation of 3D graphene by cold casting: (a) GO dispersion; (b) partially reduced GO dispersion; (c) filtration and assembly into a wet film; (d) freezing of the wet membrane, using ice crystals as template to form a porous structure; (e) the formation of graphene membranes with three-dimensional holes. (f) A corresponding enlarged SEM image [241], reproduced with permission © John Wiley and Sons.

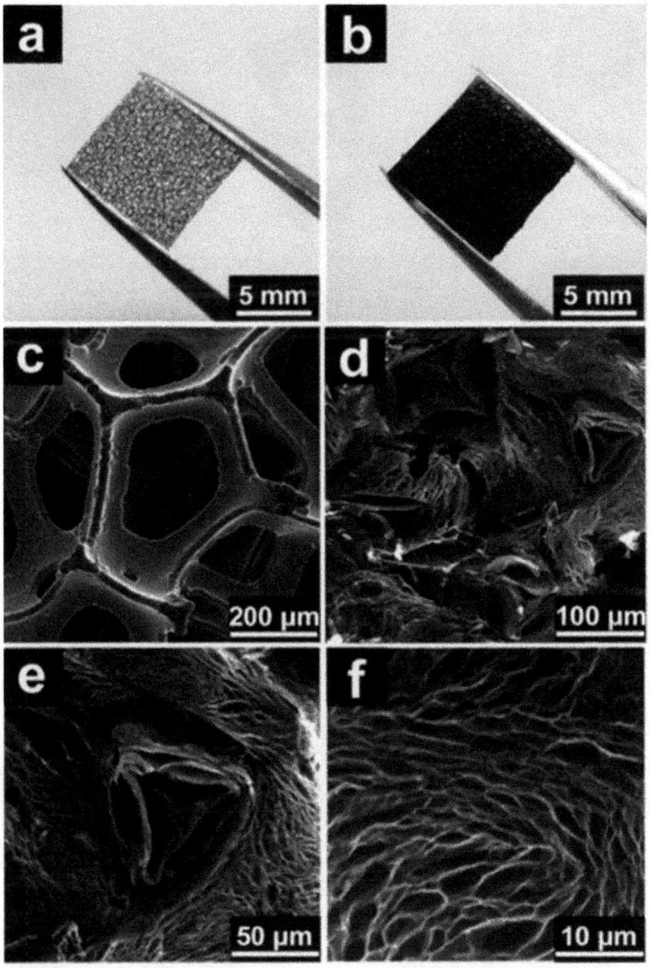

Figure 1.23. Photographs of a piece of NF (a) and an as-formed G-Gel/NF electrode (b). (c) SEM images of a 3D graphene gel prepared directly on 3D nickel foam fluid collector; (d and e) SEM images of the cross-section of the 3D graphene gel; (f) an enlarged 3D SEM image of the graphene on the nickel foam [242], reproduced with permission © John Wiley and Sons.

materials. Duan Jianfeng *et al* prepared 3D graphene materials by hydrogen peroxide oxidation etching and drilling holes and obtained high-density graphene films by compression (figure 1.24). The resulting films had a volume specific capacitance and a specific energy density of 212 F cm^{-3} and 49 Wh l^{-1}, respectively [244]. Of course, the preparation of porous graphene–carbon nanotube sandwich materials should also lead to high volume performance. For example, the volume ratio capacitance of a hybrid film made from open graphene and a carbon tube reached as much as 331 F cm^{-3} [245]. Other hybrid materials made from graphene and carbon fiber or porous carbon have also been used as supercapacitor materials, achieving higher volume performance [246, 247].

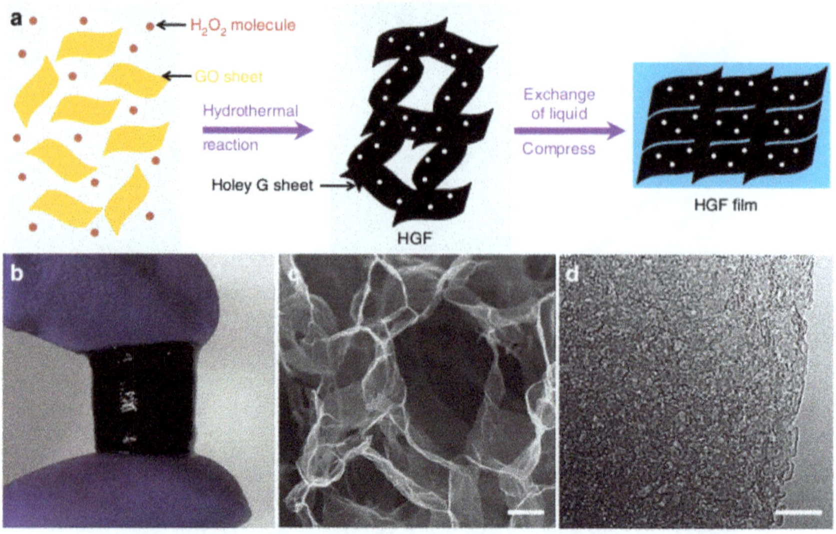

Figure 1.24. (a) The preparation and characterization of porous graphene-based electrodes; (b) a schematic diagram of the preparation route; (c) a photo of an integrated unsupported hole fossil ink electrode; (d) an SEM image of the internal microstructure of the electrode; (e) high-resolution TEM images of graphene sheets with internal holes in the electrode [244], reproduced with permission © Springer Nature.

Figure 1.25. A comparison between the preparation of 3D graphene materials high density porous graphene macroform (HPGM) by the capillary action of solvent evaporation and by freeze-drying [249], reprinted with permission © Elsevier.

Using the graphene–water interface tension method [248], a high-density 3D graphene (1.58 g cm^{-3}) was prepared by controlling the self-evaporation rate of the water content in the graphene self-assembled gel, which achieved a volume ratio of 376 F cm^{-3} (figure 1.25) [249]. Higher-density (1.15 g cm^{-3}) porous graphene materials were obtained by vacuum-assisted self-assembly compared to those obtained using mechanical compression (0.75 g cm^{-3}) [250]. This is because the self-assembly of graphene sheets was more orderly when assisted by a vacuum, which greatly reduced the chance of large holes forming between disordered graphene sheets. In another example, crumpled graphene spheres prepared by a simple spray-drying method had large SSAs and fast ion transport structures and also showed high capacitance performance when used as electrode materials for EDLCs [251]. Other strategies have included template etching/activation of tightly packed graphene and hybridization with other carbon materials. For example, the introduction of $Mg(OH)_2$ nanoparticles prevented the face-to-face stacking of graphene sheets. Due to the low processing temperature used, the resulting graphene still had many stable oxygen-containing groups on it. The contribution of these pseudocapacitors resulted in a volume ratio capacitance of 470 F cm^{-3} [252].

1.3.4.4 Template carbon

Depending on the template morphology used, the preparation methods of template carbon can be divided into the hard template method, the self-template method, and the soft template method [253]. The principle of the hard template method is to deposit carbon material on the surface of the hard template; the template effect endows the carbon material with a specific morphology and microstructure. The significant advantage of this approach is that after the hard template is removed, most of the carbon material retains the form of the template. Common templates include zeolite [254, 255], magnesium oxide [256–259], alumina [206, 260], calcium carbonate [261], etc. Zeolite templates with uniform and ordered microstructures are widely used to prepare micro/mesoporous carbon materials. For the typical preparation process and the typical template morphology (figure 1.26), the carbon layer is first deposited by CVD and then the template is etched to obtain a few layers of ordered mesoporous template carbon [262]. Based on its excellent pore structure, the mass specific capacitance of carbon-based materials was greatly improved.

In fact, due to the mesoscopic ordering of zeolite templates, mesoporous carbon materials can often be obtained. Although the mesopore size is adjustable within a certain range, fewer macropores limit the preparation of ultrahigh-power EDLC

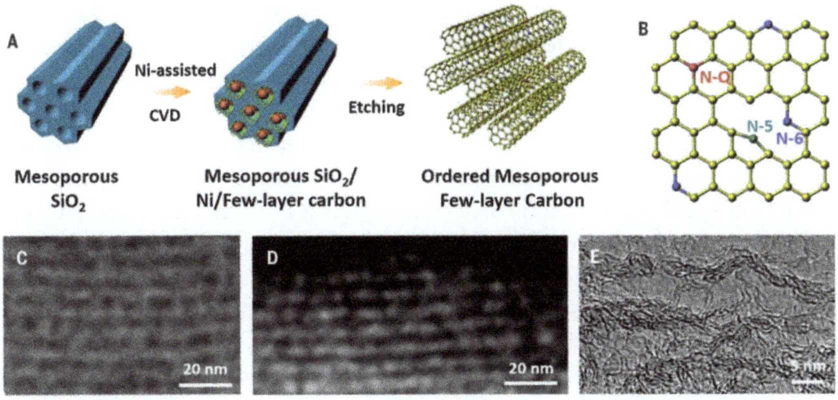

Figure 1.26. The preparation and characterization of ordered mesoporous carbon using a mesoporous silica template. (a) The preparation route used to obtain ordered and less layered mesoporous carbon from a mesoporous silica template; (b) the mesoporous carbon model after nitrogen doping; (c) a TEM image of ordered mesoporous carbon; (d) a TEM image of ordered mesoporous carbon with a few layers; (e) high-resolution TEM images of ordered mesoporous carbon with a few layers [262], reprinted with permission © American Association for the Advancement of Science.

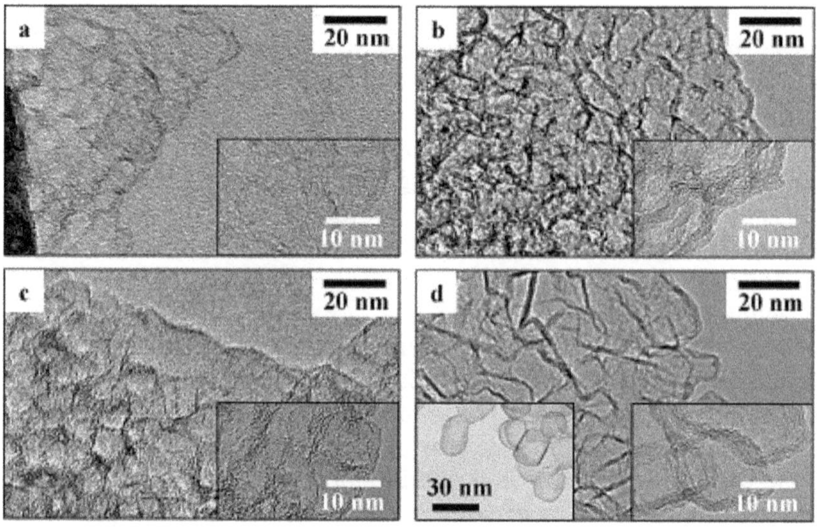

Figure 1.27. TEM characterization of carbon nanocages synthesized at different temperatures. (a) CNC670, (b) CNC700, (c) CNC800, and (d) CNC900 (after the removal of the MgO template) [257], reproduced with permission © John Wiley and Sons.

electrode materials. High SSA carbon nanocages (CNCs) with coexisting microstructure, mesopores, and macropores were prepared using an *in situ* MgO template [257]. As shown in figure 1.27, the micropores mainly originated from cracks larger than the carbon layer on the cage wall, the mesopores originated from the space inside the cage, and the macropores mainly originated from the gap between hierarchical structure sheets. These pores were connected with each other, which was very conducive to the diffusion of electrolytes. In an acidic electrolyte, the specific capacitance of the optimized sample reached 260 F g^{-1}; even at a high current density of 100 A g^{-1}, the specific capacitance was still able to reach 112 F g^{-1}. The high capacitance and rate performance of the CNCs were closely related to the pore structure consisting of connected micropores, mesopores, and macropores.

The self-template method is used to obtain porous carbon with a specific morphology and structure and self-supporting precursors through high-temperature pyrolysis; the most typical use of this method is based on metal–organic framework materials [263–265]. For example, Xu *et al* selected MOF-5 as a template and carbonized it at 530 °C–1000 °C to synthesize porous carbon with an SSA of 1140–3040 m^2 g^{-1} [266]. The sample activated at 600 °C had a high SSA and a narrow mesoporous distribution (3.9 nm), and its mass specific capacitance was greater than 100 F g^{-1} at low scanning speeds. However, the sample with the highest SSA only had a capacity of 12 F g^{-1} after carbonization, which might have been caused by its low conductivity at a low carbonization temperature. The carbonized porous zeolite metal–organic framework (MOF) ZIF-8, which has an SSA of up to 3405 m^2 g^{-1}, was used as the electrode material of a capacitor, and its specific capacitance only reduced from 188 to 160 F g^{-1} when the scanning speed was increased from 5.0 to 50 mV s^{-1} (figure 1.28). The combination of these results with the characterization of its pore structure showed that its specific capacitance mainly

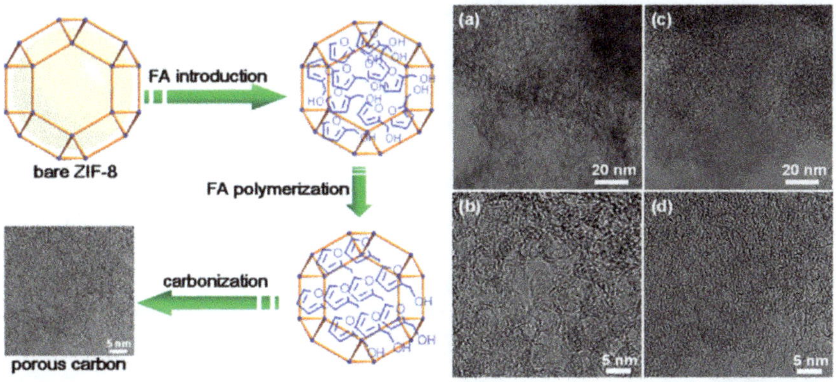

Figure 1.28. The preparation route and morphology of porous carbon synthesized from ZIF-8. TEM images of samples of (a, b) C800 and (c, d) C1000 at different magnifications [267], reproduced with permission © American Chemical Society.

originated from micropores at low scanning speeds, while it mainly originated from mesopores and macropores at high scanning speeds [267]. In fact, although traditional MOF materials have high SSAs and uniformly interconnected pore structures, their conductivity is inferior and they cannot be directly used as SC electrode materials, which is the main reason for the carbonization of the MOF.

The soft template method is mainly used to promote self-assembly with a regular pore structure via a specific interaction between materials and to obtain porous carbon materials after carbonization [268–270]. Compared to the other two methods, this method has poor controllability of the pore structure, results in a low SSA, and requires an activator for secondary activation. The low SSA is mainly caused by fewer micropores and more macropores; the micropores are conducive to improving capacitance, while the macropores are conducive to improving the penetration of the electrolyte. Porous carbon synthesized by the soft template method often has large pores and high rate performance. For example, phenolic resin was used as the carbon source to prepare 3D porous carbon materials with macroporous structures. Macropores function as electrolyte reservoirs and good graphitized pore walls, accelerating the transmission of ions and electrons, which also results in high energy and power density when these materials are used as electrode materials for EDLCs [271]. In a study that used the block copolymer F127 as a soft template, the prepared mesoporous carbon material with a regular pore structure and a high graphitization degree had high-speed performance in an organic electrolyte, but its specific capacity was about 50 F g^{-1} [269]. High-performance N-doped mesoporous carbon was easily prepared using a mixed template of soluble phenolic resin and F127. At a low current density, the specific capacitance of the N-doped carbon material in an acid electrolyte was 262 F g^{-1} (figure 1.29) [270].

In general, the template method is an effective way of preparing 3D carbon materials with controllable pore structures. It provides a good platform with which to study the dependence of the transmission of ions and electrons on pore structure, size distribution, channel shape, and other parameters. High-capacitance carbon can

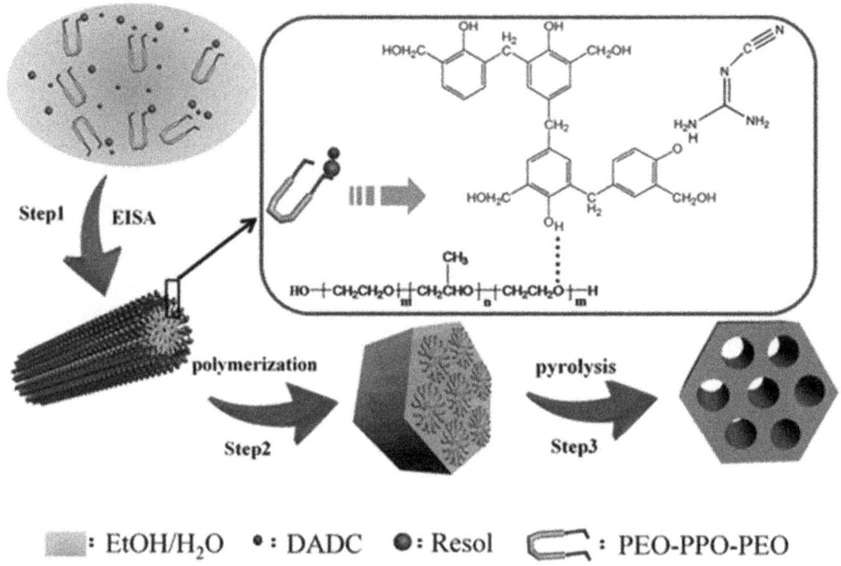

Figure 1.29. The formation process of ordered N-doped mesoporous carbon in a one-pot assembly method using dicyandiamide (DCDA) as a nitrogen source [270], reproduced with permission © John Wiley and Sons.

be obtained by regulating and optimizing the template, carbon source, and preparation parameters, but this approach has the disadvantages of high synthesis cost, low yield, and pollutant emission.

1.4 Conclusions and outlook

This chapter provided general information about the electrodes and structural configurations of EDLCs, which have been recognized as an important energy storage technology. In particular, the recent progress in carbon materials with different dimensions (zero, one, two, and three) was comprehensively summarized. Carbon nanomaterials have various merits of large surface areas, highly porous structures, good conductivities, and useful mechanical properties; as a result, they are promising candidates for use as EDLC electrodes. Moreover, surface modification and dopant-based engineering of carbon nanomaterials can significantly enhance the overall performance of EDLCs. Although great efforts have been devoted to developing carbon-based electrodes for high-performance EDLCs, some limitations still hinder their potential applications in future energy storage devices. The increasing market for electrochemical energy storage will further stimulate the study and development of EDLCs. In our opinion, future research into carbon-based EDLCs can be carried out in the following areas:

(I) EDLCs generally possess high power densities and long cyclic lives, however, they are still limited by the low energy densities caused by their relatively low specific capacitances and narrow operating voltages. To date, most EDLCs have been operated within a potential window of less than 2 V. Thus, novel electrolyte designs are urgently required in order to expand the cell voltage. Moreover, the surface area, porous structure, particle size, conductivity, surface functionalities, and available activated sites of carbon-based electrodes need to be optimized, as these ultimately determine the capacitance. In addition, optimizing the mass ratios of different electrodes is crucial to improve the energy density of EDLCs.

(II) Carbon-based electrodes still suffer from high charge transfer resistance caused by poor electrical conductivity and structural damage that occurs during cycling. Thus, the rational design of nanostructured carbon materials can promote electrochemical kinetics and reduce charge transfer resistance. For example, (i) the production of porous carbon that has a uniform pore size distribution would reduce the ion transport length, while a suitable pore size would improve contact with the electrolyte and ion migration; (ii) heteroatomic doping (B, N, F, P, and S) is beneficial to improving the chemical reactivity, electrical conductivity, and stability of carbon materials, while also providing an opportunity to enhance the rate and cyclic performance of EDLCs; (iii) combining carbon nanomaterials with different dimensions can simultaneously improve electrical conductivity and protect electrodes from structural damage and dissolution in the electrolyte.

(III) Flexible EDLCs are urgently required to support the expansion of the electronic product market. New flexible packaging and electrodes should be implemented to meet the needs of flexible and wearable technology. The development of thinner, moisture-impermeable, and flexible packages that prevent the electrolyte from flowing out of energy storage systems would help to increase their volumetric energy density. More importantly, active electrode materials with high electronic conductivity are required to eliminate current collectors. The use of conductive 1D and 2D carbon nanomaterials could eliminate the need for resistive polymer binders. However, it is still very challenging to guarantee the electrochemical performance of flexible EDLCs for practical use.

(IV) Although great progress has been achieved in the development of EDLCs, these devices still exhibit poor performance in some extreme environments, such as high temperatures, low temperatures, strong acids, and strong bases. The development of novel electrodes, electrolytes, separators, and binders is necessary to meet the requirements of harsh conditions. Additionally, the use of molecular design to obtain high-performance electrolytes, separators, and binders could be a suitable path in the future. For example, ionic liquids and antifreeze agents could be good options for electrolytes in ultralow temperatures, while high-performance polymers such as polyimides (PI) could be the ideal candidate for use in separators and binders.

(V) There are still challenges for the industrialization of novel carbon-based electrode materials due to the high cost and energy consumption of the carbonization and activation processes. Therefore, it is important to develop new technologies for the industrial production of carbon electrode materials, which will benefit industry and society.

Acknowledgements

The financial support of this work by the China Postdoctoral Science Foundation (2022M711381) and the National Natural Science Foundation of China (21706103, 22075109, and 22250410272) is gratefully acknowledged.

References

[1] Goodenough J B and Park K-S 2013 The Li-ion rechargeable battery: a perspective *J. Am. Chem. Soc.* **135** 1167–76
[2] Yoshino A 2012 The birth of the lithium-ion battery *Angew. Chem. Int. Edit.* **51** 5798–800
[3] El-Khodary S A, Abomohra A E-F, El-Enany G M, Aboalhassan A A, Ng D H L, Wang S and Lian J 2019 Sonochemical assisted fabrication of 3D hierarchical porous carbon for high-performance symmetric supercapacitor *Ultrason. Sonochem.* **58** 104617
[4] El-Kady M F, Strong V, Dubin S and Kaner R B 2012 Laser scribing of high-performance and flexible graphene-based electrochemical capacitors *Science* **335** 1326–30
[5] El-Khodary S A, Subburam G, Zou B-B, Wang J, Qiu J-X, Liu X-H, Ng D H L, Wang S and Lian J-B 2021 Mesoporous silica anchored on reduced graphene oxide nanocomposite as anode for superior lithium-ion capacitor *Rare Metals* **41** 368–77
[6] Ramachandran K, El-Khodary S A, Subburam G, Cui Y, Li S, Li J, Wang J, Liu X, Lian J and Li H 2021 Optimizing the microstructure of carbon nano-honeycombs for high-energy sodium-ion capacitor *Electrochim. Acta* **403** 139675
[7] Staffell I, Scamman D, Abad A V, Balcombe P, Dodds P E, Ekins P, Shah N and Ward K R 2019 The role of hydrogen and fuel cells in the global energy system *Energy Environ. Sci.* **12** 463–91
[8] Jamesh M I, Moni P, Prakash A and Harb M 2021 ORR/OER activity and zinc–air battery performance of various kinds of graphene-based air catalysts *Mater. Sci. Energy Technol.* **4** 1–22
[9] Sun W, Wang F, Zhang B, Zhang M, Küpers V, Ji X, Theile C, Bieker P, Xu K, Wang C and Winter M *et al* 2021 A rechargeable zinc–air battery based on zinc peroxide chemistry *Science* **371** 46–51
[10] Imanishi N and Yamamoto O 2019 Perspectives and challenges of rechargeable lithium–air batteries *Mater. Today Adv.* **4** 100031
[11] Simon P and Gogotsi Y 2010 Charge storage mechanism in nanoporous carbons and its consequence for electrical double layer capacitors *Phil. Trans. R. Soc.* A **368** 3457–67
[12] Shao Y, El-Kady M F, Sun J, Li Y, Zhang Q, Zhu M, Wang H, Dunn B and Kaner R B 2018 Design and mechanisms of asymmetric supercapacitors *Chem. Rev.* **118** 9233–80
[13] Lukatskaya M R, Kota S, Lin Z, Zhao M-Q, Shpigel N, Levi M D, Halim J, Taberna P-L, Barsoum M W, Simon P and Gogotsi Y *et al* 2017 Ultra-high-rate pseudocapacitive energy storage in two-dimensional transition metal carbides *Nat. Energy* **2** 17105
[14] Simon P and Gogotsi Y 2008 Materials for electrochemical capacitors *Nat. Mater.* **7** 845–54
[15] Simon P and Gogotsi Y 2020 Perspectives for electrochemical capacitors and related devices *Nat. Mater.* **19** 1151–63
[16] Jiang Y and Liu J 2019 Definitions of pseudocapacitive materials: a brief review *Energy Environ. Mater.* **2** 30–37
[17] Lee H Y and Goodenough J B 1999 Supercapacitor behavior with KCl electrolyte *J. Solid State Chem.* **144** 220–3

[18] Conway B E, Birss V and Wojtowicz J 1997 The role and utilization of pseudocapacitance for energy storage by supercapacitors *J. Power Sources* **66** 1–14

[19] Augustyn V, Come J, Lowe M A, Kim J W, Taberna P-L, Tolbert S H, Abruña H D, Simon P and Dunn B 2013 High-rate electrochemical energy storage through Li^+ intercalation pseudocapacitance *Nat. Mater.* **12** 518–22

[20] Cook J B, Kim H-S, Yan Y, Ko J S, Robbennolt S, Dunn B and Tolbert S H 2016 Mesoporous MoS_2 as a transition metal dichalcogenide exhibiting pseudocapacitive Li and Na-ion charge storage *Adv. Energy Mater* **6** 1501937

[21] Kim J W, Augustyn V and Dunn B 2012 The effect of crystallinity on the rapid pseudocapacitive response of Nb_2O_5 *Adv. Energy Mater.* **2** 141–8

[22] Liu Y, Jiang S P and Shao Z 2020 Intercalation pseudocapacitance in electrochemical energy storage: recent advances in fundamental understanding and materials development *Mater. Today Adv.* **7** 100072

[23] Augustyn V, Simon P and Dunn B 2014 Pseudocapacitive oxide materials for high-rate electrochemical energy storage *Energy Environ. Sci.* **7** 1597–614

[24] Lukatskaya M R, Dunn B and Gogotsi Y 2016 Multidimensional materials and device architectures for future hybrid energy storage *Nat. Commun.* **7** 12647

[25] Wang J, Polleux J, Lim J and Dunn B 2007 Pseudocapacitive contributions to electrochemical energy storage in TiO_2 (anatase) nanoparticles *J. Phys. Chem.* C **111** 14925–31

[26] Brezesinski K, Wang J, Haetge J, Reitz C, Steinmueller S O, Tolbert S H, Smarsly B M, Dunn B and Brezesinski T 2010 Pseudocapacitive contributions to charge storage in highly ordered mesoporous group V transition metal oxides with iso-oriented layered nanocrystalline domains *J. Am. Chem. Soc.* **132** 6982–90

[27] Eftekhari A and Fan Z 2017 Ordered mesoporous carbon and its applications for electrochemical energy storage and conversion *Mate. Chem. Front.* **1** 1001–27

[28] Zhao Z, Liu S, Zhu J, Xu J, Li L, Huang Z, Zhang C and Liu T 2018 Hierarchical nanostructures of nitrogen-doped porous carbon polyhedrons confined in carbon nanosheets for high-performance supercapacitors *ACS Appl. Mater. Interfaces* **10** 19871–80

[29] Zhang L L and Zhao X S 2009 Carbon-based materials as supercapacitor electrodes *Chem. Soc. Rev.* **38** 2520–31

[30] Mathis T S, Kurra N, Wang X, Pinto D, Simon P and Gogotsi Y 2019 Energy storage data reporting in perspective-guidelines for interpreting the performance of electrochemical energy storage systems *Adv. Energy Mater.* **9** 1902007

[31] El-Kady M F, Shao Y and Kaner R B 2016 Graphene for batteries, supercapacitors and beyond *Nat. Rev. Mater.* **1** 16033

[32] Shao Y, El-Kady M F, Wang L J, Zhang Q, Li Y, Wang H, Mousavi M F and Kaner R B 2015 Graphene-based materials for flexible supercapacitors *Chem. Soc. Rev.* **44** 3639–65

[33] Eftekhari A 2018 The mechanism of ultrafast supercapacitors *J. Mater. Chem.* A **6** 2866–76

[34] Sharma K, Arora A and Tripathi S K 2019 Review of supercapacitors: materials and devices *J. Energy Storage* **21** 801–25

[35] El-Kady M F, Ihns M, Li M, Hwang J Y, Mousavi M F, Chaney L, Lech A T and Kaner R B 2015 Engineering three-dimensional hybrid supercapacitors and microsupercapacitors for high-performance integrated energy storage *Proc. Natl Acad. Sci.* **112** 4233–8

[36] Chang J, Jin M, Yao F, Kim T H, Le V T, Yue H, Gunes F, Li B, Ghosh A and Xie S 2013 Asymmetric supercapacitors based on graphene/MnO_2 nanospheres and graphene/MoO_3 nanosheets with high energy density *Adv. Funct. Mater.* **23** 5074–83

[37] Gogotsi Y and Penner R M 2018 Energy storage in nanomaterials—capacitive, pseudocapacitive, or battery-like? *ACS Nano* **12** 2081–3

[38] Wang X, Mathis T S, Sun Y, Tsai W-Y, Shpigel N, Shao H, Zhang D, Hantanasirisakul K, Malchik F and Balke N 2021 Titanium carbide MXene shows an electrochemical anomaly in water-in-salt electrolytes *ACS Nano* **15** 15274–84

[39] Wang X, Bak S-M, Han M, Shuck C E, McHugh C, Li K, Li J, Tang J and Gogotsi Y 2021 Surface redox pseudocapacitance of partially oxidized titanium carbide MXene in water-in-salt electrolyte *ACS Energy Lett.* **7** 30–5

[40] Zhong C, Deng Y, Hu W, Qiao J, Zhang L and Zhang J 2015 A review of electrolyte materials and compositions for electrochemical supercapacitors *Chem. Soc. Rev.* **44** 7484–539

[41] Duay J, Gillette E, Liu R and Lee S B 2012 Highly flexible pseudocapacitor based on freestanding heterogeneous MnO_2/conductive polymer nanowire arrays *Phys. Chem. Chem. Phys.* **14** 3329–37

[42] Wang X, Li M, Chang Z, Wang Y, Chen B, Zhang L and Wu Y 2015 Orientated Co_3O_4 nanocrystals on MWCNTs as superior battery-type positive electrode material for a hybrid capacitor *J. Electrochem. Soc.* **162** A1966

[43] Alghunaim N S, El-Khodary S A, Ibrahim M and El-Enany G M 2019 Spectroscopic analyses of iron doped protonated polyaniline/graphene oxide system *Spectrochim. Acta* **216** 349–58

[44] El-Khodary S, Yahia I, Zahran H and Ibrahim M 2019 Preparation of polypyrrole-decorated MnO_2/reduced graphene oxide in the presence of multi-walled carbon nanotubes composite for high performance asymmetric supercapacitors *Physica* B **556** 66–74

[45] Galiński M, Lewandowski A and Stępniak I 2006 Ionic liquids as electrolytes *Electrochim. Acta* **51** 5567–80

[46] El-Khodary S A, El-Enany G M, El-Okr M and Ibrahim M 2014 Preparation and characterization of microwave reduced graphite oxide for high-performance supercapacitors *Electrochim. Acta* **150** 269–78

[47] Lee Y-H, Chang K-H and Hu C-C 2013 Differentiate the pseudocapacitance and double-layer capacitance contributions for nitrogen-doped reduced graphene oxide in acidic and alkaline electrolytes *J. Power Sources* **227** 300–8

[48] Wang D W, Fang H T, Li F, Chen Z G, Zhong Q S, Lu G Q and Cheng H M 2008 Aligned titania nanotubes as an intercalation anode material for hybrid electrochemical energy storage *Adv. Funct. Mater.* **18** 3787–93

[49] Bonsoa J S, Rahya A, Perera S D, Nour N, Seitz O, Chabal Y J, Balkus K J Jr, Ferraris J P and Yang D J 2012 Exfoliated graphite nanoplatelets–V_2O_5 nanotube composite electrodes for supercapacitors *J. Power Sources* **203** 227–32

[50] Sivaraman P, Bhattacharrya A R, Mishra S P, Thakur A P, Shashidhara K and Samui A B 2014 Asymmetric supercapacitor containing poly (3-methyl thiophene)-multiwalled carbon nanotubes nanocomposites and activated carbon *Electrochim. Acta* **94** 182–91

[51] Amatucci G G, Badway F, Pasquier A D and Zheng T 2001 An asymmetric hybrid nonaqueous energy storage cell *J. Electrochem. Soc.* **148** A930

[52] Naoi K 2010 'Nanohybrid capacitor': the next generation electrochemical capacitors *Fuel Cells* **10** 825–33

[53] Simon P and Gogotsi Y 2013 Capacitive energy storage in nanostructured carbon–electrolyte systems *Accounts Chem. Res.* **46** 1094–103

[54] Armand M, Endres F, MacFarlane D R, Ohno H and Scrosati B 2011 Ionic-liquid materials for the electrochemical challenges of the future *Nat. Mater.* **8** 129–37

[55] Huddleston J E, Visser A E, Reichert W M, Willauer H D, Broker G A and Rogers R D 2001 Characterization and comparison of hydrophilic and hydrophobic room temperature ionic liquids incorporating the imidazolium cation *Green Chem.* **3** 156–64

[56] Fan Q Y, Wang H Q, Song Y X, Zhang W and Yun S 2020 Five carbon allotropes from Squaroglitter structures *Comp. Mater. Sci.* **178** 109634

[57] Geim A K and Novoselov K S 2007 The rise of graphene *Nat. Mater.* **6** 183–91

[58] Mc-Donough J K and Gogotsi Y 2013 Carbon onions: synthesis and electrochemical applications *Electrochem. Soc. Interface* **22** 61

[59] Pech D, Brunet M, Durou H, Peih H, Mochalin V, Gogotsi Y, Taberna P L and Simon P 2010 Ultrahigh-power micrometre-sized supercapacitors based on onion-like carbon *Nat. Nanotechnol* **5** 651–4

[60] Van Aken K L, Maleski K, Mathis T S, Breslin J P and Gogotsi Y 2017 Processing of onion-like carbon for electrochemical capacitors *ECS J. Solid State Sc* **6** M3103

[61] Mykhailiv O, Zubyk H and Plonska Brzezinska M E 2017 Carbon nano-onions: unique carbon nanostructures with fascinating properties and their potential applications *Inorg. Chim. Acta* **468** 49–66

[62] Van Aken K L, McDonough J K, Song L, Guang F, Chathoth S M, Mamontov E, Fulvio P F, Cummings P T, Dai S and Gogotsi Y 2014 Effect of cation on diffusion coefficient of ionic liquids at onion-like carbon electrodes *J. Phys. Condens. Matter* **26** 284104

[63] Tomita S, Fujii M, Hayashi S and Yamamoto K 2000 Transformation of carbon onions to diamond by low-temperature heat treatment in air *Diam. Relat. Mater.* **9** 856–60

[64] Mykhailiv O, Lapinski A, Molina-Ontoria A, Regulska E, Echegoyen L, Dubis A T and Plonska-Brzezinska M E 2015 Influence of the synthetic conditions on the structural and electrochemical properties of carbon nano-onions *Chem. Phys. Chem.* **16** 2182–91

[65] Portet C, Yushin G and Gogotsi Y 2007 Electrochemical performance of carbon onions, nanodiamonds, carbon black and multiwalled nanotubes in electrical double layer capacitors *Carbon* **45** 2511–8

[66] McDonough J K, Frolov A I, Presser V I, Junj N, Miller C H, Ubieto T, Fedorov M V and Gogotsi Y 2015 Influence of the structure of carbon onions on their electrochemical performance in supercapacitor electrodes *Carbon* **50** 3298–309

[67] Moussa G, Ghimbeu C M, Taberna P-L, Simon P and Vix-Guterl C 2016 Relationship between the carbon nano-onions (CNOs) surface chemistry/defects and their capacitance in aqueous and organic electrolytes *Carbon* **105** 628–37

[68] Gao Y, Zhou Y S, Qian M, He X N, Redepenning J, Goodman P, Li H M, Jiang L and Lu F L 2013 Chemical activation of carbon nano-onions for high-rate supercapacitor electrodes *Carbon* **51** 52–8

[69] Anjos D M, McDonough J K, Perre E, Brown G M, Overbury S H, Gogotsi Y and Presser V 2013 Pseudocapacitance and performance stability of quinone-coated carbon onions *Nano Energy* **2** 702–12

[70] Hu L G, Wang S M, Zhang B Z and Zeng Y W 2006 Structural changes in soot particles induced by diode laser irradiation *Carbon* **44** 1725–9

[71] Inoue A, Seto T and Otani Y 2012 Onion-like carbon nanoparticles generated by multiple laser irradiations on laser-ablated particles *Carbon* **50** 1116–22

[72] Cabioc'h T, Thune E, Riviere J, Camelio S, Girard J, Guerin P, Jaouen M, Henrard L and Lambin P 2002 Structure and properties of carbon onion layers deposited onto various substrates *J. Appl. Phys.* **91** 1560–7

[73] Han F D, Yao B and Bai Y J 2011 Preparation of carbon nano-onions and their application as anode materials for rechargeable lithium-ion batteries *J. Phys. Chem.* C **115** 8923–7
[74] Jin C H, Suenaga K and Iijima S 2009 In situ formation and structure tailoring of carbon onions by high-resolution transmission electron microscopy *J. Phys. Chem.* C **113** 5043–6
[75] Choucair M and Stride J A 2012 The gram-scale synthesis of carbon onions *Carbon* **50** 1109–15
[76] Shen J H, Zhu Y H, Yang X L and Li C Z 2012 Graphene quantum dots: emergent nanolights for bioimaging, sensors, catalysis and photovoltaic devices *Chem. Commun.* **48** 3686–99
[77] Cayuela A, Soriano M, Carrillo-Carrión C and Valcárcel M 2016 Semiconductor and carbon-based fluorescent nanodots: the need for consistency *Chem. Commun.* **52** 1311–26
[78] Zhang S, Sui L N, Dong H Z, He W B, Dong L F and Yu L Y 2018 High-performance supercapacitor of graphene quantum dots with uniform sizes *ACS Appl. Mater. Interfaces* **10** 12983–91
[79] Xiao J F, Momen R and Liu C 2021 Application of carbon quantum dots in supercapacitors: a mini review *Electrochem. Commun.* **132** 107143
[80] Zhang S, Li Y, Song H, Chen X, Zhou J, Hong S and Huang M 2016 Graphene quantum dots as the electrolyte for solid state supercapacitors *Sci. Rep.* **6** 19292
[81] Facure M H, Schneider R, Mercante L A and Correa D S 2020 A review on graphene quantum dots and their nanocomposites: from laboratory synthesis towards agricultural and environmental applications *Environ. Sci. Nano* **7** 3710–34
[82] Li Y, Zhao Y, Cheng H H, Hu Y, Shen G Q, Dai L M and Qu L T 2012 Nitrogen-doped graphene quantum dots with oxygen-rich functional groups *J. Am. Chem. Soc.* **134** 15–8
[83] Liu W W, Feng Y Q, Yan X B, Chen J T and Xue Q J 2013 Superior micro-supercapacitors based on graphene quantum dots *Adv. Funct. Mater.* **23** 4111–22
[84] Lee K, Lee H, Shin Y, Yoon Y, Kimm D and Lee H 2016 Highly transparent and flexible supercapacitors using graphene–graphene quantum dots chelate *Nano Energy* **26** 746–54
[85] Chen Q, Hu Y, Hu C, Cheng H, Zhang Z, Shao and Qu L 2014 Graphene quantum dots–three-dimensional graphene composites for high-performance supercapacitors *Phys. Chem. Chem. Phys.* **16** 19307–13
[86] Li Z, Bu F, Wei J, Yao W, Wang L, Chen Z, Pan D and Wu M 2018 Boosting the energy storage densities of supercapacitors by incorporating N-doped graphene quantum dots into cubic porous carbon *Nanoscale* **10** 22871–83
[87] Niu C M, Sichel E K, Hoch R, Moy D and Tennent H 1997 High power electrochemical capacitors based on carbon nanotube electrodes *Appl. Phys. Lett.* **70** 1480–2
[88] Dahal B, Mukhiya T, Ojha G P, Muthurasu A, Chae S-H, Kim T, Kang D and Kim H Y 2019 In-built fabrication of MOF assimilated B/N co-doped 3D porous carbon nanofiber network as a binder-free electrode for supercapacitors *Electrochim. Acta* **301** 209–19
[89] El-Khodary S A, El-Enany G M, El-Okr M and Ibrahim M 2017 Modified iron doped polyaniline/sulfonated carbon nanotubes for all symmetric solid-state supercapacitor *Synthetic Met* **233** 41–51
[90] Yue Y and Liang H 2015 Hierarchical micro-architectures of electrodes for energy storage *J. Power Sources* **284** 435–45
[91] Andrade N F, Martinez D S T, Paula A J, Silveira J V, Alves O L and Souza Filho A G 2013 Temperature effects on the nitric acid oxidation of industrial grade multiwalled carbon nanotubes *J. Nanopart. Res.* **15** 1761

[92] Zhang Y and Zheng L 2010 Towards chirality-pure carbon nanotubes *Nanoscale* **2** 1919–29
[93] Arora N and Sharma N 2014 Arc discharge synthesis of carbon nanotubes: comprehensive review *Diam. Relat. Mater* **50** 135–50
[94] Magrez A, Seo J W, Smajda R, Mionić M and Forró L 2010 Catalytic CVD synthesis of carbon nanotubes: towards high yield and low temperature growth *Materials* **3** 4871–91
[95] Chrzanowska J, Hoffman J, Małolepszy A, Mazurkiewicz M, Kowalewski T A, Szymanski Z and Stobinski L 2015 Synthesis of carbon nanotubes by the laser ablation method: effect of laser wavelength *Phys. Status Solidi* **252** 1860–7
[96] Segawa Y, Yagi A, Matsui K and Itami K 2016 Design and synthesis of carbon nanotube segments *Angew. Chem. Int. Ed.* **55** 5136–58
[97] Zhu S, Ni J and Li Y 2020 Carbon nanotube-based electrodes for flexible supercapacitors *Nano Res* **13** 1825–41
[98] Yu L, Shearer C and Shapter J 2016 Recent development of carbon nanotube transparent conductive films *Chem. Rev.* **116** 13413–53
[99] Journet C, Picher M and Jourdain V 2012 Carbon nanotube synthesis: from large-scale production to atom-by-atom growth *Nanotechnology* **23** 142001
[100] He R, Pei X, Pan L, Tian L, Luo F, Sui L, Wan Q and Wang J 2013 Effects of ultrasonic radiation intensity on the oxidation of single-walled carbon nanotubes in a mixture of sulfuric and nitric acids *Nano* **8** 1350040
[101] Zhang Y and Xie E 2021 Functionalized and tip-open carbon nanotubes for high-performance symmetric supercapacitors *Dalton Trans* **50** 12982–9
[102] Ragavan R and Pandurangan A 2017 Facile synthesis and supercapacitor performances of nitrogen doped CNTs grown over mesoporous Fe/SBA-15 catalyst *New J. Chem.* **41** 11591–9
[103] Yang Z Y, Wang Y H, Dai Z, Lu Z W, Gu X Y, Zhao H, Sun G Z, Lan W, Zhang Z X and Pan X J 2019 Nature of improved double-layer capacitance by KOH activation on carbon nanotube–carbon nanofiber hierarchical hybrids *Carbon* **146** 610–7
[104] Zhu W, Zhang Y, Zhou X, Xu J, Liu Z, Yuan N and Ding J 2017 Miniaturized stretchable and high-rate linear supercapacitors *Nanoscale Res. Lett.* **12** 448
[105] Zhang D, Miao M, Niu H and Wei Z 2014 Core-spun carbon nanotube yarn supercapacitors for wearable electronic textiles *ACS Nano* **8** 4571–9
[106] Zhang D, Wu Y, Li T, Huang Y, Zhang A and Miao M 2015 High performance carbon nanotube yarn supercapacitors with a surface-oxidized copper current collector *ACS Appl. Mater. Interfaces* **7** 25835–42
[107] Ma W, Song L, Yang R, Zhang T, Zhao Y, Sun L, Ren Y, Liu D, Liu L and Shen J 2007 Directly synthesized strong, highly conducting, transparent single-walled carbon nanotube films *Nano Lett.* **7** 2307–11
[108] Jiang S, Hou P-X, Chen M-L, Wang B-W, Sun D-M, Tang D-M, Jin Q, Guo Q-X, Zhang D-D and Du J-H 2018 Ultrahigh-performance transparent conductive films of carbon-welded isolated single-wall carbon nanotubes *Sci. Adv.* **4** eaap9264
[109] Ruzicka B, Degiorgi L, Gaal R, Thien-Nga L, Bacsa R, Salvetat J-P and Forro L 2000 Optical and dc conductivity study of potassium-doped single-walled carbon nanotube films *Phys. Rev.* **61** R2468
[110] Geng H-Z, Kim K K, So K P, Lee Y S, Chang Y and Lee Y H 2007 Effect of acid treatment on carbon nanotube-based flexible transparent conducting films *J. Am. Chem. Soc.* **129** 7758–9

[111] Mirri F, Ma A W, Hsu T T, Behabtu N, Eichmann S L, Young C C, Tsentalovich D E and Pasquali M 2012 High-performance carbon nanotube transparent conductive films by scalable dip coating *ACS Nano* **6** 9737–44

[112] Feng C, Liu K, Wu J S, Liu L, Cheng J S, Zhang Y, Sun Y, Li Q, Fan S and Jiang K 2010 Flexible, stretchable, transparent conducting films made from superaligned carbon nanotubes *Adv. Funct. Mater.* **20** 885–91

[113] Koo Y, Shanov V N, Yarmolenko S, Schulz M, Sankar J and Yun Y 2015 Inverse-ordered fabrication of free-standing CNT sheets for supercapacitor *Langmuir* **31** 7616–22

[114] Moreira J V S, Corat E J, May P W, Cardoso L D R, Lelis P A and Zanin H 2016 Freestanding aligned multi-walled carbon nanotubes for supercapacitor devices *J. Electron. Mater.* **45** 5781–8

[115] Shi Q Q, Zhan H, Zhang Y and Wang J N 2021 Highly flexible and free-standing carbon nanotube/hollow carbon nanocage hybrid films for high-performance supercapacitors *RSC Adv* **11** 6655–61

[116] Wen Y, Kok M D, Tafoya J P V, Sobrido A B J, Bell E, Gostick J T, Herou S, Schlee P, Titirici M-M and Brett D J 2021 Electrospinning as a route to advanced carbon fibre materials for selected low-temperature electrochemical devices: a review *J. Energy Chem.* **59** 492–529

[117] Kim Y K, Cha S I, Hong S H and Jeong Y J 2012 A new hybrid architecture consisting of highly mesoporous CNT/carbon nanofibers from starch *J. Mater. Chem.* **22** 20554–60

[118] Hérou S, Bailey J J, Kok M, Schlee P, Jervis R, Brett D J, Shearing P R, Ribadeneyra M C and Titirici M 2021 High-density lignin-derived carbon nanofiber supercapacitors with enhanced volumetric energy density *Adv. Sci.* **8** 2100016

[119] Aboalhassan A A, Zhang Y, Zhang H, Yan J, Yu J and Ding B 2021 Facile fabrication of flexible carbon nanofiber electrodes with both high packing density and capacity for Li-ion batteries *Adv. Energy Sustain. Res* **2** 2100020

[120] Chen L-F, Lu Y, Yu L and Lou X W D 2017 Designed formation of hollow particle-based nitrogen-doped carbon nanofibers for high-performance supercapacitors *Energy Environ. Sci.* **10** 1777–83

[121] Zhou X, Liu B, Chen Y, Guo L and Wei G 2020 Carbon nanofiber-based three-dimensional nanomaterials for energy and environmental applications *Mater. Adv* **1** 2163–81

[122] Wang L, Wang J, Ng D H, Li S, Zou B, Cui Y, Liu X, El-Khodary S A, Qiu J and Lian J 2021 Operando mechanistic and dynamic studies of N/P co-doped hard carbon nanofibers for efficient sodium storage *Chem. Commun.* **57** 9610–3

[123] Yan J, Dong K, Zhang Y, Wang X, Aboalhassan A A, Yu J and Ding B 2019 Multifunctional flexible membranes from sponge-like porous carbon nanofibers with high conductivity *Nat. Commun.* **10** 5584

[124] Zhang L, Aboagye A, Kelkar A, Lai C and Fong H 2014 A review: carbon nanofibers from electrospun polyacrylonitrile and their applications *J. Mater. Sci.* **49** 463–80

[125] Aboalhassan A A, Yan J, Zhao Y, Dong K, Wang X, Yu J and Ding B 2019 Self-assembled porous-silica within N-doped carbon nanofibers as ultra-flexible anodes for soft lithium batteries *iScience* **16** 122–32

[126] Barranco V, Lillo-Rodenas M, Linares-Solano A, Oya A, Pico F, Ibañez J, Agullo-Rueda F, Amarilla J M and Rojo J 2010 Amorphous carbon nanofibers and their activated carbon nanofibers as supercapacitor electrodes *J. Phys. Chem.* **114** 10302–7

[127] Yun Y S, Im C, Park H H, Hwang I, Tak Y and Jin H-J 2013 Hierarchically porous carbon nanofibers containing numerous heteroatoms for supercapacitors *J. Power Sources* **234** 285–91

[128] Zhi M, Liu S, Hong Z and Wu N 2014 Electrospun activated carbon nanofibers for supercapacitor electrodes *RSC Adv* **4** 43619–23

[129] Zhao J, Zhu J, Li Y, Wang L, Dong Y, Jiang Z, Fan C, Cao Y, Sheng R and Liu A 2020 Graphene quantum dot reinforced electrospun carbon nanofiber fabrics with high surface area for ultrahigh rate supercapacitors *ACS Appl. Mater. Interfaces* **12** 11669–78

[130] Zhang R, Shen W, Zhong M, Zhang J and Guo S 2020 Carbon nanofibers cross-linked and decorated with graphene quantum dots as binder-free electrodes for flexible supercapacitors *J. Phys. Chem.* C **125** 143–51

[131] Guo Q, Zhou X, Li X, Chen S, Seema A, Greiner A and Hou H 2009 Supercapacitors based on hybrid carbon nanofibers containing multiwalled carbon nanotubes *J. Mater. Chem.* **19** 2810–6

[132] Deng L, Young R J, Kinloch I A, Abdelkader A M, Holmes S M, De Haro-Del Rio D A and Eichhorn S J 2013 Supercapacitance from cellulose and carbon nanotube nanocomposite fibers *ACS Appl. Mater. Interfaces* **5** 9983–90

[133] Cheng L, He J, Jin Y, Chen H and Chen M 2015 Single-walled carbon nanotube embedded porous carbon nanofiber with enhanced electrochemical capacitive performance *Mater. Lett.* **144** 123–6

[134] Zhou Z and Wu X-F 2013 Graphene-beaded carbon nanofibers for use in supercapacitor electrodes: synthesis and electrochemical characterization *J. Power Sources* **222** 410–6

[135] Liu Y, Xu X, Lu T, Sun Z, Chua D H and Pan L 2015 Nitrogen-doped electrospun reduced graphene oxide–carbon nanofiber composite for capacitive deionization *RSC Adv* **5** 34117–24

[136] Hsu H, Wang C and Chang Y 2015 Graphene oxides and carbon nanotubes embedded in polyacrylonitrile-based carbon nanofibers used as electrodes for supercapacitor *J. Phys. Chem. Solids* **85** 62–8

[137] Yu Z, Tetard L, Zhai L and Thomas J 2015 Supercapacitor electrode materials: nanostructures from 0 to 3 dimensions *Energy Environ. Sci.* **8** 702–30

[138] Gogotsi Y 2014 What nano can do for energy storage *ACS Nano* **8** 5369–71

[139] Velasco A, Ryu Y K, Boscá A, Ladrón-de-Guevara A, Hunt E, Zuo J, Pedrós J, Calle F and Martinez J 2021 Recent trends in graphene supercapacitors: from large area to microsupercapacitors *Sustain. Energy Fuels* **5** 1235–54

[140] Su F and Wu Z 2021 A perspective on graphene for supercapacitors: current status and future challenges *J. Energy Chem.* **53** 354–7

[141] Li X, Cai W, An J, Kim S, Nah J, Yang D, Piner R, Velamakanni A, Jung I, Tutuc E, Banerjee S K, Colombo L and Ruoff R S *et al* 2009 Large-area synthesis of high-quality and uniform graphene films on copper foils *Science* **324** 1312–4

[142] Cai M, Outlaw R A, Premathilake D, Butler S M and Miller J R 2014 Vertically oriented graphene nanosheet electric double layer capacitors synthesized from C_2H_2 *ACS Nano* **8** 5873–82

[143] Yang H, Yang J, Zheng B, Zhang S, Yan J and Cen K 2016 Edge effects in vertically-oriented graphene based electric double-layer capacitors *J. Power Sources* **324** 309–16

[144] Faraezi S, Khan M S and Ohba T 2020 Dehydration of cations inducing fast ion transfer and high electrical capacitance performance on graphene electrode in aqueous electrolytes *Ind. Eng. Chem. Res.* **59** 5768–74

[145] Miller J R, Outlaw R A and Holloway B C 2011 Graphene electric double layer capacitor with ultra-high-power performance *Electrochim. Acta* **56** 10443–9
[146] Cai M, Outlaw R A, Butler S M and Miller J R 2012 A high density of vertically-oriented graphenes for use in electric double layer capacitors *Carbon* **50** 5481–8
[147] Jia Z, Wang J, Wang Y, Qi T and Li C 2015 Mechano-chemically reduced macro-mesoporous hierarchical graphene for high-performance electric double-layer capacitor applications *Adv. Electron. Mater* **1** 1500123
[148] Kil H-J, Yun K, Yoo M-E, Kim S and Park J-W 2020 Solution-processed graphene oxide electrode for supercapacitors fabricated using low temperature thermal reduction *RSC Adv.* **10** 22102–11
[149] Du X, Guo P, Song H and Chen X 2010 Graphene nanosheets as electrode material for electric double-layer capacitors *Electrochim. Acta* **55** 4812–9
[150] Mitani S, Sathish M, Rangappa D, Unemoto A, Tomai T and Honma I 2012 Nanographene derived from carbon nanofiber and its application to electric double-layer capacitors *Electrochim. Acta* **68** 146–52
[151] Li J, Ye F, Vaziri S, Muhammed M, Lemme M C and Östling M 2013 Efficient inkjet printing of graphene *Adv. Mater.* **25** 3985–92
[152] Tamgadge R M and Shukla A 2019 A pH-dependent partial electrochemical exfoliation of highly oriented pyrolytic graphite for high areal capacitance electric double layer capacitor electrode *Electrochim. Acta* **325** 134933
[153] Lee N E, Lee J, Jeong H Y, Lee S Y and Cho S O 2022 Vertically aligned graphene prepared by the electrochemical anodization of graphite foil for supercapacitor electrodes *J. Power Sources* **526** 231137
[154] Fu X Y, Chen Z D, Han D D, Zhang Y L, Xia H and Sun H B 2020 Laser fabrication of graphene-based supercapacitors *Photon. Res.* **8** 577–88
[155] Fu X Y, Cai Q, Ma J N, Zhu L, Han D D and Zhang Y L 2021 Free-standing and flexible graphene supercapacitors of high areal capacitance fabricated by laser holography reduction of graphene oxide *Appl. Phys. Lett.* **118** 071601
[156] Syed A W and Mohammad M A 2021 Laser scribed graphene-based flexible micro-supercapacitors with fractal design *IEEE Access* **9** 154957–154964
[157] Bhorkar K, Samartzis N, Athanasiou M, Sygellou L, Boukos N, Dracopoulos V, Ioannides T and Yannopoulos S N 2022 Laser-assisted explosive synthesis and transfer of turbostratic graphene-related materials for energy conversion applications *NPJ 2D Mater. Appl* **6** 56
[158] Li S M, Yang S Y, Wang Y S, Tsai H P, Tien H W, Hsiao S T, Liao W H, Chang C L, Ma C C M and Hu C C 2015 N-doped structures and surface functional groups of reduced graphene oxide and their effect on the electrochemical performance of supercapacitor with organic electrolyte *J. Power Sources* **278** 218–229
[159] Youn H C, Bak S M, Kim M S, Jaye C, Fischer D A, Lee C W, Yang X Q, Roh K C and Kim K B 2015 High-surface-area nitrogen-doped reduced graphene oxide for electric double-layer capacitors *ChemSusChem.* **8** 1875–1884
[160] Haque E, Islam M M, Pourazadi E, Hassan M, Faisal S N, Roy A K, Konstantinov K, Harris A T, Minettc A I and Gomes V G 2015 Nitrogen doped graphene via thermal treatment of composite solid precursors as a high performance supercapacitor *RSC Adv.* **5** 30679–86
[161] Wu Y, Liu X, Xia D, Sun Q, Yu D, Sun S, Liu X, Teng Y, Zhang W and Zhao X 2020 Synthesis of few-layer N-doped graphene from expandable graphite with melamine and its application in supercapacitors *Chinese Chem. Lett.* **31** 559–564

[162] Zhang X and Zhang R 2019 Outstanding long-term cycling stability of a sulfur-doped graphene electrode for supercapacitors obtained by post-tailoring the chemical states of doped-sulfur *Appl. Surf. Sci.* **479** 1039–1047

[163] Thirumal V, Pandurangan A, Jayavel R and Ilangovan R 2016 Synthesis and characterization of boron doped graphene nanosheets for supercapacitor applications *Synthetic Met.* **220** 524–532

[164] Balaji S S, Karnan M, Anandhaganesh P, Tauquir S M and Sathish M 2019 Performance evaluation of B-doped graphene prepared via two different methods in symmetric supercapacitor using various electrolytes *Appl. Surf. Sci.* **491** 560–569

[165] Xia K S, Huang Z Y, Zheng L, Han B, Gao Q, Zhou C G, Wang H Q and Wu J P 2017 Facile and controllable synthesis of N/P co-doped graphene for high-performance supercapacitors *J. Power Sources* **365** 380–388

[166] Samani M N, Shishavan S H, Saman M N, Kim M S, Cho B W, Oh S H, Bozorg S F K and Kim K B 2017 *J. Power Sources* **372** 286–96 Rational hybrid modulation of P, N dual-doped holey graphene for high-performance supercapacitors

[167] Yu X, Pei C G and Feng L G 2019 Surface modulated hierarchical graphene film via sulfur and phosphorus dual-doping for high performance flexible supercapacitors *Chinese Chem. Lett.* **30** 1121–5

[168] Cheng L L, Hu Y Y, Qiao D D, Zhu Y, Wang H and Jiao Z 2018 One-step radiolytic synthesis of heteroatom (N and S) co-doped graphene for supercapacitors *Electrochim. Acta* **259** 587–97

[169] He Y, Zhuang X, Lei C, Lei L, Hou Y, Mai Y and Feng X 2019 Porous carbon nanosheets: synthetic strategies and electrochemical energy related applications *Nano Today* **24** 103–19

[170] He L, Zhang X Q and Lu A H 2017 Two-dimensional carbon-based porous materials: synthesis and applications *Acta Phys.-Chim. Sin* **33** 709–28

[171] Wang J, Xu Y, Ding B, Chang Z, Zhang X, Yamauchi Y and Wu K C W 2018 Confined self-assembly in two-dimensional interlayer space: monolayered mesoporous carbon nanosheets with in-plane orderly arranged mesopores and a highly graphitized framework *Angew Chem. Int. Ed.* **57** 2894–8

[172] Huang B, Liu Y and Xie Z 2021 Two dimensional nanocarbons from biomass and biological molecules: synthetic strategies and energy related applications *J. Energy Chem.* **54** 795–814

[173] Hou J, Cao C, Idrees F and Ma X 2015 Hierarchical porous nitrogen-doped carbon nanosheets derived from silk for ultrahigh-capacity battery anodes and supercapacitors *ACS Nano* **9** 2556–64

[174] Qian W, Sun F, Xu Y, Qiu L, Liu C, Wang S and Yan F 2014 Human hair-derived carbon flakes for electrochemical supercapacitors *Energy Environ. Sci.* **7** 379–86

[175] Zheng X, Luo J, Lv W, Wang D W and Yang Q H 2015 Two-dimensional porous carbon: synthesis and ion-transport properties *Adv. Mater.* **27** 5388–95

[176] Li S, Liu P, Zheng X and Wu M 2022 High-performance dual carbon lithium-ion capacitors based on nitrogen-doped 2D carbon nanosheets as both anode and cathode *Electrochim. Acta* **428** 140921

[177] Taer E, Apriwandi A, Taslim R, Agutino A and Yusra D A 2020 Conversion *Syzygium oleana* leaves biomass waste to porous activated carbon nanosheet for boosting supercapacitor performances *J. Mater. Res. Technol* **9** 13332–40

[178] Taer E, Apriwandi A, Mardiah M A and Taslim R 2022 Synthesis free-template highly micro-mesoporous carbon nanosheet as electrode materials for boosting supercapacitor performances *Int. J. Energy Res* **46** 18740–56
[179] Ling Z, Wang Z, Zhang M, Yu C, Wang G, Dong Y, Liu S, Wang Y and Qiu J 2016 Sustainable synthesis and assembly of biomass-derived B/N co-doped carbon nanosheets with ultrahigh aspect ratio for high-performance supercapacitors *Adv. Funct. Mater.* **26** 111–9
[180] Li Y, Wang G, Wei T, Fan Z and Yan P 2016 Nitrogen and sulfur co-doped porous carbon nanosheets derived from willow catkin for supercapacitors *Nano Energy* **19** 165–75
[181] Jin J, Qiao X, Zhou F, Wu Z S, Cui L and Fan H 2017 Interconnected phosphorus and nitrogen codoped porous exfoliated carbon nanosheets for high-rate supercapacitors *ACS Appl. Mater. Interfaces* **9** 17317–25
[182] Liu R, Wang Y and Wu X 2020 Two-dimensional nitrogen and oxygen co-doping porous carbon nanosheets for high volumetric performance supercapacitors *Micropor. Mesopor. Mater.* **295** 109954
[183] Subramanian V, Luo C, Stephan A, Nahm K, Thomas S and Wei B 2007 Supercapacitors from activated carbon derived from banana fibers *J. Phys. Chem.* C **111** 7527–31
[184] Wei L and Yushin G 2011 Electrical double layer capacitors with activated sucrose-derived carbon electrodes *Carbon* **49** 4830–8
[185] Wei L and Yushin G 2011 Electrical double layer capacitors with sucrose derived carbon electrodes in ionic liquid electrolytes *J. Power Sources* **196** 4072–9
[186] Wei L, Sevilla M, Fuertes A B, Mokaya R and Yushin G 2011 Hydrothermal carbonization of abundant renewable natural organic chemicals for high-performance supercapacitor electrodes *Adv. Energy Mater* **1** 356–61
[187] Balathanigaimani M, Shim W G, Lee M J, Kim C, Lee J W and Moon H 2008 Highly porous electrodes from novel corn grains-based activated carbons for electrical double layer capacitors *Electrochem. Commun.* **10** 868–71
[188] Rufford T E, Hulicova-Jurcakova D, Khosla K, Zhu Z and Lu G Q 2010 Microstructure and electrochemical double-layer capacitance of carbon electrodes prepared by zinc chloride activation of sugar cane bagasse *J. Power Sources* **195** 912–8
[189] Li X, Han C, Chen X and Shi C 2010 Preparation and performance of straw based activated carbon for supercapacitor in non-aqueous electrolytes *Micropor. Mesopor. Mat* **131** 303–9
[190] Zhang J, Gong L, Sun K, Jiang J and Zhang X 2012 Preparation of activated carbon from waste camellia oleifera shell for supercapacitor application *J. Solid State ElectroChem.* **16** 2179–86
[191] Li Z, Zhang L, Amirkhiz B S, Tan X, Xu Z, Wang H, Olsen B C, Holt C and Mitlin D 2012 Carbonized chicken eggshell membranes with 3D architectures as high-performance eectrode materials for supercapacitors *Adv. Energy Mater* **2** 431–7
[192] Chun S E and Whitacre J 2012 The evolution of electrochemical functionality of carbons derived from glucose during pyrolysis and activation *Electrochim. Acta* **60** 392–400
[193] Jänes A, Kurig H and Lust E 2007 Characterisation of activated nanoporous carbon for supercapacitor electrode materials *Carbon* **45** 1226–33
[194] Teng H, Chang Y J and Hsieh C T 2001 Performance of electric double-layer capacitors using carbons prepared from phenol–formaldehyde resins by KOH etching *Carbon* **39** 1981–7

[195] Kim Y J, Horie Y, Ozaki S, Matsuzawa Y, Suezaki H, Kim C, Miyashita N and Endo M 2004 Correlation between the pore and solvated ion size on capacitance uptake of PVDC-based carbons *Carbon* **42** 1491–500

[196] Alonso A, Ruiz V, Blanco C, Santamaria R, Granda M, Menendez R and De Jager S 2006 Activated carbon produced from Sasol-Lurgi gasifier pitch and its application as electrodes in supercapacitors *Carbon* **44** 441–6

[197] Huang C W, Hsieh C T, Kuo P L and Teng H 2012 Electric double layer capacitors based on a composite electrode of activated mesophase pitch and carbon nanotubes *J. Mater. Chem.* **22** 7314–22

[198] Hulicova-Jurcakova D, Puziy A M, Poddubnaya O I, Suárez-García F, Tascón J M and Lu G-Q 2009 Highly stable performance of supercapacitors from phosphorus-enriched carbons *J. Am. Chem. Soc.* **131** 5026–7

[199] Gamby J, Taberna P, Simon P, Fauvarque J and Chesneau M 2001 Studies and characterisations of various activated carbons used for carbon/carbon supercapacitors *J. Power Sources* **101** 109–16

[200] Frackowiak E and Béguin F 2001 Carbon materials for the electrochemical storage of energy in capacitors *Carbon* **39** 937–50

[201] Wei L, Sevilla M, Fuertes A B, Mokaya R and Yushin G 2012 Polypyrrole-derived activated carbons for high-performance electrical double-layer capacitors with ionic liquid electrolyte *Adv. Funct. Mater.* **22** 827–34

[202] Hulicova-Jurcakova D, Seredych M, Lu G Q and Bandosz T J 2009 Combined effect of nitrogen-and oxygen-containing functional groups of microporous activated carbon on its electrochemical performance in supercapacitors *Adv. Funct. Mater.* **19** 438–47

[203] Raymundo-Piñero E, Cadek M and Béguin F 2009 Tuning carbon materials for supercapacitors by direct pyrolysis of seaweeds *Adv. Funct. Mater.* **19** 1032–9

[204] Raymundo-Piñero E, Leroux F and Béguin F 2006 A high-performance carbon for supercapacitors obtained by carbonization of a seaweed biopolymer *Adv. Mater.* **18** 1877–82

[205] Balducci A, Dugas R, Taberna P L, Simon P, Plee D, Mastragostino M and Passerini S 2007 High temperature carbon–carbon supercapacitor using ionic liquid as electrolyte *J. Power Sources* **165** 922–7

[206] Liang Y, Schwab M G, Zhi L, Mugnaioli E, Kolb U, Feng X and Müllen K 2010 Direct access to metal or metal oxide nanocrystals integrated with one-dimensional nanoporous carbons for electrochemical energy storage *J. Am. Chem. Soc.* **132** 15030–7

[207] Shi H 1996 Activated carbons and double layer capacitance *Electrochim. Acta* **41** 1633–9

[208] Qu D and Shi H 1998 Studies of activated carbons used in double-layer capacitors *J. Power Sources* **74** 99–107

[209] Lee S W, Yabuuchi N, Gallant B M, Chen S, Kim B S, Hammond P T and Shao-Horn Y 2010 High-power lithium batteries from functionalized carbon-nanotube electrodes *Nat. Nanotechnol* **5** 531–7

[210] Portet C, Taberna P, Simon P and Flahaut E 2005 Influence of carbon nanotubes addition on carbon–carbon supercapacitor performances in organic electrolyte *J. Power Sources* **139** 371–8

[211] Emmenegger C, Mauron P, Sudan P, Wenger P, Hermann V, Gallay R and Züttel A 2003 Investigation of electrochemical double-layer (ECDL) capacitors electrodes based on carbon nanotubes and activated carbon materials *J. Power Sources* **124** 321–9

[212] Futaba D N, Hata K, Yamada T, Hiraoka T, Hayamizu Y, Kakudate Y, Tanaike O, Hatori H, Yumura M and Iijima S 2006 Shape-engineerable and highly densely packed single-walled carbon nanotubes and their application as super-capacitor electrodes *Nat. Mater.* **5** 987–94

[213] Gao L, Peng A, Wang Z Y, Zhang H, Shi Z, Gu Z, Cao G and Ding B 2008 Growth of aligned carbon nanotube arrays on metallic substrate and its application to supercapacitors *Solid State Commun* **146** 380–3

[214] Barisci J, Wallace G, MacFarlane D and Baughman R 2004 Investigation of ionic liquids as electrolytes for carbon nanotube electrodes *Electrochem. Commun.* **6** 22–7

[215] Chen J, Li W, Wang D, Yang S, Wen J and Ren Z 2002 Electrochemical characterization of carbon nanotubes as electrode in electrochemical double-layer capacitors *Carbon* **40** 1193–7

[216] Shiraishi S, Kurihara H, Okabe K, Hulicova D and Oya A 2002 Electric double layer capacitance of highly pure single-walled carbon nanotubes (HiPco™ Buckytubes™) in propylene carbonate electrolytes *Electrochem. Commun.* **4** 593–8

[217] Ma R, Liang J, Wei B, Zhang B, Xu C and Wu D 1999 Study of electrochemical capacitors utilizing carbon nanotube electrodes *J. Power Sources* **84** 126–9

[218] Lu W, Qu L, Henry K and Dai L 2009 High performance electrochemical capacitors from aligned carbon nanotube electrodes and ionic liquid electrolytes *J. Power Sources* **189** 1270–7

[219] Frackowiak E, Metenier K, Bertagna V and Beguin F 2000 Supercapacitor electrodes from multiwalled carbon nanotubes *Appl. Phys. Lett.* **77** 2421–3

[220] Yoon B J, Jeong S H, Lee K H, Kim H S, Park C G and Han J H 2004 Electrical properties of electrical double layer capacitors with integrated carbon nanotube electrodes *Chem. Phys. Lett.* **388** 170–4

[221] Lee S W, Kim B S, Chen S, Shao-Horn Y and Hammond P T 2008 Layer-by-layer assembly of all carbon nanotube ultrathin films for electrochemical applications *J. Am. Chem. Soc.* **131** 671–9

[222] An K H, Kim W S, Park Y S, Jeong H J, Choi Y C, Moon J M, Bae D J, Lim S C and Lee Y H 2001 Characterization of supercapacitors using singlewalled carbon nanotube electrodes *J. Korean Phys. Soc.* **39** 511–7

[223] Talapatra S, Kar S, Pal S, Vajtai R, Ci L, Victor P, Shaijumon M, Kaur S, Nalamasu O and Ajayan P 2006 Direct growth of aligned carbon nanotubes on bulk metals *Nat. Nanotechnol* **1** 112–6

[224] Du C and Pan N 2006 High power density supercapacitor electrodes of carbon nanotube films by electrophoretic deposition *Nanotechnology* **17** 5314

[225] Barisci J N, Wallace G G and Baughman R H 2000 Electrochemical studies of single-wall carbon nanotubes in aqueous solutions *J. Electroanal. Chem.* **488** 92–8

[226] Zhang H, Cao G, Yang Y and Gu Z 2008 Capacitive performance of an ultralong aligned carbon nanotube electrode in an ionic liquid at 60 °C *Carbon* **46** 30–4

[227] Jiang Q, Qu M, Zhou G, Zhang B and Yu Z 2002 A study of activated carbon nanotubes as electrochemical super capacitors electrode materials *Mater. Lett.* **57** 988–91

[228] Frackowiak E, Delpeux S, Jurewicz K, Szostak K, Cazorla-Amoros D and Beguin F 2002 Enhanced capacitance of carbon nanotubes through chemical activation *Chem. Phys. Lett.* **361** 35–41

[229] Vivekchand S, Rout C S, Subrahmanyam K S, Govindaraj A and Rao C 2008 Graphene-based electrochemical supercapacitors *J. Chem. Sci.* **120** 9–13

[230] Stoller M D, Park S, Zhu Y, An J and Ruoff R S 2008 Graphene-based ultracapacitors *Nano Lett.* **8** 3498–502
[231] Wang Y, Shi Z, Huang Y, Ma Y, Wang C, Chen M and Chen Y 2009 Supercapacitor devices based on graphene materials *J. Phys. Chem.* C **113** 13103–7
[232] Chen Y, Zhang X, Yu P and Ma Y 2010 Electrophoretic deposition of graphene nanosheets on nickel foams for electrochemical capacitors *J. Power Sources* **195** 3031–5
[233] An X, Simmons T, Shah R, Wolfe C, Lewis K M, Washington M, Nayak S K, Talapatra S and Kar S 2010 Stable aqueous dispersions of noncovalently functionalized graphene from graphite and their multifunctional high-performance applications *Nano Lett.* **10** 4295–301
[234] Shao Y, Wang J and Engelhard M 2010 Facile and controllable electrochemical reduction of graphene oxide and its applications *J. Mater. Chem.* **20** 743–8
[235] Chen Y, Zhang X, Zhang D, Yu P and Ma Y 2011 High performance supercapacitors based on reduced graphene oxide in aqueous and ionic liquid electrolytes *Carbon* **49** 573–80
[236] Lin Z, Liu Y, Yao Y, Hildreth O, Li Z, Moon K and Wong C 2011 Superior capacitance of functionalized graphene *J. Phys. Chem.* C **115** 7120–5
[237] Liu C, Zhenning Y, David N, Aruna Z and Bor Z J 2010 Graphene-based supercapacitor with an ultrahigh energy density *Nano Lett.* **10** 4863–8
[238] Jeong H M, Lee J W, Shin W H, Choi Y J, Shin H J, Kang J K and Choi J W 2011 Nitrogen-doped graphene for high-performance ultracapacitors and the importance of nitrogen-doped sites at basal planes *Nano Lett.* **11** 2472–7
[239] Zhao J, Lai H, Lyu Z, Jiang Y, Xie K, Wang X, Wu Q, Yang L, Jin Z and Ma Y 2015 Hydrophilic hierarchical nitrogen-doped carbon nanocages for ultrahigh supercapacitive performance *Adv. Mater.* **27** 3541–5
[240] Zhu Y, Murali S, Stoller M D, Ganesh K J, Cai W, Ferreira P J, Pirkle A, Wallace R M, Cychosz K A, Thommes M, Su D, Stach E A and Ruoff R S *et al* 2011 Carbon-based supercapacitors produced by activation of graphene *Science* **332** 1537–41
[241] Shao Y, El-Kady M F, Lin C W, Zhu G, Marsh K L, Hwang J Y, Zhang Q, Li Y, Wang H and Kaner R B 2016 3D Freeze-casting of cellular graphene films for ultrahigh-power-density supercapacitors *Adv. Mater.* **28** 6719–26
[242] Chen J, Sheng K, Luo P, Li C and Shi G 2012 Graphene hydrogels deposited in nickel foams for high-rate electrochemical capacitors *Adv. Mater.* **24** 4569–73
[243] Han X, Funk M R, Shen F, Chen Y, Li Y, Campbell C J, Dai J, Yang X, Kim J and Liao Y 2014 Scalable holey graphene synthesis and dense electrode fabrication toward high-performance ultracapacitors *ACS Nano* **8** 8255–65
[244] Xu Y, Lin Z, Zhong X, Huang X, Weiss N O, Huang Y and Duan X 2014 Holey graphene frameworks for highly efficient capacitive energy storage *Nat. Commun.* **5** 4554
[245] Jiang L, Sheng L, Long C and Fan Z 2015 Densely packed graphene nanomesh-carbon nanotube hybrid film for ultra-high volumetric performance supercapacitors *Nano Energy* **11** 471–80
[246] Yu D, Goh K, Wang H, Wei L, Jiang W, Zhang Q, Dai L and Chen Y 2014 Scalable synthesis of hierarchically structured carbon nanotube-graphene fibres for capacitive energy storage *Nat. Nanotechnol* **9** 555–62
[247] Yan J, Wang Q, Lin C, Wei T and Fan Z 2014 Interconnected frameworks with a sandwiched porous carbon layer/graphene hybrids for supercapacitors with high gravimetric and volumetric performances *Adv. Energy Mater* **4** 1294–305

[248] Shao J, Lv W and Yang Q 2014 Self-assembly of graphene oxide at interfaces *Adv. Mater.* **26** 5586–612

[249] Tao Y, Xie X, Lv W, Tang D, Kong D, Huang Z, Nishihara H, Ishii T, Li B and Golberg D 2013 Towards ultrahigh volumetric capacitance: graphene derived highly dense but porous carbons for supercapacitors *Sci. Rep.* **3** 2975

[250] Ghaffari M, Zhou Y, Xu H, Lin M, Kim T, Ruoff R and Zhang Q 2013 High-volumetric performance aligned nano-porous microwave exfoliated graphite oxide-based electrochemical capacitors *Adv. Mater.* **25** 4879–85

[251] Luo J, Jang H D and Huang J 2013 Effect of sheet morphology on the scalability of graphene-based ultracapacitors *ACS Nano* **7** 1464–71

[252] Yan J, Wang Q, Wei T, Jiang L, Zhang M, Jing X and Fan Z 2014 Template-assisted low temperature synthesis of functionalized graphene for ultrahigh volumetric performance supercapacitors *ACS Nano* **8** 4720–9

[253] Zhao X S, Su F, Yan Q, Guo W, Bao X Y, Lv L and Zhou Z 2006 Templating methods for preparation of porous structures *J. Mater. Chem.* **16** 637–48

[254] Sang L C, Vinu A and Coppens M O 2011 Ordered mesoporous carbon with tunable, unusually large pore size and well-controlled particle morphology *J. Mater. Chem.* **21** 7410–7

[255] Itoi H, Nishihara H, Kogure T and Kyotani T 2011 Three-dimensionally arrayed and mutually connected 1.2-nm nanopores for high-performance electric double layer capacitor *J. Am. Chem. Soc.* **133** 1165–7

[256] He X, Li R, Qiu J, Xie K, Ling P, Yu M, Zhang X and Zheng M 2012 Synthesis of mesoporous carbons for supercapacitors from coal tar pitch by coupling microwave-assisted KOH activation with a MgO template *Carbon* **50** 4911–21

[257] Xie K, Qin X, Wang X, Wang Y, Tao H, Wu Q, Yang L and Hu Z 2012 Carbon nanocages as supercapacitor electrode materials *Adv. Mater.* **24** 347–52

[258] Morishita T, Tsumura T, Toyoda M, Przepiórski J, Morawski A, Konno H and Inagaki M 2010 A review of the control of pore structure in MgO-templated nanoporous carbons *Carbon* **48** 2690–707

[259] Bu Y, Sun T, Cai Y, Du L, Zhuo O, Yang L, Wu Q, Wang X and Hu Z 2017 Compressing carbon nanocages by capillarity for optimizing porous structures toward ultrahigh-volumetric-performance supercapacitors *Adv. Mater.* **29** 1700470

[260] Xue Y, Ding Y, Niu J, Xia Z, Roy A, Chen H, Qu J, Wang Z L and Dai L 2015 Rationally designed graphene-nanotube 3D architectures with a seamless nodal junction for efficient energy conversion and storage *Sci. Adv* **1** e1400198

[261] Xu B, Peng L, Wang G, Cao G and Wu F 2010 Easy synthesis of mesoporous carbon using nano-$CaCO_3$ as template *Carbon* **48** 2377–80

[262] Lin T, Chen I W, Liu F, Yang C, Bi H, Xu F and Huang F 2015 Nitrogen-doped mesoporous carbon of extraordinary capacitance for electrochemical energy storage *Science* **350** 1508–13

[263] Chaikittisilp W, Hu M, Wang H, Huang H S, Fujita T, Wu K C, Chen L, Yamauchi Y and Ariga K 2012 Nanoporous carbons through direct carbonization of a zeolitic imidazolate framework for supercapacitor electrodes *Chem. Commun.* **48** 7259–61

[264] Salunkhe R, Kamachi Y, Torad N, Hwang S, Sun Z, Dou S, Kim J H and Yamauchi Y 2014 Fabrication of symmetric supercapacitors based on MOF-derived nanoporous carbons *J. Mater. Chem.* A **2** 19848–54

[265] Xu Y, Shi G and Duan X 2015 Self-assembled three-dimensional graphene macro-structures: synthesis and applications in supercapacitors *Acc. Chem. Res.* **48** 1666–75
[266] Liu B, Shioyama H, Jiang H, Zhang X and Xu Q 2010 Metal–organic framework (MOF) as a template for syntheses of nanoporous carbons as electrode materials for supercapacitor *Carbon* **48** 456–63
[267] Jiang H, Liu B, Lan Y Q, Kuratani K, Akita T, Shioyama H, Zong F and Xu Q 2011 From metal–organic framework to nanoporous carbon: toward a very high surface area and hydrogen uptake *J. Am. Chem. Soc.* **133** 11854–7
[268] Li Q, Jiang R, Dou Y, Wu Z, Huang T, Feng D, Yang J, Yu A and Zhao D 2011 Synthesis of mesoporous carbon spheres with a hierarchical pore structure for the electrochemical double-layer capacitor *Carbon* **49** 1248–57
[269] Fulvio P F, Mayes R T, Wang X, Mahurin S M, Bauer J C, Presser V, McDonough J, Gogotsi Y and Dai S 2011 'Brick-and-mortar' self-assembly approach to graphitic mesoporous carbon nanocomposites *Adv. Funct. Mater.* **21** 2208–15
[270] Wei J, Zhou D, Sun Z, Deng Y, Xia Y and Zhao D 2013 A controllable synthesis of rich nitrogen-doped ordered mesoporous carbon for CO_2 capture and supercapacitors *Adv. Funct. Mater.* **23** 2322–8
[271] Weng Z, Li F, Wang D W, Wen L and Cheng H M 2013 Controlled electrochemical charge injection to maximize the energy density of supercapacitors *Angew. Chem. Int. Ed.* **52** 3722–5

Chapter 2

Aqueous electrochemical capacitors based on porous carbon anodes and battery cathodes

Wei Ni and Ling-Ying Shi

Aqueous electrochemical capacitors, also called aqueous hybrid supercapacitors (AHSCs), are safe and eco-friendly. They bridge the gap between aqueous supercapacitors and aqueous batteries, i.e. they offer the advantages of high energy and high power densities along with the long life spans of capacitive electrodes. In this chapter, we review and summarize the recent development of AHSCs which are characterized by the mechanisms of porous carbon anodes and battery cathodes. We also compare and discuss strategies for advanced material preparation and electrode design, especially those based on intercalative compounds. Finally, we present the challenges, future research directions, and perspectives for the future breakthroughs and realistic applications of AHSCs.

2.1 Introduction

The simultaneous combination of high power density, desirable energy density, and superior cyclic stability is an ideal solution to the intrinsic shortcomings of energy storage devices. Among such devices, hybrid supercapacitors (HSCs), sometimes also referred to as asymmetric supercapacitors (ASCs) or battery–supercapacitor hybrid (BSH) devices, which have both capacitive and battery-type electrodes, are of great interest [1–3]. HSCs inherit the advantages of both SCs and batteries and are thus promising for next-generation energy storage devices/systems. As an emerging electrochemical energy storage system, the promising AHSCs or aqueous BSH devices have attracted great interest due to their high safety, low cost, and desirable high energy density and power density, which bridge the gap between electric double-layer capacitors (EDLCs) and batteries [4–7]. For electrode materials and for the cathodes of AHSCs in particular, battery-type materials and/or pseudocapacitive materials play a vital role in dominating the electrochemical performance of AHSCs [8, 9]. The focus of this chapter is 'battery–supercapacitor hybrid devices,' which is

doi:10.1088/978-0-7503-5042-6ch2 © IOP Publishing Ltd 2023

similar to the term 'supercapattery,' formed by combining 'supercapa' (from supercapacitor) and 'ttery' (from battery), which is used to refer to hybrid energy storage systems characterized by high energy and power densities, in which each device is constructed using a battery electrode and a capacitive electrode (see figure 2.1 for a visual representation) [10–15]. In this chapter, we particularly review the progress, mechanisms, and perspectives of the more popular porous carbon anode//battery cathode configuration (i.e. a non-faradic capacitive anode and a faradic cathode) for aqueous HSCs. These battery-type cathodes are mainly intercalative compounds, including $LiMn_2O_4$, $NaMnO_2$, $KMnO_2$, etc. Some pseudocapacitive materials such as (binary/ternary) transition-metal oxides (TMOs) and hydroxides are also briefly referred to for comparison and better understanding.

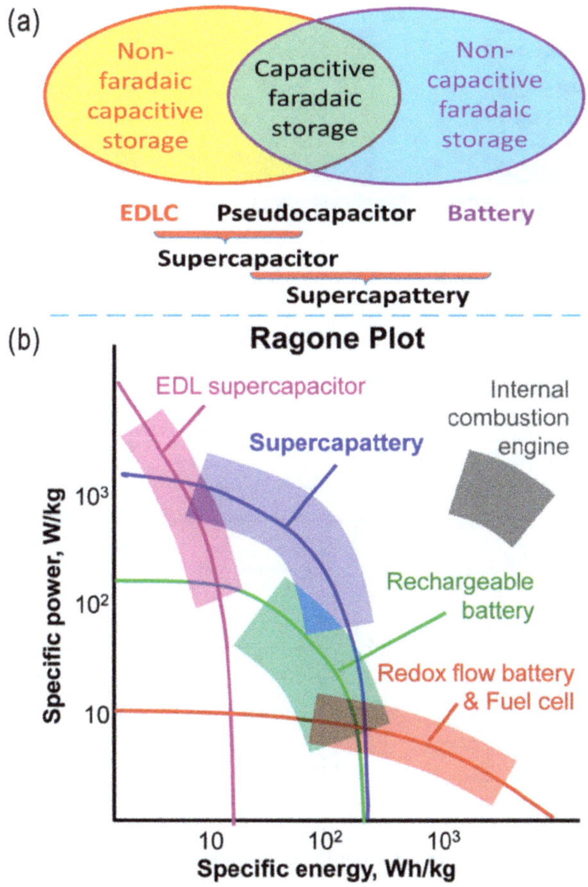

Figure 2.1. (a) A schematic of the relationships between EDLCs, pseudocapacitors, supercapacitors, batteries, and supercapatteries in terms of their capacitive and faradic charge storage processes. *Supercapattery*: a hybrid of a supercapacitor and a battery [15], reproduced with permission © Elsevier. (b) A schematic Ragone plot showing the relative positions (in terms of power density vs energy density) of supercapatteries and other energy technologies [13], reproduced with permission © Taylor & Francis.

2.2 Merits, properties, and mechanisms

AHSCs, which represent a breakthrough in energy storage devices, combine the advantages of EDLCs and batteries, showing a much higher energy density and power density than those of symmetric AC//AC capacitors as well as significantly enhanced rate performance and capacitance retention compared to conventional batteries [16]. The energy storage mechanisms of the electrode materials can be classified as follows: capacitive, pseudocapacitive, and battery-like (see figure 2.2) [9, 17].

Synchronous cyclic chronopotentiometry measurements can facilely predict the cell voltage (or potential difference) based on the potentials of the cathode and anode (vs a reference). These characteristic galvanostatic curves may reveal the distinct charge accumulation mechanisms of each electrode, i.e. the battery-type

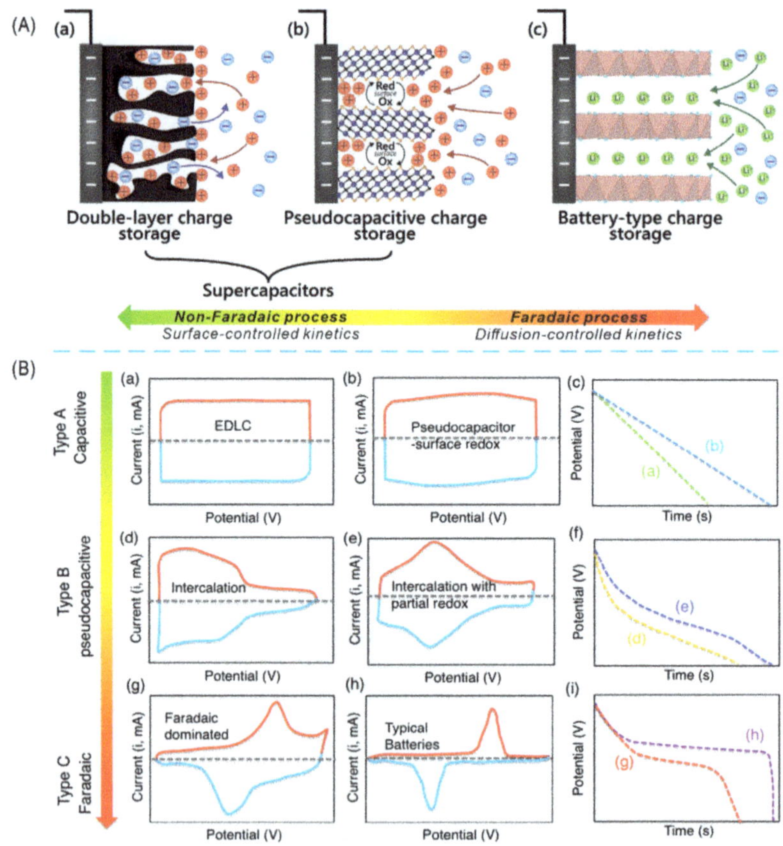

Figure 2.2. (A) An illustration of the conventional classification of electrode processes occurring in (a) electrical double-layer capacitive, (b) pseudocapacitive, and (c) faradic electrodes, as well as the corresponding charge storage mechanisms [14], reproduced with permission © John Wiley & Sons. (**B**) (a, b, d, e, g, and h) Cyclic voltammetry (CV) curves and (c, f, and i) the corresponding galvanostatic discharge curves of different kinds of energy storage material, showing the specific electrochemical characteristics of one or a combination of the abovementioned categories, of which (g–i) correspond to battery-type materials [17], reproduced with permission © American Chemical Society.

cathode has an insertion/de-insertion mechanism with typically two plateaus, while the carbonaceous anode exhibits an EDLC or non-faradic capacitive mechanism with a typical triangular profile (viz. the redox reaction mechanism of battery-type cathodes and the EDL absorption/desorption mechanism of porous carbon anodes) (figure 2.3 *center*) [5, 6, 18, 19]. At high enough current densities, AHSCs exhibit good capacitive behaviors with almost linear galvanostatic charge–discharge (GCD) profiles, i.e. the electrochemical adsorption–desorption process dominates over faradic charge-transfer reactions, although both occur simultaneously at the electrode/electrolyte interfaces at lower current densities [20]. Likewise, due to nanoscale effects, some battery materials have similar behaviors to those of pseudocapacitive materials [6]. Although these alkali-ion insertion materials and their enhanced electrochemical performances obtained via hybridization, compositing, or metal doping are well known, their advantages and characteristics are yet to be further investigated in HSCs based on these insertion-type cathode compounds (figure 2.3 *outer ring*) [6].

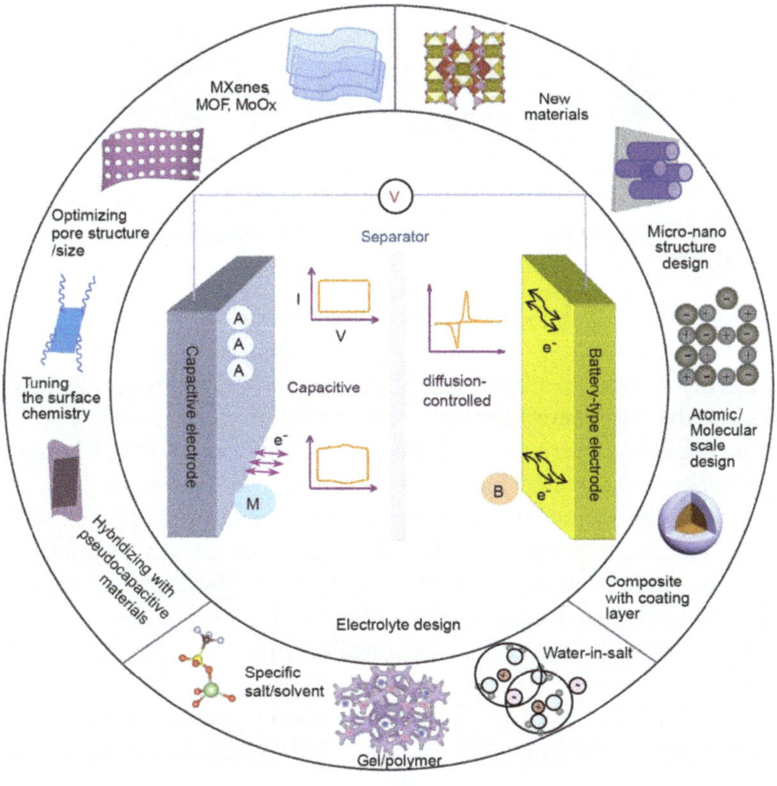

Figure 2.3. A schematic illustration of the device structure, charge storage mechanism, and materials design concepts of AHSCs. 'A (M)' and 'B' are the ions involved in the charge storage processes of capacitive electrodes (adsorption–desorption/(near-)surface faradic reaction mechanism(s)) and battery-type electrodes (bulk redox reaction mechanisms), respectively. The combination of these complementary mechanisms endows AHSCs with both high energy density and high power density [6], reproduced with permission © Springer Nature.

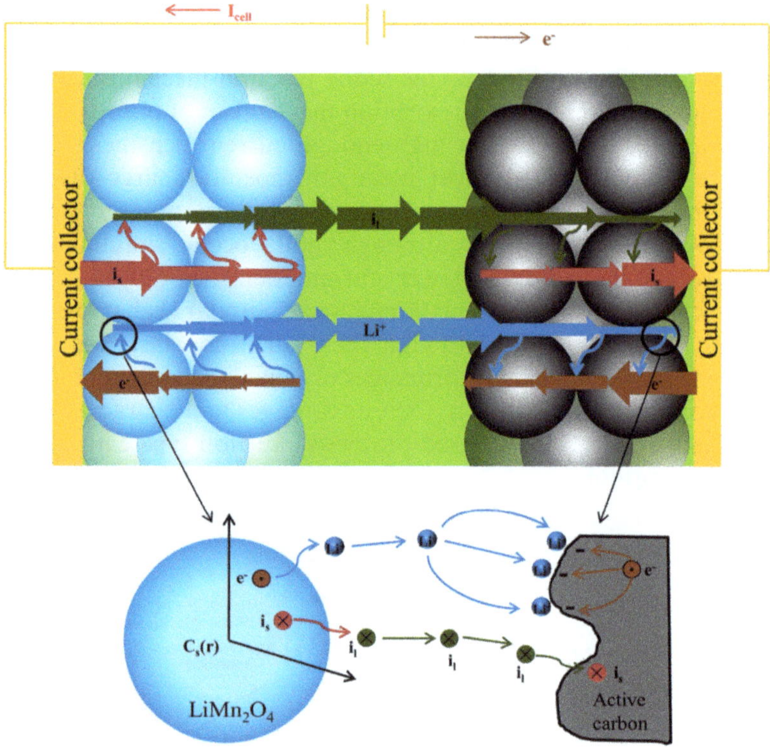

Figure 2.4. A schematic illustration of the charge flow in a typical AHSC. The red and green arrows indicate the current flow/density in the solid and electrolyte phases, respectively, while the blue and brown arrows represent the transfer of ions and electrons, respectively (e.g. Li^+/e^- transfer in a $LiMn_2O_4$ cathode//Li_2SO_4 electrolyte//AC anode full cell) [21], reproduced with permission © Springer Nature.

Taking $LiMn_2O_4$ as an example, in AHSCs, two contributions appear in modeling/simulations, i.e. the EDL (non-faradic or electrostatic) absorption/desorption and the faradic redox reactions (figure 2.4) [21]. It is worth mentioning that although it was not previously a concern, a distinct 'relaxation effect' may appear in AHSCs, even for the AC electrode, i.e. with increasing AC electrode thickness, the specific capacity, voltage drop, and lifetime can deteriorate due to rising resistance [21]. In addition, it should be noted that with a rapid increase in current density, the cathode shows a capacitive-type feature (i.e. an almost linear GCD curve rather than the nonlinear behavior resulting from pseudocapacitive redox/diffusion-controlled processes) [20, 22–24], which has also been noticed in other battery systems [25–29]. Of course, the proportion of battery-type (faradic) versus capacitive (non-faradic) materials in the cathode can be tuned to meet specific energy/power density requirements, but even for battery-type materials, it may still show a small amount of capacitive behavior [30].

2.3 AHSC configurations

AHSCs may be classified into different categories, such as aqueous Li/Na/K-ion HSCs, H/OH-ion HSCs, or Zn-ion HSCs based on the ion transport that takes place in the electrolyte; they may also be classified according to the electrode's chemical composition/structure and device configuration [6, 31–33].

2.3.1 Cathodes

The faradic or pseudocapacitive materials used as the cathodes of HSCs may be divided into the following three types: i.e. alloying/dealloying, conversion reaction, and insertion/de-insertion. Amongst these types, insertion or intercalative compounds (battery-type) are a most promising option, as they have much less volume variation and superior reaction kinetics and most of all, their high and flat voltage plateaus are of great interest for stable voltage output [18]. The typical insertion materials investigated to date for use as cathodes include lithium transition-metal oxides ($LiMn_2O_4$ [20, 21, 34, 35], $LiCoO_2$ [36, 37], $LiNi_{0.5}Mn_{1.5}O_4$, $LiCo_{1/3}Mn_{1/3}Ni_{1/3}O_2$ [38], $Li_4Ti_5O_{12}$ [39, 40], $LiFePO_4$ [41, 42], $LiMnPO_4$ [43, 44], $Li_3V_2(PO_4)_3$ [18]), Na_xMnO_2 [45], and K_xMnO_2 [46–48] as well as some transition-metal oxides such as the classic layered MnO_2. Compared to $LiCo_{1/3}Ni_{1/3}Mn_{1/3}O_4$ or $LiCoO_2$, the alkali and alkaline-earth intercalated layered manganese oxides, which represent a popular kind of insertion compound, have shown superior rate capability and enhanced cyclic life [49]. Usually, these battery-type electrodes are not considered to be pseudocapacitive electrode materials [9, 31], although as a result of nanoscale effects (i.e. highly exposed surfaces/interfaces and significantly shortened ion diffusion lengths compared to those of traditional bulk materials) they are similar in terms of their fast redox kinetics and many other features [6].

Some AHSCs based on pseudocapacitive transition-metal compounds have been studied in the last two decades. These transition-metal chalcogenides, hydroxides, phosphates/phosphides, nitrides, metal–organic frameworks (MOFs), and their hybrids/composites also show similar features, e.g. $CoCr_2O_4$/C and $MnCr_2O_4$/Mn_3O_4/C composites [50], porous $Co_3V_2O_8$ nanosheets supported on carbon cloth [51], S-Co_3O_4@$NiCo_2S_4$ hierarchical structures grown on Ni foam [52], bouquet-like cobalt phosphate [53], Ni_2P lamellas [54], porous VN@C nanobelts [55], flower/nanosheet-like NiCo–layered double hydroxides (LDHs) [56–60], Prussian blue crystals [61], hexacyanoferrate nanosheet arrays or nanoparticles [62–68], and core–shell Ni_3S_2@$Co(OH)_2$ nanowires grown on Ni foam [65]. These materials, which are used in the cathodes of AHSCs, are somewhat similar to typical battery-type materials with an ion insertion/de-insertion mechanism. However, they are not detailed in this chapter because they are better classified as pseudocapacitors (with a reversible surface faradic redox reaction mechanism, i.e. materials that present both EDL absorption/desorption and faradic redox reactions) [5, 9]. The interested reader is referred to the following recent reviews: [3, 5, 7].

2.3.2 Anodes: diverse porous carbons

Carbon materials are an ideal choice due to their high conductivity, low cost, high chemical stability, and environmental friendliness; they can be utilized as either cathode or anode materials in HSCs or ASCs. Porous or activated carbon has been extensively used in HSCs due to its excellent conductivity, superior surface area, and chemical stability [66–68]. Here, we focus on its application in the anodes (i.e. negative electrodes) of HSCs. In addition to porous/activated carbon [20], we also discuss 1D carbon nanotubes/nanofibers [69], 2D carbon nanosheets [70], and graphene [56, 71, 72] with porous morphologies for comparison and better understanding. For example, heteroatom-doped graphene usually endows the pristine graphene with an enriched surface chemistry, controlled electronic properties, and surface wettability [39, 56]. N-doped graphene oxide (NGO) usually displays a high specific capacitance of over 200 F g^{-1} at high rates and excellent cyclic stability with a capacity retention of more than 90% after thousands of cycles [39]. Biomass-derived honeycomb/porous carbon [40, 60], activated carbon [73], graphene-like porous carbon [38, 74], and nitrogen-rich carbon [75], which have been utilized as anode materials in AHSCs, are also being combined [3]; these renewable and sustainable carbon sources are comparable to petroleum and coal sources.

Since the anode materials of porous activated carbon or other forms of carbon usually deliver a moderate specific capacitance of 200–300 F g^{-1} (at 1 A g^{-1}) [37, 39–41, 75, 76], they are not specifically discussed here, as the focus of this chapter is the fabrication and modification of cathode materials. However, the reader should note that the adsorption of some alkali cations or the metal cation doping of AC anodes (e.g. K^+-adsorbed holey carbon) can be used to boost the hydrogen evolution reaction (HER) overpotential and thus increase the voltage window of AHSCs [7, 46].

2.3.3 Electrolytes

The voltages of AHSCs are usually restricted to less than 2 V as a result of the theoretical H_2O electrolysis voltage of 1.23 V, although the appropriate cathode/anode materials and electrolytes can fortunately influence/withstand the O_2/H_2 evolution overpotential [7, 68, 77, 78]. Most critically, the electrolytes themselves essentially determine the potential window for AHSCs. For example, neutral aqueous electrolytes (e.g. Li_2SO_4, Na_2SO_4) have larger and stabler potential windows than those of alkaline or acidic electrolytes (usually up to more than 2.0 V compared to 1.0 V or 0.8 V due to the depressed crucial dehydration energy of ions and the low concentration of H^+/OH^-, i.e. a high overpotential for gas evolution reactions) [8, 79]. Low-pH electrolytes may also induce capacity fading in some lithium transition-metal compounds due to potential proton intercalation [80]. Some alkali ions, such as Na^+ from a neutral aqueous Na_2SO_4 electrolyte, can act as a physical barrier that suppresses the formation of HER intermediates and thus HER activity; the use of high overpotential (viz. energy density) cathode materials such as layered Na_xMnO_2 simultaneously improves the energy density and power density of AHSCs [68]. Thus, by leveraging the overpotentials of the HER and the oxygen evolution reaction (OER), the bottleneck in the potential window can be overcome and an extended operating potential can be achieved [68]. It is worth mentioning that the optimization of the electrolyte–electrode interaction, i.e. the selection of an electrode with low HER/OER activity in a specific electrolyte is also used to raise the voltage output of AHSCs [68, 78]. Overall, electrolyte optimization and electrode modification are two major strategies for increasing the voltage windows of AHSCs beyond 2.0 V via the mechanisms of boosting the HER/OER overpotentials or introducing a higher work function difference (WFD) between the cathode and the anode [7, 77, 81–83].

In addition, reducing or eliminating oxygen in the electrolyte can also enhance the cyclic stability of aqueous Li-ion batteries (ALIBs) [84]; for AHSCs, the impact may be much smaller due to the choice of more robust water-resistant cathode materials; moreover, the elimination of the presence or evolution of O_2 in AHSC systems seems impossible. The construction of a Janus separator (e.g. a modified Nafion membrane) between the electrodes can offer a kind of mixed/hybrid electrolyte, i.e. dual alkali–neutral electrolytes on each side of the separator respectively contribute to the cyclic stability of the cathode and the high energy density of the AHSC [75].

In addition to conventional aqueous electrolytes [16, 20, 21, 39], some novel electrolytes such as ionic liquids (ILs) [18], 'water-in-salt' (WIS) electrolytes [85], and solid-state electrolytes (aqueous solid and gel electrolytes) [64, 86] are also emerging and have attracted increasing attention. Usually, these electrolytes, which have no or much lower water content, endow full cells with relatively higher voltage outputs compared to these of aqueous HSCs although lower than those of HSCs based on organic electrolytes.

2.3.4 Material and electrode design

Rational material/electrode design is essential to the high performance of electrochemical devices. In AHSCs with porous carbon anodes and battery-type cathodes, the cathodes and the corresponding devices have greater tunability [31]. Materials design and control methods include hydrothermal methods [37, 39, 47, 87], stoichiometric solid-state reactions (calcination/annealing) [20, 50, 88], electrospinning [89], ball milling, etc. while some material syntheses use integrative methods to tailor the related structures in an extended way [90], e.g. hydrothermal/calcination methods for nanotube/nanorod structures [16], ball milling/calcination for porous microparticles [45], solution synthesis/electrochemical oxidation for nanosheet structures [46].

In addition to the use of common aqueous HSCs (or ASCs) based on classic porous carbon capacitive materials [91], the incorporation of some components with faradic pseudocapacitance, e.g. metal-free π-conjugated compounds [92], conductive polymers [93], transition-metal oxides [4, 94], carbides [95], or their composites can further enhance the energy density of AHSCs. These typical examples may be detailed in other chapters.

2.4 Li-ion hybrid supercapacitors

Among the various battery-type electrodes, lithium intercalation/deintercalation materials, e.g. $LiMn_2O_4$ and $Li_4Ti_5O_{12}$, have attracted increasing interest for uses beyond traditional Li-ion batteries [80, 96]. Although these Li-ion insertion materials and their enhanced electrochemical performances via hybridization, compositing, or metal doping are well known, their advantages and characteristics should be further investigated in AHSCs based on insertion-type cathode compounds.

Lithium titanate (i.e. lithium titanium oxide, LTO, $Li_4Ti_5O_{12}$, which has a 'zero-strain' spinel structure) is an emerging electrode material of exceptional electrochemical stability, which is not only utilized in conventional organic electrolyte-based Li-ion batteries/HSCs but is also attracting increasing attention for use in aqueous HSCs [97–102]. Xing *et al* synthesized a unique kind of $Li_4Ti_5O_{12}$ hollow sphere with a chestnut-like shell via a facile hydrothermal strategy [39]. The specific hollow structural LTO spheres showed an outstanding capacitance of 653 F g^{-1} and a good retention of 88.6% over 4000 cycles (at a current density of 1 A g^{-1}). When the spheres were configured with porous N-doped graphene oxide (NGO) to form an anode for an aqueous ASC, the resulting ASC showed good specific capacitances of 73.6 F g^{-1} at 1 A g^{-1} and 40.6 F g^{-1} at 10 A g^{-1}, as well as a high energy density of 26.2 Wh kg^{-1} (at a power density of 800 W kg^{-1}); when the power density increased by 10 times, i.e. to 8.0 kW kg^{-1}, the energy density still remained at a high level of 14.5 Wh kg^{-1} (figure 2.5). Similar systems, such as a flower-shaped LTO-graphene hybrid as the cathode and biomass (pine needle)-derived honeycomb carbon as the anode have also been used and showed similar but better performance [40].

$LiMn_2O_4$ is another lithium transition-metal oxide with high capacity and a 3D tunnel structure that allows facile ion diffusion and good cyclic stability [20, 34, 87]; compared to other Li-ion intercalated compounds, such as $LiCo_{1/3}Ni_{1/3}Mn_{1/3}O_2$ and $LiCoO_2$, it exhibits simultaneous superior cyclic stability and rate capability [80]. Elaborately prepared porous (and hollow) microspheres of $LiMn_2O_4$ can be obtained via a reactive template ($MnCO_3$ microsphere)-assisted solid-state thermal reaction

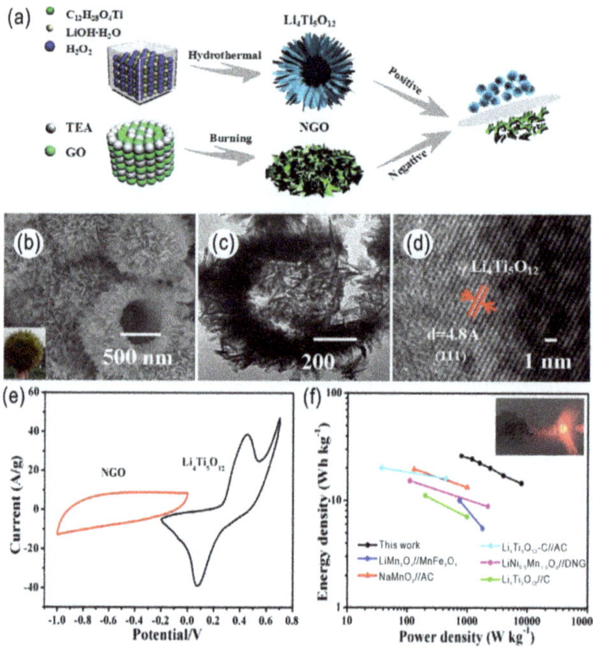

Figure 2.5. (a) A schematic illustration of an AHSC based on hollow $Li_4Ti_5O_{12}$ and N-doped graphene oxide (NGO). (b–d) Scanning electron microscopy (SEM), transmission electron microsocopy (TEM) and high-resolution transmission electron microsocopy (HRTEM) images of the $Li_4Ti_5O_{12}$ hollow spheres. (e) A plot of the CV results for $Li_4Ti_5O_{12}$ and NGO at 50 mV s^{-1}. (f) A Ragone plot (energy density vs power density) of the $Li_4Ti_5O_{12}$//NGO AHSC compared with other typical AHSC devices [39], reproduced with permission © Elsevier.

(followed by selective HCl etching to obtain hollow structures) [20]. When configured with activated carbon (AC), the AC//porous $LiMn_2O_4$ microsphere HSC devices with an aqueous electrolyte exhibited a high specific energy of 33.1 Wh kg^{-1} at a power density of 90 W kg^{-1}, which remained at 25.0 Wh kg^{-1} at 900 W kg^{-1}, which was slightly higher than those of their hollow-structured counterparts and much higher than the energy/power densities of EDLCs. Similar performances can also be achieved through the use of other nanostructures, e.g. a $LiMn_2O_4$ nanohybrid consisting of nanotubes, nanorods, and nanoparticles prepared from a hydrothermally synthesized α-MnO_2 nanotube precursor that was subsequently calcinated (figure 2.6) [16]. AHSCs

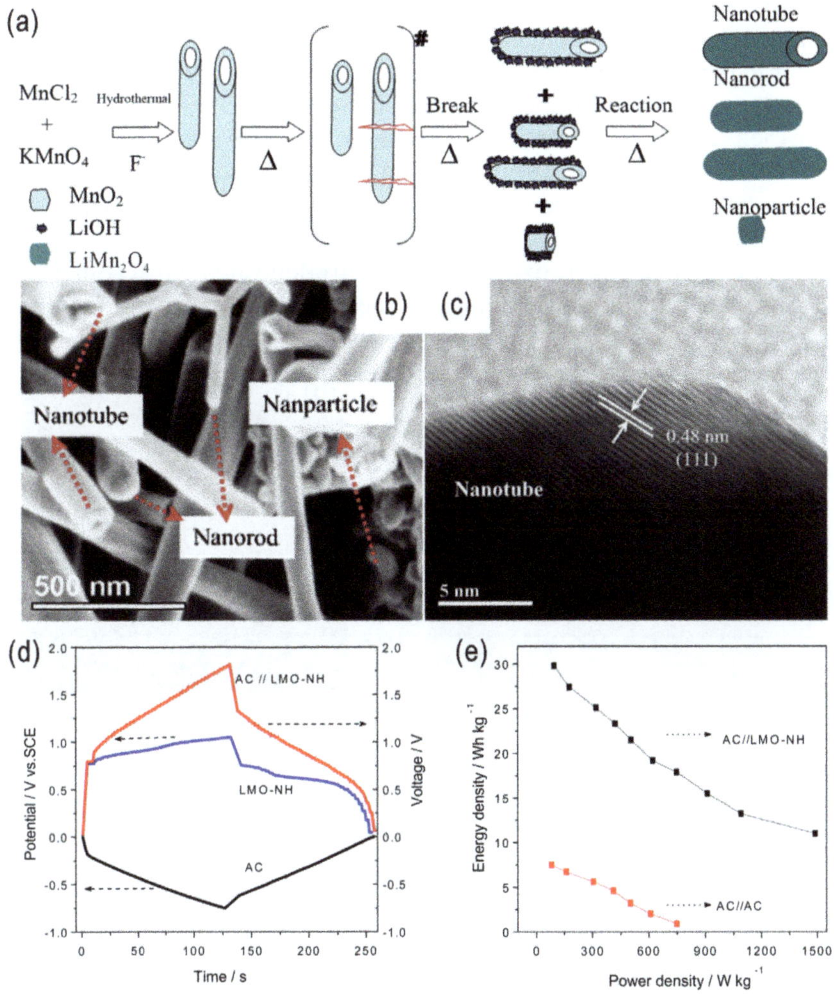

Figure 2.6. (a) A schematic illustration of the synthesis of the $LiMn_2O_4$ nanohybrid (LMO-NH) and (b and c) the corresponding field emission scanning electron microscopy (FESEM) and HRTEM images. (d) Potential–time curves of the individual electrodes (vs the saturated calomel electrode (SCE)) and the voltage–time profile of the AC//LMO-NH AHSC (at a current density of 3 A g^{-1}). (e) Ragone plot for the AC//LMO-NH AHSC compared to the Ragone plot of an AC//AC EDLC [16], reproduced with permission © Elsevier.

are somewhat similar to the previously reported aqueous rechargeable lithium batteries (ARLBs, first referred to in 1994) or post-lithium aqueous batteries [103, 104], which usually have the same configuration but are rated by specific capacity rather than by capacitance. For example, porous $LiMn_2O_4$ was prepared using a polystyrene (PS) colloidal template; following calcination, outstanding rate performance and a superior capacity retention of 93% were obtained after 10 000 cycles in a 0.5 M Li_2SO_4 aqueous electrolyte (figure 2.7) [105]. It is interesting to note that an AHSC reached an even higher voltage output of up to 2.3 V and a superior energy density of 50 Wh kg^{-1} at a power density of 571 W kg^{-1} via a double electrolyte separated by a K^+ exchange membrane. The specific alkaline($KOH+K_2SO_4$)-neutral ($Li_2SO_4+K_2SO_4$) electrolyte took full advantage of electrolyte–electrode reactions and exhibited an energy density about twice that of common AHSCs. Overall, considering its capacitance, performance, cost-effectiveness, and environmental issues, $LiMn_2O_4$ is one of the most promising cathode materials for AHSCs.

Lithium-rich $Li_4Mn_5O_{12}$, usually used as a cathode material in lithium-ion batteries (LIBs) with organic electrolytes, may also demonstrate a high potential or a large voltage window even in aqueous electrolytes and AHSCs. Zhao *et al* designed a carbon cloth that supported $Li_4Mn_5O_{12}$ nanoflakes that was synthesized via the *in situ* electrochemical oxidation of electrodeposited Mn_3O_4 nanowalls [8]. The $Li_4Mn_5O_{12}$ electrode showed higher specific capacitance (627 F g^{-1}) than

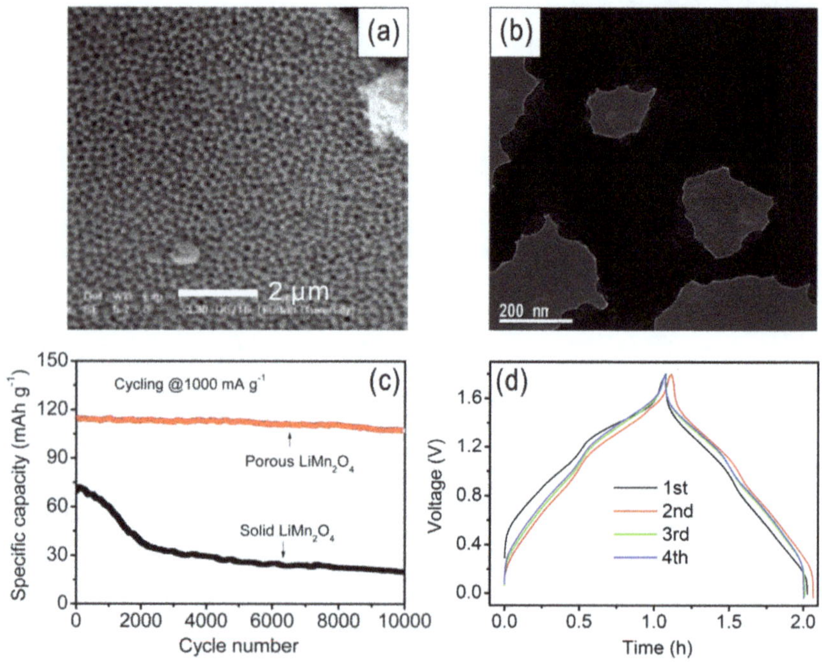

Figure 2.7. (a and b) SEM and TEM images of the prepared porous $LiMn_2O_4$. (c) The cyclic performance of a porous $LiMn_2O_4$ cathode (compared to a solid one) configured with AC as the anode. (d) Typical GCD curves of an AC//porous $LiMn_2O_4$ AHSC (at a current density of 100 mA g^{-1}) [105], reproduced with permission © The Royal Society of Chemistry.

Mn_3O_4 (527 F g^{-1}) at 1 mA cm^{-2}, and the large capacitance was attributed to the diffusion-controlled contribution rather than the surface capacitive contribution of Mn_3O_4. As a result of the the enlarged voltage window, AC//$Li_4Mn_5O_{12}$ AHSCs with Li_2SO_4 electrolytes exhibited a wide voltage window of 2.2 V and a superior energy density that reached 78 Wh kg^{-1} at 0.295 kW kg^{-1} (figure 2.8). The facile synthesis and relatively low cost of the carbon-cloth-supported lithium-rich Mn-based cathode with high capacitance and higher voltage output make it a promising candidate for real-world applications.

Electrospinning is also an effective method for controlling the electrode morphology of supercapacitors, especially those with freestanding or flexible nanofiber electrodes [69, 89]. Aswathy *et al* prepared a nanofibrous $LiNi_{0.5}Mn_{1.5}O_4$ octahedral

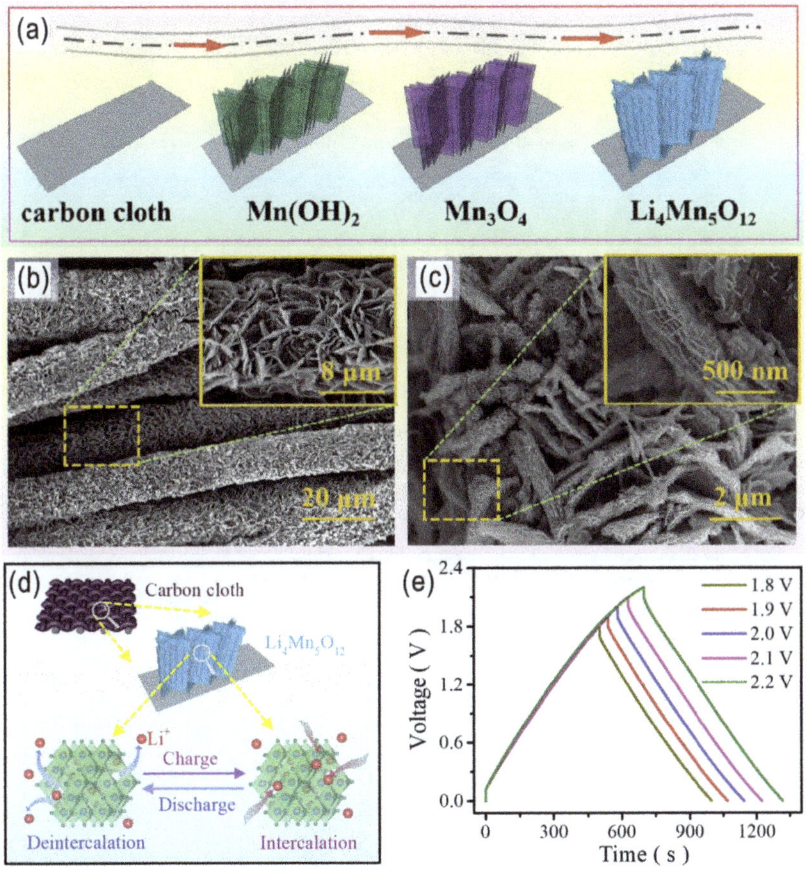

Figure 2.8. (a) A schematic illustration of the structural evolution from $Mn(OH)_2$ nanowalls to $Li_4Mn_5O_{12}$ nanoflakes via *in situ* synthesis on carbon cloth via electrochemical deposition/oxidation, and (b and c) corresponding SEM images of the resulting $Li_4Mn_5O_{12}$ nanoflakes. (d) A schematic illustration of the electrochemical reaction process of the $Li_4Mn_5O_{12}$ cathode at the atomic scale. (e) GCD curves of the assembled $Li_4Mn_5O_{12}$//AC AHSC device for different voltage windows (at 5 mA cm^{-2}) [8], reproduced with permission © Elsevier.

spinel composite via electrospinning and subsequent calcination [89]. When used as a cathode, the highly crystalline $LiNi_{0.5}Mn_{1.5}O_4$ spinel showed a high voltage of 1.1 V in a 3 M $LiNO_3$ aqueous electrolyte; when the cathode was assembled into an AHSCs with nitrogen-doped graphene (NDG) as the anode, the AHSC demonstrated superior cyclic stability and a high capacitance of 72 F g^{-1} (at a current density of 0.5 mA cm^{-2}), as well as a maximum energy density of 15 Wh kg^{-1} (at a power density of 110 W kg^{-1}) (figure 2.9). However, it should be mentioned that electrospun and calcinated nanofiber electrodes are usually brittle and the electrodes produced in this way are in fact based on ground powders, which may limit the advantages of such porous mat structures when used as flexible/freestanding electrodes.

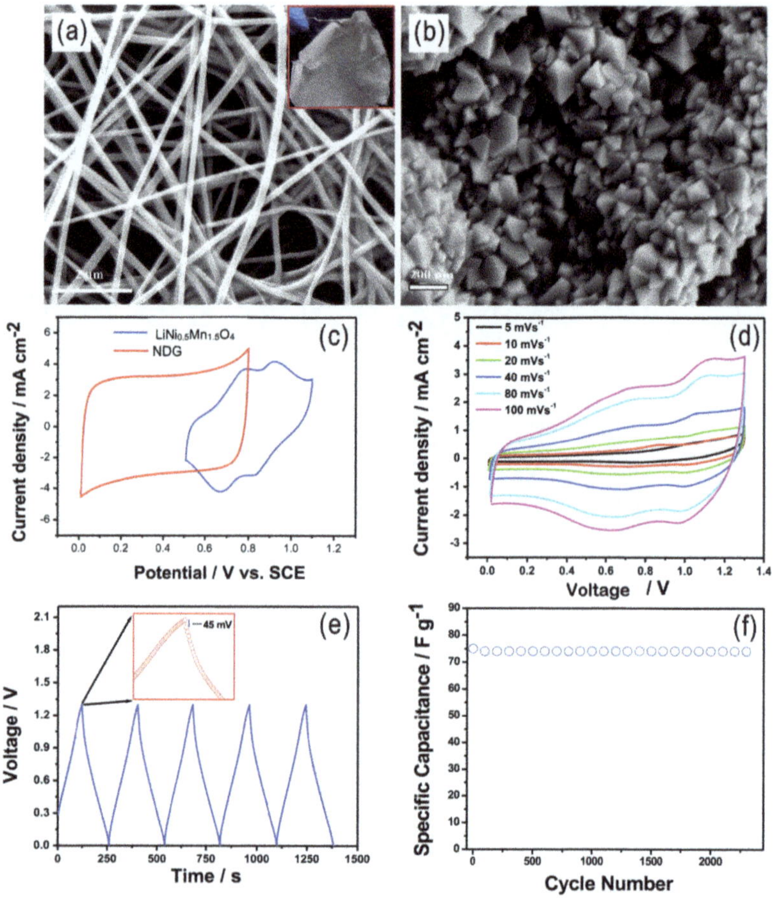

Figure 2.9. (a and b) FESEM images of the $LiNi_{0.5}Mn_{1.5}O_4$/PVP nanofibers before and after heating at 800 °C. (c) A plot of the CV results for a $LiNi_{0.5}Mn_{1.5}O_4$ cathode and a nitrogen-doped graphene (NDG) anode in a three-electrode cell (3 M $LiNO_3$ aqueous electrolyte). (d) A plot of the CV results for the $LiNi_{0.5}Mn_{1.5}O_4$/NDG AHSC device at different scan rates. (e) GCD curves of the AHSC and the corresponding cyclic performance (at a current density of 0.5 mA cm^{-2}) [89], reproduced with permission © The Royal Society of Chemistry.

Since capacitor-type electrodes usually have much higher power densities, it is essential to enhance the rate capability of the corresponding battery-type electrodes; thus, various approaches have been exploited [104]. In addition to nanostructuring or nanocrystallization, the incorporation of highly conductive materials (e.g. super P, carbon nanotubes, graphene, carbon matrices) or coatings (e.g. carbon, LaF_3) into lithium transition-metal oxides is another way of enhancing their rate capability and cyclic life [38, 91, 106–108]. Nanostructuring results in a larger surface area and a higher surface-to-volume ratio, enhanced active-material/electrolyte contact, and facilitated ion (de)intercalation due to shorter ion transport distances, while conductive coatings and carbon matrices greatly enhance the conductivity and can also act as a buffer during repeated charging/discharging [106]. As an example, ball milling can simultaneously enhance the specific capacitance, power density, and energy density of $LiMn_2O_4$ cathodes and the AHSCs made from them; the power and energy densities can be further significantly increased via the synergistic effect of ball milling and super P addition [106].

In addition, some lithium transition-metal phosphates such as $LiFePO_4$ and $LiMnPO_4$ have been investigated as cathode materials for AHSCs, although their capacity (energy density) and particularly their cyclic stability need to be further improved for competitive applications [41, 42, 44].

The electrochemical features of some typical intercalative cathode materials for Li-ion AHSCs and the Na-ion and K-ion AHSCs discussed below are summarized in table 2.1 for reference and comparison.

2.5 Na-ion hybrid supercapacitors

Layered transition-metal oxides are of great interest for facilitating ion transfer in high-power energy devices. Layer-structured birnessite (Na_xMnO_2) has attracted increasing interest due to its high theoretical capacitance, high oxygen evolution potential, low cost, and environmental benignity [45, 76]. Compared to Li-ion HSCs, the Na_xMnO_2-based cathodes of Na-HSCs can endow AHSCs with improved voltage outputs of up to more than 2.5 V [76]. For example, Xiong *et al* prepared an electrochemically oxidized inserted-Na MnO_2 cathode ($Na_{0.25}MnO_2$ nanosheets supported on carbon paper) and paired it with an electrochemically reduced porous carbon anode; 1 M of Na_2SO_4 electrolyte was added to create an AHSC that achieved a high potential of 2.7 V, a high capacitance of 82.9 F g^{-1} (at 1 A g^{-1}), excellent cyclic stability (a capacity retention of 93.7% after 10 000 cycles at 20 A g^{-1}), and an excellent energy density of 61 Wh kg^{-1} at a power density of 0.98 kW kg^{-1} (16 Wh kg^{-1} at 43 kW kg^{-1}) (figure 2.10) [68]. The use of electrode–electrolyte interaction, specifically by increasing the HER/OER overpotentials and minimizing their activity, may aid in the development of AHSCs with wide potential windows. Tong *et al* designed carbon-cloth-supported Na_xMnO_2 nanosheets wrapped in rGO to further enhance the conductivity [76]. The resulting rGO/V-Na_xMnO_2/CC showed a high capacitance of 363 F g^{-1} at 1 A g^{-1}, a high capacitance retention of 56.6% at current densities of up to 24 A g^{-1}, as well as a superior cyclic stability of 90.2% over 10 000 charge–discharge cycles.

Table 2.1. The electrochemical performances of typical porous carbon anode//battery-type cathode devices.

Cathode materials	Synthesis methods	Capacitance [F g^{-1}@A g^{-1}](V [V])[(a)]	Capacitance retention [%] (cycles)	E/P[(b)][Wh kg^{-1}@kW kg^{-1}] (V [V])[(a)]	Anode	Electrolytes	Year/ Reference
$LiMn_2O_4$ powder	Calcination	/	Cathode: 95(20k)	35@0.1 10@2 (0.8–1.8)	AC	1 M Li_2SO_4	2006/[35]
Spinel $LiMn_2O_4$ nanoparticles (NPs)	Hydrothermal	409@2 mA cm^{-2}348@100 mA cm^{-2}(0–1.1)	Cathode: 93(2k)	44.3@0.60 24@12.4 (0–1.8)	N-GC	0.5 M Li_2SO_4	2015/[37]
$LiMn_2O_4$ microspheres (porous)	Precipitation, calcination	536@1 mV s^{-1} (0–1.2)	Cathode: 82(0.2k)	33.1@0.09 25.0@0.9 (0–1.8)	AC	1 M Li_2SO_4	2018/[20]
$LiMn_2O_4$ nanohybrids (nanotubes/nanorods/ nanoparticles)	Hydrothermal, calcination	415@0.5 359@5 208@10 (0–1.05)	Full cell: 91(1k)	29.8@0.09 11.2@1.5 (0–1.8)	AC	0.5 M Li_2SO_4	2014/[16]
Nanostructured $LiMn_2O_4$/super P composite	Ball milling	306@1 229@40 (0.5–1.1)	Full cell: 85(4.5k)	21.6@0.29 13@5.2 (0–1.3)	AC	1 M Li_2SO_4	2018/ [106]
$Li_4Mn_5O_{12}$nanoflakes @carbon cloth	Electrochemical oxidation	627@1 mA cm^{-2} (0–1.2)		78@0.295(0–2.2)	AC	2 M Li_2SO_4	2020/[8]

(Continued)

Table 2.1. *(Continued)*

Cathode materials	Synthesis methods	Capacitance [F g^{-1}@A g^{-1}](V [V])[(a)]	Capacitance retention [%] (cycles)	$E/P^{(b)}$[Wh kg^{-1}@kW kg^{-1}] (V [V])[(a)]	Anode	Electrolytes	Year/ Reference
$Li_4Ti_5O_{12}$ hollow spheres (3D chestnut shell-like)	Hydrothermal, calcination	653@1 420@5 375@10 298@40 (0–0.45)	Cathode: 88.6(4k) 92.9(2k) Full cell: 91.8(1k)	26.2@0.8 14.5@8 (0–1.6)	NGO	2 M KOH	2018/[39]
$Li_4Ti_5O_{12}$-graphene hybrid (3D flower-shaped)	Hydrothermal, annealing	707@1 566@2 424@10 (0–0.45)	Cathode: 90.3(4k) Full cell: 90.2(2k)	35.1@0.8 20.8@8 (0–1.6)	PNDCN	2 M KOH	2018/[40]
Octahedral $LiNi_{0.5}Mn_{1.5}O_4$ spinel nanofibers	Electrospinning, sintering	75@0.5 (mA cm^{-2}) (0.5–1.1)	Full cell: 100(2.3k)	15@0.11 8.8@2.2 (0–1.3)	NDG	3 M $LiNO_3$	2015/[89]
$LiNi_{0.5}Mn_{1.5}O_4$-rGO	Hydrothermal, calcination	572@3.75 351@37.5 (0–1.1)	Full cell: 88.9(1k)	60@1.2 23@13 (0–2.0)	AC	5 M $LiNO_3$	2020/[96]
Carbon-coated $LiFePO_4$	*In situ* polymerization, calcination	506@1 mV s^{-1} (−0.6 to 1.0)	Full cell: 73(1k)	11.5@0.1 (0–1.0)	ARGO	1 M Li_2SO_4	2019/[41]
Graphene-functionalized Ni-doped $LiMnPO_4$	Solvothermal, calcination	138@5 mV s^{-1} 60@1 (0.2–1.2)	Full cell: 83(0.75k)	14@5.5 5@19 (0–2.0)	AC	1 M Li_2SO_4	2020/[44]

$LiMn_2O_4$ nanorods	Hydrothermal, annealing	440@0.5 330@20 (0–1.2)	Full cell: 88(5k)	50@0.57 17@6.9 (0–2.3)	NBC	Dual (Li_2SO_4 + K_2SO_4// KOH + K_2SO_4)	2019/[75]
$NaMnO_2$ (porous microparticles)	Ball milling, calcination	140@2C (0–1.1)	Full cell: 97(10k)	38.9@0.02 19.5@0.13 13.2@1 (0–1.9)	AC	0.5 M Na_2SO_4	2009/[45]
Carbon-paper-supported $Na_{0.25}MnO_2$ nanosheets	Electrochemical oxidization	295@2 182@10 80@50 (0–1.3)	Cathode: 95.8(10k) Full cell: 93.7/82.6(10k/20k)	61.1@0.98 16.3@42.9 (0–2.7)	ERPC	1 M Na_2SO_4	2018/[68]
rGO/V-Na_xMnO_2/CC	Deposition, spraying, reduction	363@1 205@24 (0–1.2)	Cathode: 90.2(10k) Full cell: 88.5(10k)	72.2@1.25 29.3@10 (0–2.5)	ACMK-3	1 M Na_2SO_4	2022/[76]
$K_{1.04}Mn_8O_{16}$ (KMO, nanoparticle/nanorod assembly)	Calcination	363@0.5 mV s^{-1} 115@100 mV s^{-1} (0–1.1)	Full cell: 97(10k)	128@0.10 14.5@2.3 (0–2.4)	AC	1 M $LiNO_3$	2020/[109]

(*Continued*)

Table 2.1. (*Continued*)

Cathode materials	Synthesis methods	Capacitance [F g^{-1}@A g^{-1}](V [V])$^{(a)}$	Capacitance retention [%] (cycles)	$E/P^{(b)}$[Wh kg^{-1}@kW kg^{-1}] (V [V])$^{(a)}$	Anode	Electrolytes	Year/ Reference
		397@0.5103@100 mV s^{-1}(0–1.1)	Full cell:85 (10k)	171@0.1319.9@3 (0–2.4)		1 M Na_2SO_4	
$K_{0.19}MnO_2$ nanowires	Hydrothermal	148@0.2(0–1.0)	Cathode: ~100(2.5k)	41.3@0.16(0–1.7)	AC	0.5 M K_2SO_4	2014/[47]
Carbon-paper-supported $K_{0.6}MnO_2$ nanosheets	Electrochemical oxidization	254@1 154@10 72@50 (0–1.2)	Full cell: 95(10k)	52.8@0.5 7@58.4 (0–2.4)	KHC	1 M KTFSI	2018/[46]

Note: $^{(a)}$(Voltage window [V]); $^{(b)}$Energy/power density. *Abbr.* AC: activated carbon; ACMK-3: activated CMK-3; ARGO: activated reduced graphene oxide; CC: carbon cloth; ERPC: electrochemically reduced porous carbon; KHC: K^+-adsorbed holey carbon; KTFSI: potassium bis(trifluoromethanesulfonyl)imide; NBC: nitrogen-rich biomass carbon; NDG: nitrogen-doped graphene; N-GC: nitrogen-doped graphene/porous carbon composite; NGO: N-doped graphene oxide; NPs: nanoparticles; PNDCN: pine-needle-derived carbon nanopores; rGO: reduced graphene oxide.

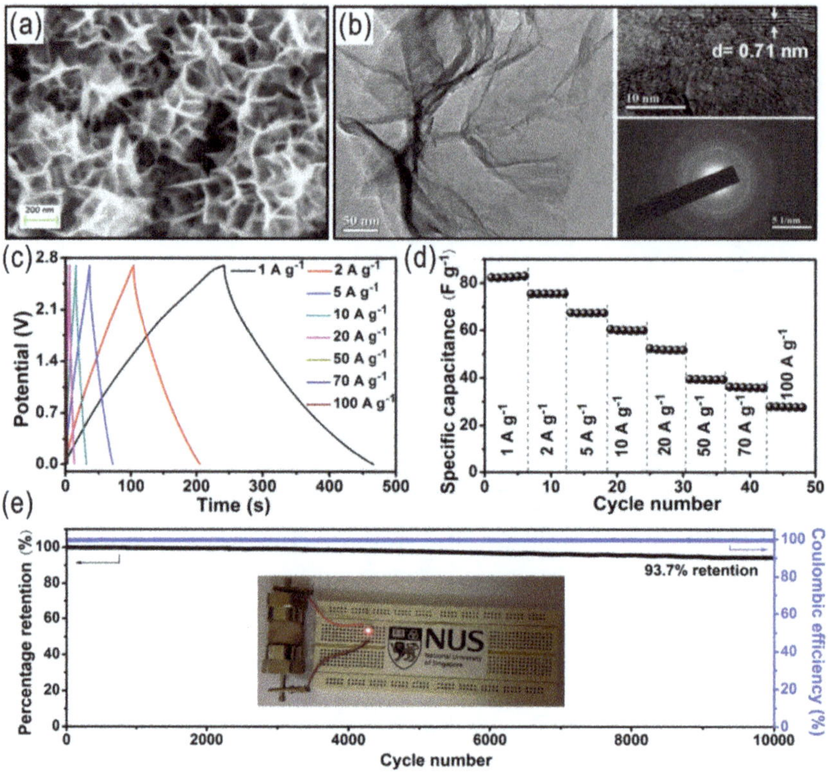

Figure 2.10. (a and b) SEM and TEM images (insets: an HRTEM image and a selected area electron diffraction (SAED) pattern) of the $Na_{0.25}MnO_2$ nanosheets grown on carbon paper. (c and d) GCD curves and rate performances at various current densities. (e) The cyclic stability and coulombic efficiency of the $Na_{0.25}MnO_2$//electrochemically reduced porous carbon AHSC (at 20 A g^{-1}) [68], reproduced with permission © John Wiley & Sons.

When this material was assembled with mesoporous carbon (CMK-3) as an anode and 1 M Na_2SO_4 as an electrolyte to form an AHSC, the resulting device exhibited a wider potential window of up to 2.5 V, and according to the classic energy density formula $E=0.5CV^2$, it unprecedentedly revealed outstanding energy densities of 72.2 Wh kg^{-1} at 1.25 kW kg^{-1} and 29.3 Wh kg^{-1} at 10.0 kW kg^{-1}; in addition, an excellent capacitance retention of 88.5% was obtained over 10 000 cycles. Furthermore, if the carbon anode is replaced by a carbon hybrid of transition-metal compounds [110], the capacitance of the anode and the energy density of AHSCs can be further elevated. Overall, these layered Na_xMnO_2-based AHSCs show great promise for applications.

In addition, some other alkali-ion pre-inserted transition-metal oxides such as $K_{1.04}Mn_8O_{16}$ (KMO) nanostructures prepared by a facile calcination/annealing process also exhibited excellent electrochemical properties of high specific capacitance (397 F g^{-1} for Na and 363 F g^{-1} for Li storage at a scan rate of 0.5 mV s^{-1}), superior energy/power densities (171 Wh kg^{-1} at 134 W kg^{-1} for Na-ion AHSCs and

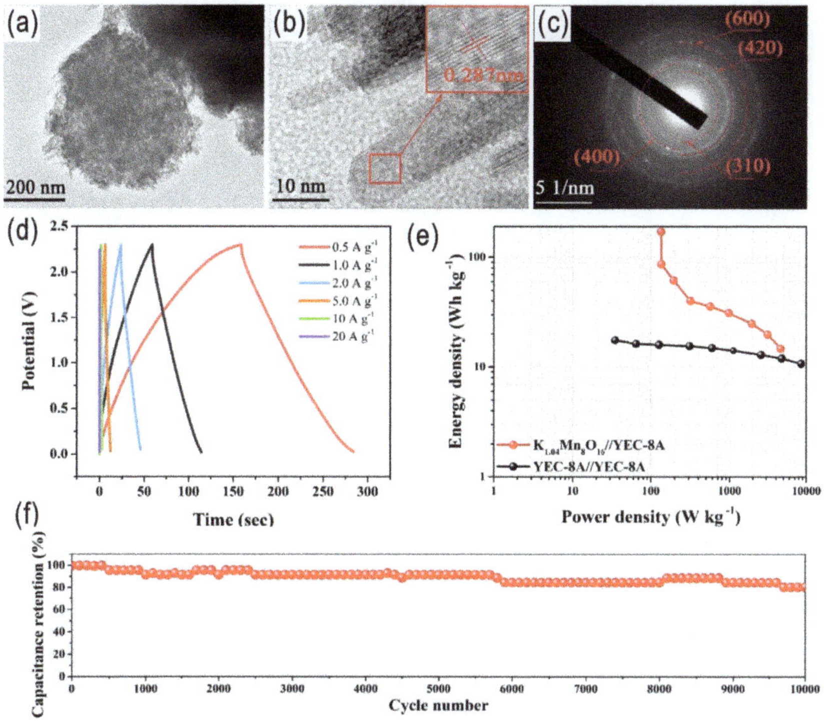

Figure 2.11. (a–c) TEM and HRTEM images and the SAED pattern of the synthesized $K_{1.04}Mn_8O_{16}$ (KMO) nanorods. The electrochemical performance of the KMO//YEC Na-ion AHSC: (d) GCD curves at different current densities; (e) Ragone plots, and (f) cyclic life (anode: commercial activated carbon, YEC-8A) [109], reproduced with permission © Elsevier.

128 Wh kg^{-1} at 99.9 W kg^{-1} for Li-ion AHSCs), and cyclic stability (85% for Na-ion AHSCs and 97% for Li-ion AHSCs after 10 000 cycles); such oxides have good application prospects (figure 2.11) [109].

2.6 K-ion hybrid supercapacitors

K^+ has been recently chosen as an intercalating cation due to its higher mobility and conductivity compared to those of alkali-metal ions such as Li^+ and Na^+ [46–48, 61, 111]. For example, a layered K^+-intercalated MnO_2 compound, namely $K_{0.6}MnO_2$, has exhibited competitive performance as a cathode material. When configured with K^+-adsorbed holey carbon as an anode and 1 M of KTFSI as an aqueous electrolyte, an AHSC produced a voltage output of up to 2.4 V and delivered an outstanding energy density of 52.8 Wh kg^{-1} (at 0.5 kW kg^{-1}), a superior power density of 58.4 kW kg^{-1} (at 7 Wh kg^{-1}), and a high capacitance retention of 95% after 10 000 cycles (at a current density of 20 A g^{-1}), viz. high rate performance and outstanding cyclic stability (figure 2.12) [46]. K^+-intercalated MnO_2 is another promising cathode material for wide-voltage-window and high-performance AHSCs.

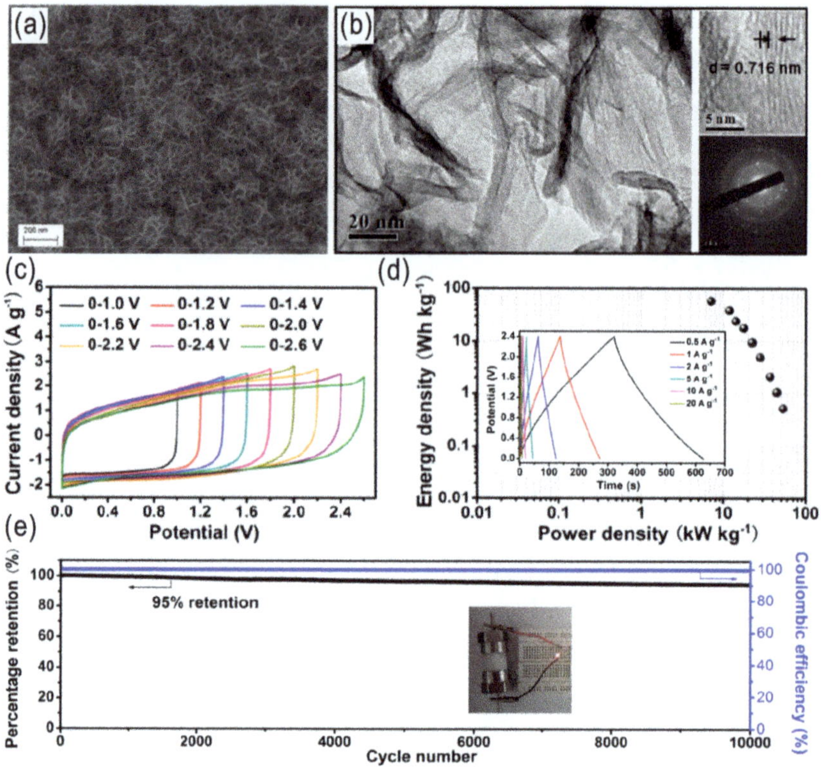

Figure 2.12. (a and b) SEM and TEM images (insets: HRTEM image and SAED pattern) of synthesized $K_{0.6}MnO_2$ nanosheets grown on carbon paper. The electrochemical performance of the $K_{0.6}MnO_2$//KHC K-ion AHSC. (c) A plot of the CV results for the $K_{0.6}MnO_2$//KHC at various operating voltage windows. (d) A Ragone plot (inset: GCD curves at different current densities). (e) The cyclic performance (at 20 A g^{-1}) [46], reproduced with permission © American Chemical Society. *KHC*: K^+-adsorbed holey carbon.

2.7 Conclusions and perspectives

AHSCs with a battery-type cathode and a capacitor-type cathode have been regarded as one of the most promising energy storage devices, as they are safe and offer high energy densities and high power densities. Although they are promising advanced energy storage systems, some challenges have to be overcome before their realistic application.

(i) Compared to HSCs with organic electrolytes, HSCs with aqueous electrolytes have higher ionic conductivity, better safety, and usually larger capacitance, although their output voltages are somewhat lower.

(ii) The combination of two or multiple strategies, i.e. synergistic effects, in order to enhance the performance of battery-type cathodes has shown great promise in advanced HSCs. The choice of electrolytes and electrodes, as well as the electrolyte–electrode interaction, impacts the practical voltage output of AHSCs. For example, the presence of additional higher potential redox peaks and an additional insertion process that requires a higher potential for the insertion of alkaline metal cations into the cathodes of layered structures can increase the OER overpotential; the simultaneous formation of such cations on the surface of an AC anode (or a cation-adsorbed AC anode) increases the H^+ adsorption barrier and reduces HER activity, thus leading to an AHSC with a higher voltage window.

(iii) The conductivity of the battery-type cathode dominates the reaction/diffusion kinetics, and several strategies (including synergistic methods) have been adopted to address critical concerns about electronic conductivity, the voltage window, rate performance, and cyclic stability (or lifetime). Some essential progress has been made toward meeting real-world application requirements. Compared with transition-metal oxides (TMOs) and transition-metal chalcogenides (TMCs), alkali-metal cations with pre-inserted TMOs (especially MnO_2) are of great interest, as they exhibit enhanced ion diffusion kinetics and conductivity and inhibited volume expansion; they thus show the most promise for use in advanced and practical AHSCs.

(iv) Although nanostructured materials have many advantages as excellent electrode materials for electrochemical energy storage devices, including higher capacitance and/or superior rate capability compared to conventional microsized materials, the tapped-density-induced energy density decrease of the whole device should be taken into consideration; thus, a balance or trade-off between rate capability/capacitance and energy density should be reached for cost-effective real-world applications.

Acknowledgments

This work was supported by National Natural Science Foundation of China (Grant Nos. 51403193, 51403132), Pangang Group Institute Fund (2021P4Y1E03Q).

List of abbreviations

3D	three-dimensional
AC	activated carbon
AHSCs	aqueous hybrid supercapacitors
ASCs	asymmetric supercapacitors
BSH	battery–supercapacitor hybrid
CV	cyclic voltammetry
EDLCs	electric double-layer capacitors
(FE)SEM	(field emission) scanning electron microscopy
GCD	galvanostatic charge–discharge
HRTEM	high-resolution transmission electron microscope
HSCs	hybrid supercapacitors
LDHs	layered double hydroxides
SAED	selected area electron diffraction

References

[1] Wang Q, Wang S L, Guo X H, Ruan L M, Wei N, Ma Y, Li J Y, Wang M, Li W Q and Zeng W 2019 *Adv. Electron. Mater.* **5** 1900537

[2] Shao Y, El-Kady M F, Sun J, Li Y, Zhang Q, Zhu M, Wang H, Dunn B and Kaner R B 2018 *Chem. Rev.* **118** 9233–80

[3] Ruiz-Montoya J G, Quispe-Garrido L V, Gomez J C C, Baena-Moncada A M and Goncalves J M 2021 *Sustain. Energy Fuels* **5** 5332–65

[4] Guan S D, Fu X L, Zhang B, Lei M and Peng Z J 2020 *J. Mater. Chem.* A **8** 3300–10

[5] Wang F, Xiao S, Hou Y, Hu C, Liu L and Wu Y 2013 *RSC Adv.* **3** 13059–84

[6] Gui Q, Ba D, Li L, Liu W, Li Y and Liu J 2022 *Sci. China Mater.* **65** 10–31

[7] Liu J-C, Huang Z-H and Ma T-Y 2020 *Small Struct.* **1** 2000020

[8] Zhao N, Fan H, Zhang M, Ma J, Wang C, Yadav A K, Li H, Jiang X and Cao X 2020 *Nano Energy* **71** 104626

[9] Fleischmann S, Mitchell J B, Wang R, Zhan C, Jiang D-e, Presser V and Augustyn V 2020 *Chem. Rev.* **120** 6738–82

[10] Numan A, Zhan Y, Khalid M and Hatamvand M 2021 *Advances in Supercapacitor and Supercapattery* ed N Arshid, M Khalid and A N Grace (Amsterdam: Elsevier) 3 45–61 pp

[11] Krishnan S G, Arunachalam A and Jagadish P 2021 *Advances in Supercapacitor and Supercapattery* ed N Arshid, M Khalid and A N Grace (Amsterdam: Elsevier) 12 311–48

[12] Yu L and Chen G Z 2016 *J. Power Sources* **326** 604–12

[13] Chen G Z 2017 *Int. Mater. Rev.* **62** 173–202

[14] Mathis T S, Kurra N, Wang X, Pinto D, Simon P and Gogotsi Y 2019 *Adv. Energy Mater.* **9** 1902007

[15] Guan L, Yu L and Chen G Z 2016 *Electrochim. Acta* **206** 464–78

[16] Wang F X, Xiao S Y, Zhu Y S, Chang Z, Hu C L, Wu Y P and Holze R 2014 *J. Power Sources* **246** 19–23

[17] Gogotsi Y and Penner R M 2018 *ACS Nano* **12** 2081–3

[18] dos Santos G A Jr, Fortunato V D S, Bastos G A A, Silva G G, Ortega P F R and Lavall R L 2020 *ACS Appl. Energy Mater.* **3** 9028–39

[19] Cao J and Wang W 2022 Pseudocapacitive Materials *Handbook of Energy Materials* ed R Gupta (Singapore: Springer Nature Singapore) 1–29

[20] Xiao H, Wang Y, Xie K, Cheng S and Cheng X 2018 *J. Alloys Compd.* **738** 25–31

[21] Hao C, Wang X, Yin Y and You Z 2016 *J. Electron. Mater.* **45** 515–26

[22] Liao J, Ni W, Wang C and Ma J 2020 *Chem. Eng. J.* **391** 123489

[23] Ni W and Khan A 2020 *Metal-Organic Framework Nanocomposites: From Design to Application* (Boca Raton, FL: CRC Press) 75–131

[24] Ni W and Shi L 2019 *Metal-Organic-Framework Composites* 58 (Millersville, PA: Materials Research Forum LLC) 177–238

[25] Ni W, Li X, Shi L-Y and Ma J 2022 *Nanoscale* **14** 9609–35

[26] Ni W and Shi L-Y 2022 *Handbook of Energy Materials* ed R Gupta (Singapore: Springer Nature Singapore) 1–25

[27] Wu M, Ni W, Hu J and Ma J 2019 *Nano-Micro Lett.* **11** 44

[28] Huang J, Wei Z, Liao J, Ni W, Wang C and Ma J 2019 *J. Energy Chem.* **33** 100–24

[29] Wei Z, Wang L, Zhuo M, Ni W, Wang H and Ma J 2018 *J. Mater. Chem.* A **6** 12185–214

[30] Shellikeri A, Yturriaga S, Zheng J S, Cao W, Hagen M, Read J A, Jow T R and Zheng J P 2018 *J. Power Sources* **392** 285–95

[31] Choudhary N, Li C, Moore J, Nagaiah N, Zhai L, Jung Y and Thomas J 2017 *Adv. Mater.* **29** 1605336
[32] Xia L, Tang B, Wei J and Zhou Z 2021 *Batter. Supercaps* **4** 1108–21
[33] Liu X, Sun Y, Tong Y, Wang X, Zheng J, Wu Y, Li H, Niu L and Hou Y 2021 *Nano Energy* **86** 106070
[34] Fan J, Yu L-L, Fan G-D, Xu W-L, Xing J-J and Zhao J-T 2021 *J. Materiomics* **7** 488–97
[35] Wang Y-g and Xia Y-y 2006 *J. Electrochem. Soc.* **153** A450–4
[36] Xu Y, Ding L, Zhong T, Han X, Jiao L, Yuan H and Wang Y 2015 *J. Energy Chem.* **24** 193–8
[37] Yang X, Qu F, Niu H, Wang Q, Yan J and Fan Z 2015 *Electrochim. Acta* **180** 287–94
[38] Ramkumar B, Yuvaraj S, Surendran S, Pandi K, Ramasamy H V, Lee Y S and Kalai Selvan R 2018 *J. Phys. Chem. Solids* **112** 270–9
[39] Xing L-L, Huang K-J, Cao S-X and Pang H 2018 *Chem. Eng. J.* **332** 253–9
[40] Xing L-L, Wu X and Huang K-J 2018 *J. Colloid Interface Sci.* **529** 171–9
[41] Gao H, Wang J, Zhang R, Wu L, Tao C-a and Deng G 2019 *Mater. Res. Express* **6** 045509
[42] Qiu S, Zhang X, Li Y, Sun T, Wang C and Qin C 2016 *J. Mater. Sci., Mater. Electron.* **27** 7255–64
[43] Prabaharan S R S, Anslin Star R, Kulkarni A R and Michael M S 2015 *Curr. Appl Phys.* **15** 1624–33
[44] Hlongwa N W, Ikpo C O, Ndipingwi M M, Nolly C, Raleie N, Dywili N and Iwuoha E I 2020 *Electroanalysis* **32** 2812–26
[45] Qu Q T, Shi Y, Tian S, Chen Y H, Wu Y P and Holze R 2009 *J. Power Sources* **194** 1222–5
[46] Xiong T, Lee W S V and Xue J 2018 *ACS Appl. Energy Mater.* **1** 5619–26
[47] Zhang B H, Liu Y, Chang Z, Yang Y Q, Wen Z B and Wu Y P 2014 *Electrochim. Acta* **130** 693–8
[48] Qu Q, Li L, Tian S, Guo W, Wu Y and Holze R 2010 *J. Power Sources* **195** 2789–94
[49] Chomkhuntod P, Ma N, Kosasang S, Duangdangchote S, Phattharasupakun N, Jangsan C and Sawangphruk M 2020 *ACS Appl. Energy Mater.* **3** 1402–9
[50] Zhou Q, Wang J, Zheng R, Gong Y and Lin J 2018 *Electrochim. Acta* **283** 197–211
[51] Tang A C, Wan C B, Meng X H, Li X C, Hu X Y, Huang M F and Ju X *Nano Res.* **15** 5123–33
[52] Ouyang Y, Ye H, Xia X, Jiao X, Li G, Mutahir S, Wang L, Mandler D, Lei W and Hao Q 2019 *J. Mater. Chem.* A **7** 3228–37
[53] Lu P, Chen Y, Zhou R, Guo C, Liu X, Yang F and Zhu Y 2022 *Sci. China Mater.* **65** 1503–11
[54] Ma D, Chen L, Wang G, Shen X, Yuan A, Li H, Yu H, Premlatha S and Ji Z 2022 *Appl. Surf. Sci.* **590** 153121
[55] Ji M R *et al* 2022 *Appl. Surf. Sci.* **575** 151734
[56] Cao J, Mei Q, Wu R and Wang W 2019 *Electrochim. Acta* **321** 134711
[57] Shahrokhian S, Rahimi S and Mohammadi R 2018 *Int. J. Hydrog. Energy* **43** 2256–67
[58] Hou Z-Q, Tian F-S, Gao Y-P, Wu W, Yang L-X, Jia X-L and Huang K-J 2018 *J. Alloys Compd.* **753** 525–31
[59] Li X, Zhao Y, Yu J, Liu Q, Chen R, Zhang H, Song D, Li R, Liu J and Wang J 2019 *J. Colloid Interface Sci.* **557** 691–9
[60] Ouyang Y, Xing T, Chen Y, Zheng L, Peng J, Wu C, Chang B, Luo Z and Wang X 2020 *J. Energy Storage* **30** 101454
[61] Zhou L, Zhang M, Wang Y, Zhu Y, Fu L, Liu X, Wu Y and Huang W 2017 *Electrochim. Acta* **232** 106–13

[62] Yao H, Zhang F, Zhang G and Yang Y 2019 *Electrochim. Acta* **294** 286–96
[63] Pazhamalai P, Krishnamoorthy K, Sahoo S and Kim S J 2018 *J. Alloys Compd.* **765** 1041–8
[64] Liu R, Wang Y, Sun S, Chen C and Wu X 2020 *J. Electroanal. Chem.* **878** 114571
[65] Chen F, Wang H, Ji S, Linkov V and Wang R 2018 *Chem. Eng. J.* **345** 48–57
[66] Gu X Y, Hong Y, Ai G, Wang C Y and Mao W F 2018 *Acta Chim. Sin.* **76** 644–8
[67] Zhao P, Yang B J, Chen J T, Lang J W, Zhang T Y and Yan X B 2020 *Acta Phys.-Chim. Sin.* **36** 7
[68] Xiong T, Tan T L, Lu L, Lee W S V and Xue J 2018 *Adv. Energy Mater.* **8** 1702630
[69] Liang J, Zhao H, Yue L, Fan G, Li T, Lu S, Chen G, Gao S, Asiri A M and Sun X 2020 *J. Mater. Chem.* A **8** 16747–89
[70] Ni W and Shi L 2019 *J. Vac. Sci. Technol.* A **37** 040803
[71] Lang J, Zhang X, Liu B, Wang R, Chen J and Yan X 2018 *J. Energy Chem.* **27** 43–56
[72] Pazhamalai P, Krishnamoorthy K, Sudhakaran M S P and Kim S J 2017 *Chem. Electro. Chem.* **4** 396–403
[73] Rong C, Chen S, Han J, Zhang K, Wang D, Mi X and Wei X 2015 *J. Renew. Sustain. Energy* **7** 023104
[74] Athanasiou M, Yannopoulos S N and Ioannides T 2022 *Chem. Eng. J.* **446** 137191
[75] Li C, Wu W, Zhang S, He L, Zhu Y, Wang J, Fu L, Chen Y, Wu Y and Huang W 2019 *J. Mater. Chem.* A **7** 4110–8
[76] Tong H, Gong D, Liu J, Xiao J, Chen X, Wu Y, Zhou Y, Shen L and Zhang X 2022 *J. Alloys Compd.* **901** 163543
[77] Zang X, Shen C, Sanghadasa M and Lin L 2019 *Chem. Electro. Chem.* **6** 976–88
[78] Hu T, Wang Y, Li J, Chen S, Shi Y, Li Z, Ruan D, Xu Q and Li F 2021 *Batter. Supercaps* **4** 717–32
[79] Hanna O, Luski S, Brousse T and Aurbach D 2017 *J. Power Sources* **354** 148–56
[80] Wang Y-g, Luo J-y, Wang C-x and Xia Y-y 2006 *J. Electrochem. Soc.* **153** A1425–31
[81] Gou Q, Zhao S, Wang J, Li M and Xue J 2020 *Nano-Micro Lett.* **12** 98
[82] Guo T, Zhou D, Pang L, Sun S, Zhou T and Su J 2022 *Small* **18** 2106360
[83] Dong C, Xu F, Chen L, Chen Z and Cao Y 2021 *Small Struct.* **2** 2100001
[84] Luo J-Y, Cui W-J, He P and Xia Y-Y 2010 *Nat. Chem.* **2** 760–5
[85] Tian X, Zhu Q and Xu B 2021 *Chem. Sus. Chem.* **14** 2501–15
[86] Bashir S, Chong M Y, Hina M, Kamran K, Ramesh S and Ramesh K 2021 *Advances in Supercapacitor and Supercapattery* ed N Arshid, M Khalid and A N Grace (Amsterdam: Elsevier) pp 271–310
[87] Wang F X, Xiao S Y, Gao X W, Zhu Y S, Zhang H P, Wu Y P and Holze R 2013 *J. Power Sources* **242** 560–5
[88] Xiang J, Zhang P, Lv S, Ma Y, Zhao Q, Sui Y, Ye Y and Qin C 2021 *RSC Adv.* **11** 14891–8
[89] Aswathy R, Kesavan T, Kumaran K T and Ragupathy P 2015 *J. Mater. Chem.* A **3** 12386–95
[90] Hussain I, Sahoo S, Sayed M S, Ahmad M, Sufyan Javed M, Lamiel C, Li Y, Shim J-J, Ma X and Zhang K 2022 *Coord. Chem. Rev.* **458** 214429
[91] Mu C *et al* 2019 *J. Alloys Compd.* **805** 624–30
[92] Zhang Y *et al* 2019 *J. Mater. Chem.* A **7** 19668–75
[93] Li C, Li R and Zhou Y 2022 *Energies* **15** 3877
[94] Khalafallah D, Li X Y, Zhi M J and Hong Z L 2021 *ACS Appl. Nano Mater.* **4** 14258–73
[95] Tan Q Y *et al* 2020 *J. Power Sources* **448** 10
[96] Gajraj V, Jose R and Mariappan C R 2020 *Mater. Res. Bull.* **124** 110742

[97] Thackeray M M and Amine K 2021 *Nat. Energy* **6** 683–3
[98] Ni W and Shi L 2022 *Curr. Appl. Mater.* **1** e200521193451
[99] Zhao B, Ran R, Liu M and Shao Z 2015 *Mater. Sci. Eng.* R **98** 1–71
[100] Lee B-G, Lee S-H, Ahn H-J and Yoon J-R 2018 *J. Alloys Compd.* **748** 882–8
[101] Ruan D, Huang Y, Li L, Yuan J and Qiao Z 2017 *J. Alloys Compd.* **695** 1685–90
[102] Ni J, Yang L, Wang H and Gao L 2012 *J. Solid State Electrochem.* **16** 2791–6
[103] Li W, Dahn J R and Wainwright D S 1994 *Science* **264** 1115–8
[104] Zhang H, Liu X, Li H, Hasa I and Passerini S 2021 *Angew. Chem. Int. Ed.* **60** 598–616
[105] Qu Q, Fu L, Zhan X, Samuelis D, Maier J, Li L, Tian S, Li Z and Wu Y 2011 *Energy Environ. Sci.* **4** 3985–90
[106] Chen L *et al* 2018 *J. Power Sources* **392** 116–22
[107] Tang M, Yuan A, Zhao H and Xu J 2013 *J. Power Sources* **235** 5–13
[108] Zhu Q, Zheng S, Lu X, Wan Y, Chen Q, Yang J, Zhang L-z and Lu Z 2016 *J. Alloys Compd.* **654** 384–91
[109] Zhang Y, Chen L, Hao C, Zheng X, Guo Y, Chen L, Lai K, Zhang Y and Ci L 2020 *J. Energy Chem.* **46** 53–61
[110] Jabeen N, Hussain A, Xia Q, Sun S, Zhu J and Xia H 2017 *Adv. Mater.* **29** 1700804
[111] Wu Y, Sun Y, Tong Y, Liu X, Zheng J, Han D, Li H and Niu L 2021 *Energy Storage Mater.* **41** 108–32

Chapter 3

Aqueous electrochemical capacitors based on battery anodes and porous carbon cathodes

Qinghong Wang

Aqueous electrochemical capacitors, which bridge the gap between aqueous supercapacitors (SC) and aqueous batteries, have the characteristics of safety and eco-friendliness. In this chapter, we review and summarize the recent development of aqueous electrochemical capacitors of the type characterized by the mechanisms of battery anodes and porous carbon cathodes. We also compare and discuss strategies for advanced material preparation and electrode design. Moreover, the opportunities and challenges for these aqueous electrochemical capacitors are also highlighted.

3.1 Introduction

Pseudocapacitor materials generally deliver higher specific capacitances than those of electric double-layer capacitors (EDLCs) due to their battery-like energy storage mechanism, which is based on reversible faradic reactions [1]. They also exhibit fast redox reactions that occur at rates comparable to those of EDLC materials, i.e. timescales of a few minutes. As a consequence, pseudocapacitors generally charge much faster than rechargeable batteries, and have capacities that are 10–100 as high as those of EDLCs. Because of the way in which the capacitance is generated, enhancements in electrochemical activity and surface area are main approaches used to improve pseudocapacitor materials. In recent decades, RuO_2 has dominated the field of pseudocapacitive materials [2]. Recent investigations have revealed that various low-cost, highly efficient, and environmentally friendly metallic oxides, such as MnO_2, NiO, Co_3O_4, CuO, Fe_2O_3, TiO_2, MoO_3 etc. also display exceptionally elevated theoretical capacitances, making them tremendously useful electrode materials for use in the design of high energy SCs [3]. Among these materials, iron-based, titanium-based, and molybdenum-based oxides are considered to be the main anode materials for use in aqueous SCs due to their advantages of high theoretical capacitance and suitable negative potential window. With the rapid

doi:10.1088/978-0-7503-5042-6ch3

© IOP Publishing Ltd 2023

development of nanotechnology and composite material engineering, these materials have been widely investigated. Herein, the energy storage mechanisms of different kinds of anode material are reviewed. Moreover, the strategies for improving the electrochemical performance of these materials are summarized.

3.2 Iron-based anode materials for aqueous supercapacitors

3.2.1 Iron-based oxides and hydroxides

Various iron-based oxides and hydroxides, such as Fe_2O_3, Fe_3O_4, and FeOOH, have been widely investigated as anodes for pseudocapacitors due to their high theoretical specific capacitance, suitable negative potential window, rich abundance, and environmental benignity [4, 5]. Among these materials, Fe_2O_3 displays the highest theoretical specific capacitance of 3625 F g^{-1} and has a voltage window of 1.0 V [6]. Fe_2O_3 mainly exists in four crystallographic forms, α-Fe_2O_3, β-Fe_2O_3, γ-Fe_2O_3 and ε-Fe_2O_3, based on the atomic arrangements of its Fe^{3+} and O^{2-} ions. Among these various crystallographic forms, α-Fe_2O_3 has proved to be the most stable material for pseudocapacitors due to its more stable hexagonal corundum-like structure [7–9]. In α-Fe_2O_3, O_2 ions lie on the (001) plane of a nearly ideal hexagonal close-packed lattice, whereas the cation Fe^{3+} occupies two-thirds of the octahedral interstices in the (001) basal planes. This cationic arrangement forms a pair of FeO_6 octahedrons with edges shared by three neighboring octahedrons in the same plane. Generally, the pseudocapacitive performance of Fe_2O_3 originates from the reversible conversion between Fe(III) and Fe(II) [10]. However, it exhibits different energy storage mechanisms in different electrolytes. In a KOH electrolyte, the oxidation/reduction reaction can be described as follows [11]:

$$Fe_2O_3 + 2K^+ + 2e^- \leftrightarrow K_2Fe_2O_3.$$

When used with a Na_2SO_4 electrolyte, no characteristic redox peaks are observed, indicating ideal capacitive characteristics [12].

The theoretical specific capacitance of Fe_3O_4 is 2299 F g^{-1} and its voltage window is 1.45 V. The capacitive behavior of Fe_3O_4 was first investigated in 2003 by Wu *et al.* It was revealed that Fe_3O_4 in a Na_2SO_3 electrolyte delivered a high capacitance of 510 F g^{-1} at an operating potential of 1.2 V. Its specific capacity may originate from the oxidation/reduction reaction between Fe(II) and Fe(III) or the surface redox reaction of sulfur in the form of sulfate and sulfite anions [13]. In 2006, they further confirmed the energy storage mechanism of Fe_3O_4 in Na_2SO_3, Na_2SO_4, and KOH electrolytes. In a Na_2SO_3 electrolyte, the specific capacity of Fe_3O_4 arises from the combination of both EDLC-type and pseudocapacitive behaviors, which involve the reduction/oxidation of adsorbed sulfite anions [14]:

$$2SO_3^{2-} + 3H_2O + 4e^- \leftrightarrow S_2O_3^{2-} + 6OH^-$$

$$S_2O_3^{2-} + 3H_2O + 8e^- \leftrightarrow 2S^{2-} + 6OH^-.$$

In Na_2SO_4 solutions, the capacitive behavior of Fe_3O_4 can be attributed to the double layer formed. The small capacitance value observed for the KOH electrolyte was attributed to the surface oxidization of the oxide electrode.

Due to its suitable negative potential window and high theoretical specific capacitance (2606 F g^{-1} at a voltage interval of 1.25 V), FeOOH has emerged as a promising anode material in recent years. Depending on the arrangement of the basic FeO_6 octahedra, FeOOH can exhibit one of four different crystal structures: (1) orthorhombic α-FeOOH is constructed from edge-shared FeO_6 octahedra, which form 2×1 tunnels; (2) monoclinic β-FeOOH consists of double chains of edge-shared FeO_6 octahedra, which form 2×2 tunnels [15]; (3) orthorhombic γ-FeOOH possesses two-dimensional channels constructed from layers of FeO_6 octahedra; (4) δ-FeOOH is a metastable phase with a hexagonal structure, in which oxygen ions form a hexagonal close-packed lattice and half of the octahedral interstices are occupied by ferric ions [16]. The charge storage mechanism of FeOOH still remains unclear. Chen's group investigated the charge/discharge processes of γ-FeOOH nanosheets in Li_2SO_4 electrolytes via *in situ* x-ray absorption spectroscopy and electrochemical measurement [17]. They found that Li^+ can reversibly insert/de-insert into/from the 2D channels of γ-FeOOH, accompanied by a Fe^{2+}/Fe^{3+} redox transition during the charge/discharge process, thus producing the superpseudocapacitive behavior of the γ-FeOOH electrode.

Many efforts have been made to explore various iron-based materials for SCs in recent decades. However, their electrochemical performance and cyclic stability are limited by their intrinsic features of poor electrical conductivity and unsatisfactory structural stability [4, 18]. Therefore, the construction of electrodes that possess good structural stability and have smooth pathways for both electrons and electrolytes that can support extremely rapid charge/discharge processes is crucial for high-performance iron-based anode materials for aqueous SCs.

3.2.2 Defect engineering

It has been demonstrated that defects can change the atomic configuration and electronic states at the surfaces of materials and can vary their surface activity [19, 20]. Intrinsic defect engineering based on, for example, oxygen or cation vacancies, can significantly alter the physiochemical properties of materials and take advantage of low-cost material modification routes. This approach has been progressively employed to enhance the supercapacitive performance of iron oxides and hydroxide materials [3]. As shown in figures 3.1(a)–(c), oxygen-deficient Fe_2O_3 quantum dots anchored on graphene nanosheets (V-FG) have been fabricated successfully [21]. The surface oxygen vacancies of Fe_2O_3 resulted in tunable adsorption of redox--active SO_3^{2-} in a Na_2SO_3 electrolyte. Therefore, the highly efficient dual-ion redox reactions involving both the cation and the anion at the surface facilitated the surficial redox reactions and contributed extra pseudocapacitance to the electrode (figures 3.1(d) and (e)). The V-FG electrode achieved a high specific capacity of up to 749 C g^{-1} in a 1 M Na_2SO_3 electrolyte and retained 290 C g^{-1} at an ultrahigh scan rate of 3.2 V s^{-1}. The density functional theory (DFT) calculations shown in figures 3.1(f) and (g) revealed that the bond lengths of S(sulfite)–O(Fe_2O_3) (1.541 Å) and O(sulfite)–Fe(Fe_2O_3) (2.011 Å) in Fe_2O_3 (001):Vo/Na_2SO_3 were shorter than those of Fe_2O_3 (001)/Na_2SO_3 (1.596 Å, 2.040 Å), indicating stronger interaction

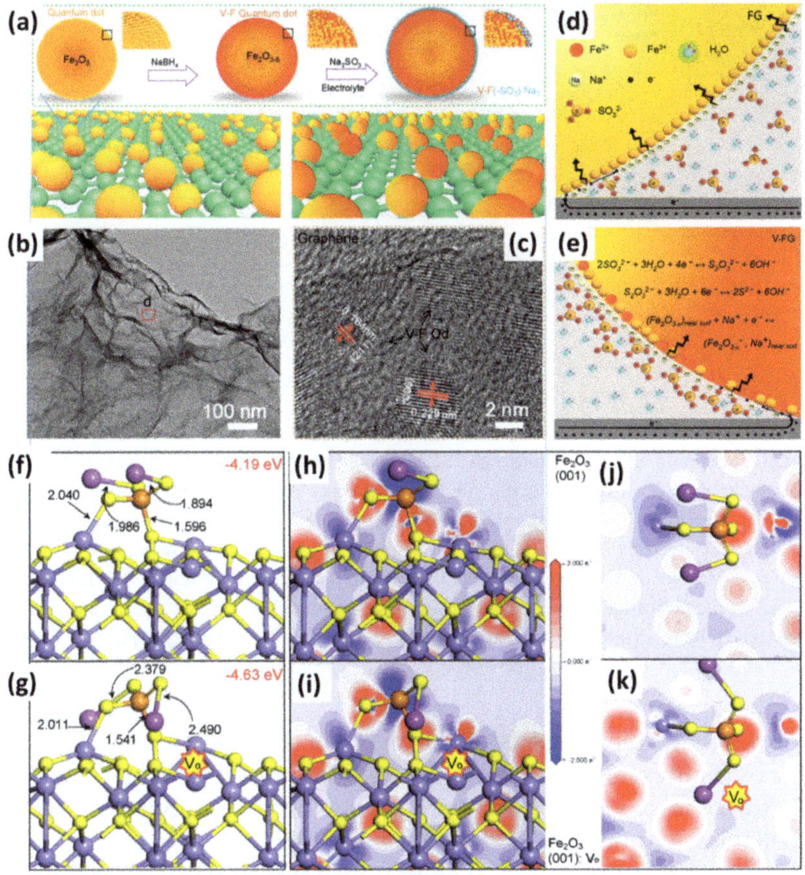

Figure 3.1. (a) A schematic diagram illustrating the fabrication procedure used to obtain V-FG and the adsorption of SO_3^{2-} on the V-FG surface. (b) Transmission electron microsocopy (TEM) and (c) high-resolution transmission electron microsocopy (HRTEM) images of the V-FG sample. The arrows mark the Fe_2O_3 quantum dots on graphene. The charge storage mechanisms of the FG/Na_2SO_3 (d) and V-FG/Na_2SO_3 (e) systems. (f–k) Local minima of the Fe_2O_3 (001)/Na_2SO_3 configuration optimized via DFT, including: (f) a perfect Fe_2O_3 (001) surface and (g) a nonstoichiometric Fe_2O_3 (001) surface with Vo. The arrows mark the bond length in Å. Adsorption energies (eV) are given with respect to Na_2SO_3 on an Fe_2O_3 (001) surface. Charge density differences collected for –Fe–O–S–O– bonds: surface atoms of (h and j) Fe_2O_3 (001)/Na_2SO_3 and (i and k) Fe_2O_3 (001):Vo/Na_2SO_3. Blue and red zones denote charge depletion and charge accumulation, respectively. Blue, Fe atoms; yellow, O atoms; orange, S atoms; purple, Na atoms. Source: [21], reproduced with permission © John Wiley and Sons.

between Na_2SO_3 and Fe_2O_3 (001) after the introduction of Vo. On both the perfect and the nonstoichiometric Fe_2O_3 (001) surfaces, the adsorptions of Na_2SO_3 were exothermic by −4.19 eV and −4.63 eV, respectively, indicating spontaneous chemisorption. Charge density differences taken along the axis of the –Fe–O–S–O– bonds (figures 3.1(h) and (i)) and at the surface (figures 3.1(j) and (k)) also confirmed the enhanced binding energy between the adsorbates and the oxygen-deficient Fe_2O_3 surface. Lu *et al* also prepared oxygen-deficient α-Fe_2O_3 nanorods anchored

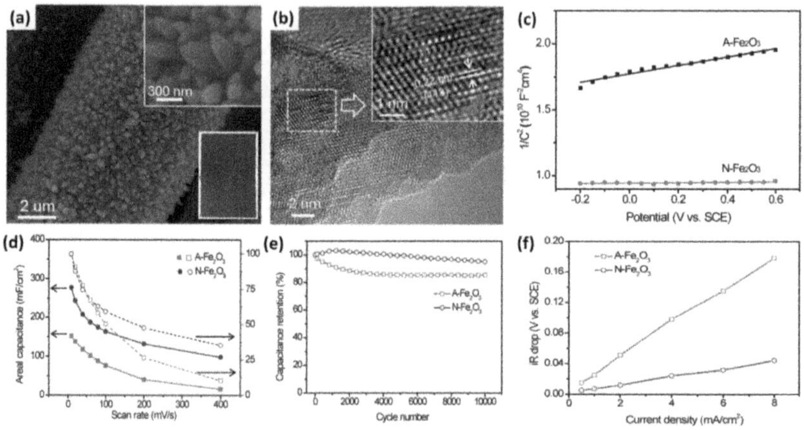

Figure 3.2. (a) A scanning electron microscopy (SEM) image and (b) an HRTEM image of oxygen-deficient Fe_2O_3 nanorods anchored on carbon cloth. (c) Mott–Schottky plots of Fe_2O_3 treated in air and oxygen-deficient Fe_2O_3 treated in an N_2 atmosphere obtained at at different potentials at a frequency of 10 kHz in the dark in a 1 M KOH electrolyte. (d) The areal capacitances and capacitance retentions of oxygen-deficient Fe_2O_3 treated in air and in an N_2 atmosphere calculated from the cyclic voltammetry (CV) curves as a function of scan rate. (e) Cyclic performance of the Fe_2O_3 nanorod electrodes collected at a scan rate of 200 mV s^{-1} for 10 000 cycles and (f) the rate capability measured at different current densities [22], reproduced with permission © John Wiley and Sons.

on carbon cloth through thermal treatment of FeOOH nanorods under a N_2 atmosphere (figures 3.2(a) and (b)). According to Mott–Schottky plots (figure 3.2(c)), the oxygen-deficient Fe_2O_3 nanorods exhibited n-type semiconductive behavior with increased donor densities from 5.5×10^{18} to 8.1×10^{19} cm^{-3} after oxygen vacancy formation. The obtained electrode displayed a high areal capacitance of 382.7 mF cm^{-2} at 0.5 mA cm^{-2} with good rate capability, excellent stability (95.2% capacitance loss after 10 000 cycles), and good rate performance (figures 3.2(d)–(f)) [22].

It has been found that less crystalline or amorphous iron metal oxides are capable of achieving better cyclic stability than their highly crystalline counterparts because of their increased structural defects and disorder [23]. Amorphous fish-scale-like FeOOH was synthesized on a Ni foam substrate through a one-step electrodeposition method [24]. The unique nanostructures of the hydroxides provided a large number of surface active sites, while the amorphous nature of the material system facilitated the diffusion and reaction of electrolyte ions and provided isotropic charging discharging processes. As a result, the FeOOH electrode exhibited a high pseudocapacitance of 1.11 F cm^{-2}. High rate capabilities and superior cyclabilities were achieved using the FeOOH as an anode and Co–Ni double hydroxides as a cathode. The aqueous asymmetric SC delivered a high energy density of 86.4 Wh kg^{-1} and a power density of 11.6 kW kg^{-1}, which were superior to those of most reported asymmetric SCs. Mai's group [25] also prepared a low-crystallinity FeOOH nanoparticle anode via a novel strategy involving the hydrothermal growth and subsequent electrochemical transformation of α-Fe_2O_3

nanoparticles. The FeOOH nanoparticles anchored on carbon fiber cloth presented capacitances of 1066 and 716 F g^{-1} at mass loadings of 1.6 and 9.1 mg cm^{-2}, respectively, a rate capability in which 74.6% of the capacitance was retained at 30 A g^{-1}, and a cyclic stability that retained 91% of the capacitance after 10 000 cycles. Oxygen-deficient Fe_2O_3-δ nanorod arrays with a crystalline core/amorphous shell heterostructure were prepared via a facile NaBH4 treatment and displayed a high specific capacitance of 701 F g^{-1} at a current density of 1 A g^{-1} [26]. This proved that enhanced electron transport and ion transport could be achieved simultaneously in the unique heterostructure, in which the tunable oxygen vacancy greatly improved the electronic conductivity while the amorphous shell facilitated fast Li^+ diffusion at the surface region. Moreover, the crystalline/amorphous interface provided extra Li^+ storage sites for the electrode, which significantly increased the charge storage capability of the pseudocapacitive electrode material.

3.2.3 Doping-based engineering

Doping a small quantity of extrinsic atoms into materials can clearly regulate their electronic structure, charge carrier concentration, and phase [3]. Doping has been widely employed to control the electrical conductivity, chemical potential, and surface activity of energy storage materials. As shown in figures 3.3(a)–(c), β-FeOOH doped with fluorine was prepared on carbon cloth via a hydrothermal route [27]. The carbon cloth modified by β-FeOOH with 5.06% of fluorine dopant exhibited an areal electrical resistance of 2.211 Ω cm^{-2}, which was much smaller than the value of 4.096 Ω cm^{-2} for the original carbon cloth (figure 3.3(d)). DFT calculation results revealed that β-FeOOH is a semiconductor with a direct bandgap of 1.05 eV, which led to the relative low conductivity of β-FeOOH. Fluorine doping significantly reduced the bandgap to about 0.2 eV, and there were finite densities of

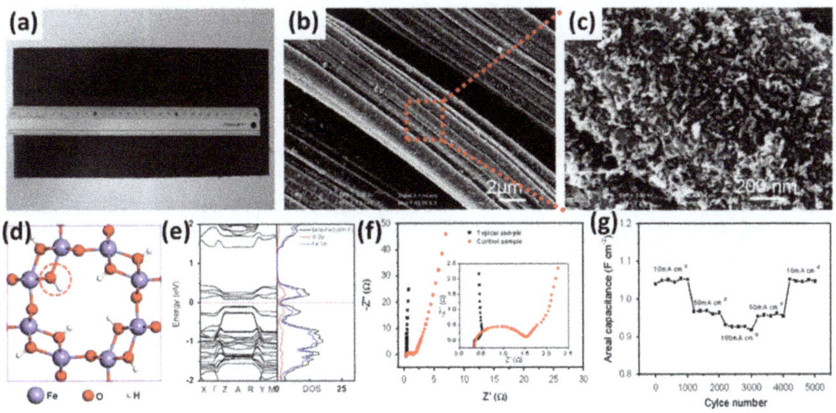

Figure 3.3. (a)–(d) The computational model used for the β-FeOOH. The marked –OH ion is used to illustrate the fluorine-doped site. (e) Calculated density of states plots for β-FeO (OH, F). (f) Electrochemical impedance spectroscopy (EIS) of SCs based on a typical sample (β-FeOOH) and a control sample (fluorine-doped β-FeOOH) performed between 0.01 and 100 kHz. (g) Cyclic performances of the fluorine-doped β-FeOOH-based SC at various current densities [27], reproduced with permission © Elsevier.

states on both sides of the Fermi level (figures 3.3(e) and (f)). Fluorine-doped β-FeOOH delivered a high pseudocapacitive response (1.12 F cm^{-2} at 1.0 mA cm^{-2}), excellent rate capability (83.0% of the capacitance was retained at a high current of 100 mA cm^{-2}), and good capacitance retention after 5000 cycles (figure 3.3(g)). Ti–Fe_2O_3 was prepared by a hydrothermal method. The donor density of Fe_2O_3 was significantly enhanced by the Ti doping, and the resulting material displayed a remarkably large areal capacitance of 1.15 F cm^{-2} with ultrahigh cyclic durability and a capacitance retention of more than 96% after 30 000 cycles [28].

3.2.4 Architecture engineering

Architecture engineering is a common strategy for optimizing the charge storage performance of aqueous SCs due to its effects of increasing the specific surface area and improving the electrolyte diffusion of electrode materials. Strategies for the synthesis of iron-based anode materials mainly include hydrothermal/solvothermal methods, electrospinning, and template-assisted methods [5]. Fe_2O_3, Fe_3O_4, and FeOOH have been formed into various nanostructures, such as 0D nanoparticles and hollow structures [29–31]. 1D nanostructures such as nanowires, nanorods, and nanotubes [32, 33], 2D nanostructures such as layered nanosheets and nanoplatelets, and 3D nanostructures such as nanoflowers [34, 35] have been fabricated and have delivered good charge storage performance. These nanostructures have also been directly grown on flexible current collectors without the use of binders, forming 3D flexible electrode architectures with high energy density and good cyclic and rate performances for wearable energy storage applications [36–39]. For example, a flexible porous Fe_2O_3@ACC electrode was fabricated by anchoring Fe_2O_3 nanospheres to activated carbon cloth (ACC, figure 3.4(a)); the resulting material displayed high specific capacitances of about 2775 mF cm^{-2} at 1 mA cm^{-2} and

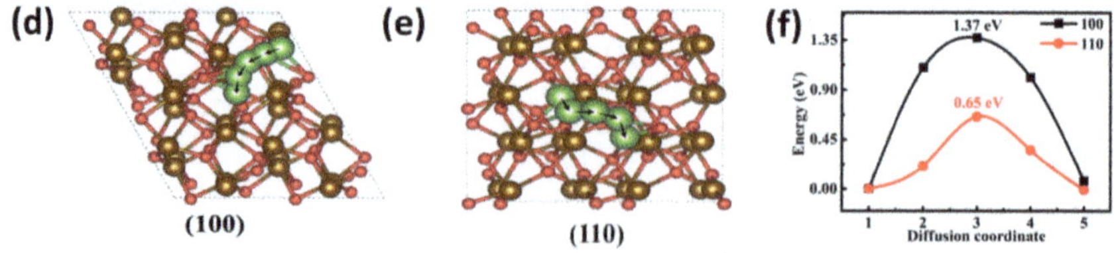

Figure 3.4. (a) A schematic illustration of the Fe_2O_3@ACC electrode and the structure of the symmetric SC. (b) The areal and specific capacitances of the Fe_2O_3@ACC electrode versus current density. (c) CV curves of the Fe_2O_3@ACC, pure Fe_2O_3 nanospheres (P-Fe_2O_3), and ACC at 10 mV s^{-1}. A top view of the lithium-ion diffusion path on the (d) 100 and (e) 110 facets of Fe_2O_3. (f) The corresponding energy barriers in the diffusion paths of the 100 and 110 facets, calculated by the nudged elastic-band method [40], reproduced with permission © Elsevier.

4755 mF cm^{-2} at 2 mV s^{-1} (figures 3.4(b) and (c)). Li-ion adsorption and diffusion mechanisms on the (100) and (110) facets of Fe_2O_3 have been investigated by the nudged elastic-band method based on the top view (figures 3.4(d) and (e)). This revealed that it was easier for lithium ions to move on the 110 plane (figure 3.4(f)). Therefore, the excellent performance of the Fe_2O_3@ACC electrode was attributed to the synergetic effect of porous Fe_2O_3 nanospheres, highly conductive ACC, relatively lower adsorption and diffusion barriers on the facets of Fe_2O_3, and the electronegativity of the exposed surface [40].

3.2.5 Compositing and hybridizing

Iron-based oxides and hydroxides are wide-bandgap materials with low electron conductivity. This property leads to a decrease in the redox-active sites due to the inefficient electron transport to the electrode/electrolyte interface. To solve this problem, carbon modification is usually employed to improve the electrochemical performance of iron-based anode materials because of the high conductivity and ultralarge surface area of carbon materials, such as carbon nanotubes (CNTs), carbon fiber and graphene. Generally, strategies for synthesizing carbon-modified iron-based materials include the reduction method, the chemical coprecipitation method, the solvothermal/hydrothermal–calcination/sintering method, the electrospinning technique, the solvent–thermal method, and the microwave method [5]. Kang *et al* [41] fabricated Fe_2O_3/CNTs nanocomposites via a one-step combustion method and obtained a high specific capacitance of 192.6 F g^{-1} at 1 A g^{-1} as well as a long cyclic life of 1000 cycles with a capacitance retention of 82%, which was superior to those of Fe_2O_3/C. The improved electrochemical performance was attributed to the enhanced electrochemical kinetics and mitigated aggregation of Fe_2O_3 nanoparticles caused by the CNTs. Wang *et al* [42] reported that Fe^{3+} cations could easily bind with oxygen-containing groups on graphene oxide sheets via electrostatic interactions and could then be transformed into an Fe_2O_3 anchored graphene composite via hydrothermal and calcination processes. Due to its improved conductivity and structural stability, the Fe_2O_3/rGO electrode delivered an ultrahigh capacity of 908 F g^{-1} at 2 A g^{-1} and a perfect rate capability (a capacity retention of 69% at 50 A g^{-1}).

Due to their good flexibility and mechanical strength, carbon nanotube fibers (CNTFs) are regarded as the most promising candidates for conductive fibrous substrates which can be used in flexible energy storage devices. As shown in figures 3.5(a)–(c), metal–organic framework (MOF)-derived spindle-like α-Fe_2O_3@C was successfully grown on oxidized carbon nanotube fiber (S-α-Fe_2O_3@C/OCNTF) [43]. Coupled with a MnO_2/CNTF cathode, the fiber-shaped asymmetric SC delivered high specific capacitance values of 201.3 and 136.4 mF cm^{-2} at current densities of 2 and 20 mA cm^{-2}, respectively (figure 3.5(d)). The device possessed high flexibility and good mechanical stability. The charge/discharge curves underwent only negligible changes when the bending angle was varied from 0° to 180° at a current density of 4 mA cm^{-2} (figure 3.5(e)). After bending at 90° for more than 4000 cycles, the device still retained 97.1% of its specific capacitance

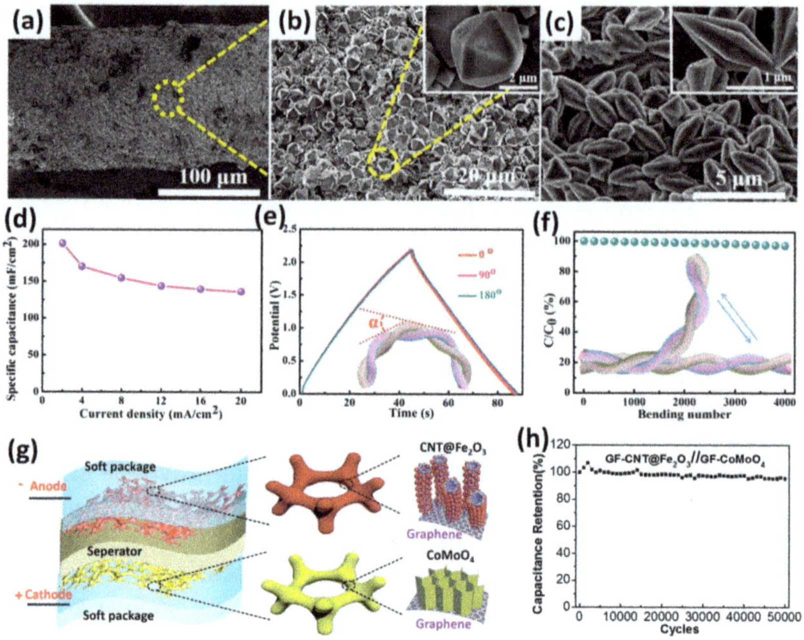

Figure 3.5. (a and b) SEM images of MIL-88-Fe/CNTFs at increasing magnification. (c) An SEM image of the MOF-derived S-α-Fe_2O_3@C/OCNTF electrode. (d) Areal specific capacitances of the S-α-Fe_2O_3@C/OCNTF//Na-MnO_2 NSs/CNTF asymmetric SC calculated from the charge–discharge curves as a function of current density. (e) Charge/discharge curves of the asymmetric SC measured at a current density of 4 mA cm^{-2} for different bending angles. (f) Normalized capacitances of the asymmetric SC at a bending angle of 90° for 4000 cycles [43], reproduced with permission © American Chemical Society. (g) A schematic of the GF/CNT@Fe_2O_3//GF/$CoMoO_4$ full-cell package and (h) its cyclic stability measured at a current density of 7 A g^{-1} [44], reproduced with permission © American Chemical Society.

(figure 3.5(f)). Guan *et al* [44] attempted to produce a hierarchical structure consisting of iron oxide nanoparticles decorated on a three-dimensional ultrathin graphite foam/CNTs (GF/CNT) forest substrate via the chemical vapor deposition method. In three electrode tests, GF/CNT/α-Fe_2O_3 presented high capacitance (~470.5 mF cm^{-2} at 20 mA cm^{-2}), excellent cyclic performance (a capacitance retention of ~93.5% after 50 000 cycles), and improved rate capacity (a capacitance retention of ~62.1% when the current density increased from 10 to 40 A g^{-1}). A full cell produced using a GF/CNT@Fe_2O_3 anode and a GF/$CoMoO_4$ cathode based on this structure gave rise to a high energy of 74.7 Wh kg^{-1} at a power of 1400 W kg^{-1}, and 95.4% of its capacitance was retained after 50 000 charge/discharge cycles at a current of 7 A g^{-1} (figures 3.5(g)–(i)).

In addition to carbon materials, some excellent conductive polymer materials, such as polyaniline (PANI) and polypyrrole (PPY) are also considered to be promising candidates for improving the electrochemical performance of metal oxides. Lu *et al* fabricated highly ordered α-Fe_2O_3@PANI core–shell nanowire arrays via a facile electrodeposition method. The highly ordered α-Fe_2O_3@PANI nanowire arrays provided a stable structure and a fast ion/electron transfer path, which all are advantageous for improving electrochemical properties [45]. As a common conductive polymer material, PPY is normally used as a coating layer to improve the structural stability of electrode materials in SCs due to its excellent chemical stability. A PPY layer can effectively prevent the severe aggregation of Fe_2O_3 nanomaterial and structural collapse during the cyclic process. Wang *et al* synthesized hierarchical α-Fe_2O_3/PPY nanoarrays grown on conductive carbon cloth, which delivered outstanding electrochemical performance (382.4 mF cm^{-2} at a relatively low current density of 0.5 mA cm^{-2}) [46].

Recently, a new family of 2D transition-metal carbides and nitrides with superior hydrophilicity, metallic conductivity, and surface redox activity, known as MXenes, has been emerging. These particular features endow MXenes with excellent high-rate pseudocapacitive energy storage properties [47–49]. As shown in figure 3.6(a), a sandwich-like MXene/Fe_3O_4/MXene film with a porous Fe_3O_4 layer was successfully prepared on a flexible Ni tape [50]. In a 1 M Li_2SO_4 aqueous electrolyte, the sandwich-like electrode exhibited an areal capacitance of 46.4 mF cm^{-2} at a current density of 0.5 mA cm^{-2}. When assembled into a symmetric all-solid-state device, the SC displayed good flexibility, which was confirmed by the almost overlapping CV curves at varying bending angles (figure 3.6(b)) and a capacitance retention of 98.4% even at the high bending angle of 180° (figure 3.6(c)). As shown in figures 3.6(d) and (e), the symmetric all-solid-state device displayed a high energy density of 0.970 μWh cm^{-2} at a power density of 0.176 mW cm^{-2} along with good cyclic stability (91.7% of its initial capacitance was preserved after 2000 cycles at a current density of 4 mA cm^{-2}). The high performance of the thin-film electrode was attributed to the anchoring effect of the upper and lower MXene thin films on the Fe_3O_4 thin film, which not only benefited electron transfer but also prevented the dissolution of Fe_3O_4 during the charge and discharge processes.

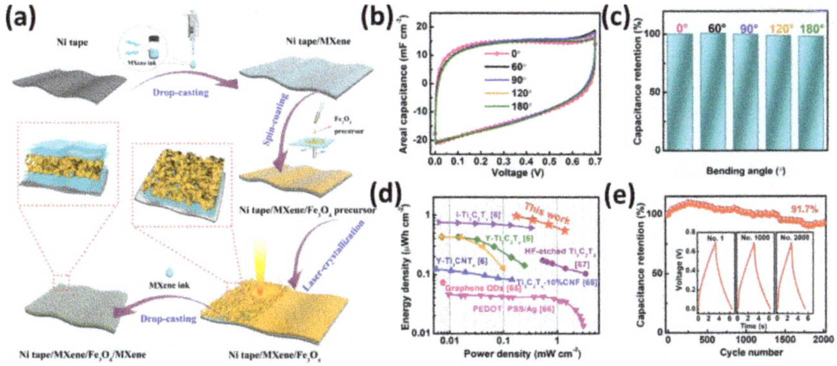

Figure 3.6. (a) A graphical illustration of the preparation of MXene/Fe_3O_4/MXene films on flexible Ni tapes. (b) The CV curves of a single symmetric all-solid-state cell at different bending angles. (c) Capacitance retention as a function of bending angle for a single device. (d) A Ragone plot of the prepared symmetric all-solid-state cell compared to previously reported results. (e) The cyclic stability of a single such device after 2000 cycles at a current density of 4 mA cm^{-2} [50], reproduced with permission © Elsevier.

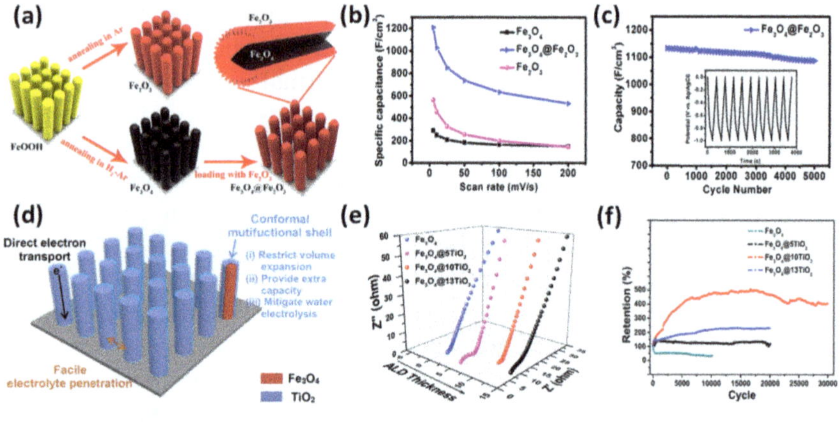

Figure 3.7. (a) A schematic illustration of the synthesis procedure used to obtain the hierarchical $Fe_3O_4@Fe_2O_3$ core–shell nanoarrays. (b) The volumetric capacitances of the Fe_3O_4, Fe_2O_3, and $Fe_3O_4@Fe_2O_3$ electrodes at different scan rates ranging from 5 to 200 mV s^{-1} and (c) the cyclic stability of the $Fe_3O_4@Fe_2O_3$ electrode collected at a current density of 1.25 mA cm^{-2} [54], reproduced with permission © American Chemical Society. (d) A schematic illustration of the structural features of the $Fe_3O_4@TiO_2$ hybrid nanorod array for electrochemical energy storage. (e) The EIS spectra and (f) cyclic performances of hybrid and pristine electrodes in a 1 M Li_2SO_4 electrolyte [55], reproduced with permission © John Wiley and Sons.

In addition to the introduction of conductive carbon and polymers, constructing hierarchical heterostructures has also proved to be an effective approach for improving the charge transport of materials [51–53]. As shown in figure 3.7(a)–(c), a porous Fe_2O_3 shell was coated on the surface of Fe_3O_4 nanoarrays via a facile electrodeposition process, followed by annealing. The hierarchical $Fe_3O_4@Fe_2O_3$ core–shell nanoarrays displayed a large volumetric capacitance of 1206 F cm^{-3} with an active mass loading of 1.25 mg cm^{-2} as well as good rate capability and cyclic performance [54]. Additionally, an $Fe_3O_4@TiO_2$ core–shell nanorod array anode exhibited a remarkable cyclic performance for more than 30 000 cycles over a wide potential window in a neutral lithium salt electrolyte. The TiO_2 shell not only buffered the internal volume expansion of the Fe_3O_4 but also contributed extra capacity through Li^+ intercalation/de-intercalation and helped to alleviate water electrolysis [55] (figure 3.7(d)–(f)).

3.2.6 Binary transition-metal oxides

Binary transition-metal oxides (TMOs) have attracted much attention for use in SCs due to their terrestrial abundance, low cost, environmental friendliness, and intriguing electrochemical nature. On the one hand, binary transition-metal oxides have many demonstrable merits because the coupling of two transition-metal elements greatly improves the chemical stability, electrical conductivity, and capacitive properties of energy storage devices [56]. On the other hand, binary transition-metal oxides may be doped with various metals, which alters the capacitive behavior of the electrode material. Metal ferrites with a universal formula of MFe_2O_4 (M = Co, Ni, Zn, Mn, Cu, etc.) show many interesting properties, such as 3D diffusion pathways, stable structure, rich redox chemistry, good electronic

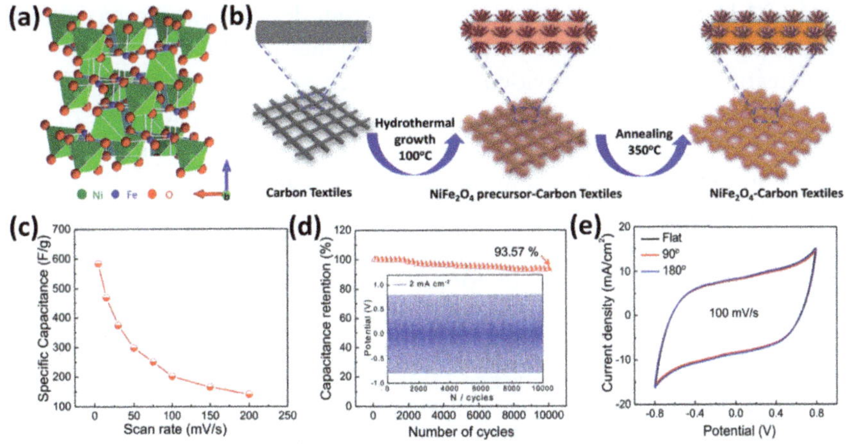

Figure 3.8. (a) The crystal structure of $NiFe_2O_4$; green tetrahedrons, blue dots, and red dots represent nickel, iron, and oxygen atoms, respectively. (b) A schematic of the growth of $NiFe_2O_4$/carbon textile (NFO/CT) by a simple hydrothermal method followed by a short post-annealing treatment. (c) The specific capacitance of the NFO/CT-based solid-state SC as a function of scan rate. (d) Capacitance retention versus number of cycles for the NFO/CT-based solid-state SC (the inset shows 10 000 cycles at a constant current density of 2.0 mA cm^{-2}). (e) The electrochemical performances of the flexible NFO/CT solid-state SC: CV curves at different bending states at a constant scan rate of 100 mV s^{-1} [59], reproduced with permission © Royal Society of Chemistry.

conductivity, eco-friendliness, and ease of synthesis [57]. It is worth noting that MFe_2O_4 has an inverse spinel cubic structure in which O adopts face-centered cubic (fcc) packing, M^{2+} occupies the octahedral (O) interstitial site, and Fe^{3+} is distributed uniformly in the O and tetrahedral (T) sites (figure 3.8(a)) [58, 59]. This interesting structure has displayed excellent electrical conductivity because of the hopping of electrons between the diverse valence states of the metals in the oxygen sites and ought to supply indispensable surface redox-active metal centers for the adsorption and activation of active substances. MFe_2O_4 have been extensively studied for many applications in SCs, lithium-ion batteries, and electrocatalysts. Nevertheless, most research has mainly focused on powdered materials or compact films for spinel-type ferrites. In addition, these materials still have intrinsic defects; for instance, ferrite nanoparticles tend to aggregate, which severely reduces the exposed active sites and leads to relatively inferior structural stability.

$NiFe_2O_4$ with an inverse spinel structure has been extensively researched. However, the poor electrochemical properties of $NiFe_2O_4$ might be due to its highly agglomerating nature and low specific surface area, which decrease the usability of the active sites in the electrode material for electrolyte ions. Therefore, conductive substrates have been introduced to improve the electrochemical properties of $NiFe_2O_4$. $NiFe_2O_4$ nanoparticles were directly anchored to a flexible carbon cloth substrate via a simple surfactant-assisted hydrothermal approach [60]. The $NiFe_2O_4$/carbon textile (NFO/CT) electrodes thus obtained exhibited high specific capacitances of 1135.5 F g^{-1} in H_2SO_4 and 922.6 F g^{-1} in KOH at a current density of 2 mA cm^{-2}, as well as superior rate performance retentions of over 80% in both 6 M

KOH and 1 M H_2SO_4 electrolytes when the current density was increased to 100 mA cm^{-2}. The carbon cloth became a three-dimensional conductive network, supplied an ion diffusion path, and provided an ultrahigh surface area for $NiFe_2O_4$ nanomaterials, thus leading to an improvement in the specific capacitance of NFO/CT. Carbon textiles possess the merits of high conductivity, flexibility, and excellent mechanical strength. As shown in figure 3.8(b), Javed *et al* [59] reported the construction of hierarchical mesoporous NFO/CT via a facile hydrothermal method and a short subsequent heat treatment. The open space between hierarchical nanocones allowed the electrolyte ions to diffuse easily and the carbon textile ensured fast electron transportation that led to outstanding electrochemical performance. Due to the benefits of this architecture, a single electrode exhibited an excellent electrochemical performance by achieving the high capacitance of 697 F g^{-1} at a scan rate of 5 mV s^{-1}, and the solid-state SC exhibited a high capacitance of 584 F g^{-1} at a scan rate of 5 mV s^{-1} (figure 3.8(c)), good cyclic stability with a capacitance retention of 93.57% after 10 000 continuous cycles (figure 3.8(d)), and good flexibility (figure 3.8(e)). The direct growth of $NiFe_2O_4$ materials on other conductive hosts such as graphene, CNTs, nickel foam, and carbon paper has also recently been reported for energy storage devices [61–65]. Nevertheless, most of the conductive hosts are inflexible and rigid.

As an essential member of the binary transition-metal oxide family, Zn_2FeO_4 has received growing attention due to its chemical stability, good redox behavior, and thermal stability. Various $ZnFe_2O_4$ nanostructures, such as nanoflakes [66], nanoparticles [67], and nanospheres [68] have been reported to be promising electrodes for SCs. Vadiyar's group [67] prepared a $ZnFe_2O_4$ nanoflake@$ZnFe_2O_4$/C nanoparticle heterostructured electrode via anchoring $ZnFe_2O_4$ and carbon nanoparticles using an *in situ* biomediated green rotational chemical bath deposition approach. The $ZnFe_2O_4$ nanoflake@$ZnFe_2O_4$/C heterostructured electrode thus prepared demonstrated an ultrahigh capacitance of 1884 F g^{-1} at 5 mA cm^{-2}. Asymmetric all-solid-state flexible SC devices designed using $ZnFe_2O_4$ nanoflake@$ZnFe_2O_4$/C heterostructures and reduced graphene oxide as electrodes displayed a high energy density of 81 Wh kg^{-1} at a power density of 3.9 kW kg^{-1}.

3.3 Molybdenum-based materials

Due to the multivalent nature of molybdenum, molybdenum oxides show high electrochemical activity and have great potential for use in the field of SCs [69, 70]. MoO_3 is a layered n-type semiconductor with three main polymorphs, namely orthorhombic α-MoO_3 (a thermodynamically stable phase), monoclinic β-MoO_3, and hexagonal h-MoO_3 (low-temperature metastable phases). α-MoO_3 in particular is deemed to be one of the best potential energy storage candidates due to its high theoretical specific capacitance, abundant reserves, large potential window, and ideal cyclic life [71, 72]. α-MoO_3 possesses an anisotropic structure that is layered in the [010] direction, which is formed by stacking bilayer sheets of MoO_6 octahedra using van der Waals forces [73]. These structures may promote the insertion/deinsertion of electrolyte ions, such as H^+ and K^+. MoO_3 possesses a high theoretical capacity of 1256 F g^{-1}. Xiao's group compared the work function difference between positive and negative electrode materials and studied its effect on the potential window of SCs. They found that carbon@$K_{0.5}Mn_2O_4$/metal-oxide asymmetric SCs with MoO_3 as the anode displayed a higher potential window (2.42 V) than those using V_2O_5 (2.4 V), Fe_2O_3 (2.3 V), or WO_3 (2.2 V) anodes. As a result, a carbon@$K_{0.5}Mn_2O_4$/MoO_3 SC displayed outstanding energy density (66.7 Wh kg^{-1}) at 5.95 kW kg^{-1} [74]. However, its low electronic conductivity and structural collapse in an aqueous electrolyte are detrimental to high-rate and long-term cyclic performance in electrochemical devices [75].

Various MoO_3 nanostructures together with their derivative materials have been synthesized for use as anode materials in SCs [70]. MoO_3 nanobelts with an average length of 200 μm were synthesized via a hydrothermal method and then assembled into a flexible transparent MoO_3 nanopaper [76]. The 3D freestanding MoO_3 nanopaper showed good ion diffusion and delivered an outstanding specific capacitance of 1198 F g^{-1} at 2 mV s^{-1}, which was just slightly below the theoretical value; it also displayed excellent long-term stability over 20 000 cycles, achieving a retention rate of 96.5%. Oxygen vacancy-enriched MoO_{3-x} nanobelts with oxygen vacancies of up to 20% (corresponding to $MoO_{2.4}$) were synthesized by Cao's group [77]. It was discovered that the coexistence of oxygen vacancies and low-valence Mo ions enhanced both the mass and charge transport kinetics and catalyzed the redox reactions. The MoO_{3-x} nanobelt electrode thus obtained possessed excellent specific capacitance, rate capacity (1230 F g^{-1} and 1220 F g^{-1} at 5 A g^{-1} and 50 A g^{-1}, respectively), and cyclic performance (100% after 38 000 cycles) in a H_2SO_4/ethylene glycol electrolyte. The MoO_{3-x}/activated carbon asymmetric SC delivered high energy and power densities of 111 Wh kg^{-1} and 803 W kg^{-1}, respectively, (alternatively, 50 Wh kg^{-1} at 27 321 W kg^{-1}) and retained 80 Wh kg^{-1} and 794 W kg^{-1}, respectively, (alternatively 17 Wh kg^{-1} at 23 565 W kg^{-1}) at a low temperature of −25°C.

Composites with carbon materials [69, 72, 78–82] (such as active carbon, graphene, carbon nanotubes, carbon cloth) conducting polymers [83, 84] (such as polyaniline, polypyrrole), and MXene [85] have been widely prepared and lead to improved supercapacitor performances. Lee *et al* [86] anchored MoO_3 nanodots on CNT films through a hydrothermal method, and the MoO_3 nanodot/CNT electrode exhibited a specific capacitance of 281 F g^{-1} at 1 mV s^{-1} and a good rate performance of 79 F g^{-2} at 200 mV s^{-1} within a potential window of −1.0 to 0 V. Ultrathin MoO_3 nanocrystals were assembled on 3D graphene oxide frameworks via a hydrothermal reaction, forming a layered structure by oxygen-bonding interactions at the interface (figures 3.9(a)–(c), (e), and (f)) [80]. As shown in figure 3.9(d), the thickness of a single layer of MoO_6 octahedra in the [010] direction was about 0.7 nm, and a MoO_3/graphene sheet/MoO_3 sandwich structure was about 1.7 nm in thickness, leading to the abundant exposed active sites of the ultrathin MoO_3 and rapid ion diffusion and electron transport in 3D graphene frameworks. The resulting hybrids possessed high specific capacitance and excellent cyclic stability in both 1 M H_2SO_4 aqueous (527 F g^{-1} at the current density of 1.0 A g^{-1}, 100% retention after 10 000 cycles in a 1 M H_2SO_4 electrolyte, figure 3.9(g)) and H_2SO_4–PVA polymer

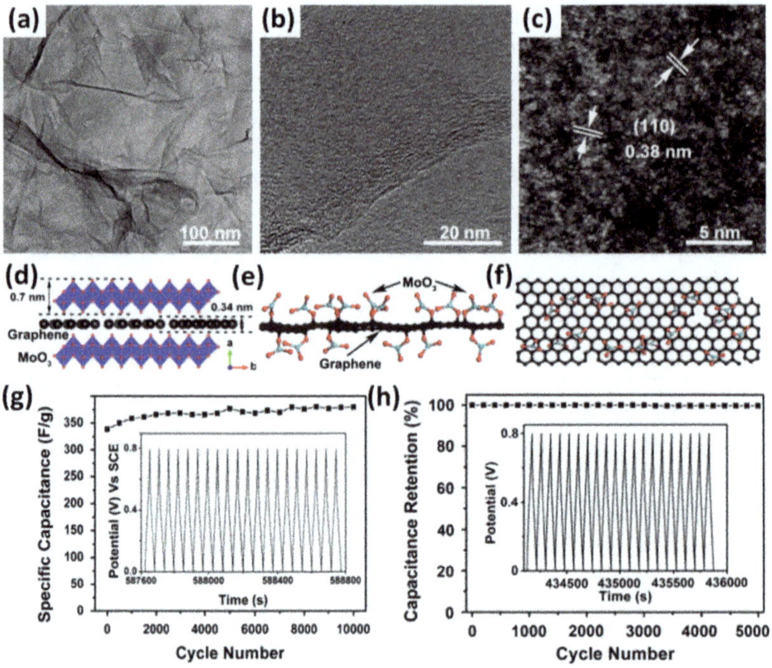

Figure 3.9. (a) Low- and (b) high-resolution TEM images of the MoO_3/graphene nanosheets. (c) A scanning transmission electron microscopy (STEM) image of the MoO_3/graphene nanosheets. (d) A schematic of self-assembled MoO_3 on graphene nanosheets. (e and f) Schematic diagrams of MoO_3 bonding with graphene. (g) The cyclic stability of the MoO_3/graphene nanosheet electrode at a current density of 10.0 A g^{-1} in a 1 M H_2SO_4 electrolyte. The inset shows the last 20 charge/discharge profiles. (h) The cyclic stability of the MoO_3/graphene nanosheet-based device at a current density of 1.0 A g^{-1} in a H_2SO_4–PVA polymer gel electrolyte. The inset shows the last 20 charge/discharge profiles [80], reproduced with permission © Elsevier.

solid electrolytes (373 F g^{-1} at 1.0 A g^{-1}, 100% retention after 5000 cycles, figure 3.9(h)). Wu *et al* [87] employed PPY as a conducting matrix and a coating agent to improve the conductivity of MoO_3 nanobelts and inhibit the dissolution of molybdenum ions during cycling. Their constructed PPy@MoO_3/activated carbon SC had a capacity retention rate of 83% after 600 cycles in a 0.5 M K_2SO_4 aqueous solution, which was much higher than the rate achieved by employing bare MoO_3 as the anode.

Heterointerfaces may facilitate electron transport and further effectively promote the electrochemical performances of electrode materials. As a consequence, core–shell structured heterostructures have been constructed to improve the supercapacitance of MoO_3 [88, 89]. A heterogeneous VO_x@MoO_3 composite was prepared via a facile electrochemical method [88]. The thin layer of MoO_3 effectively interacted with VO_x and modified its chemical environment and electronic structure. An electrode made from this composite displayed a high capacitance of 1980 mF cm^{-2} at 2 mA cm^{-2} and still retained 1166 mF cm^{-2} at a high current density of 100 mA cm^{-2}. It also supported stable long-term cycling and retained 94% of its capacitance after 10 000 cycles.

3.4 TiO_2-based materials

TiO_2 is considered to be a highly functional material with many interesting applications in diverse areas, including energy conversion and storage and solar cells due to its remarkable semiconducting characteristics and outstanding chemical stability. In an initial study, it was generally considered to be a conventional EDLC material which only contributed a very low faradic capacitance [90, 91]. Park's group [92] found that the specific capacitances of TiO_2 nanotube arrays were greatly influenced not only by their crystalline structure but also by the electrolyte composition. The volumetric capacitance of TiO_2 nanotube arrays was more than 1.8 times higher than that of activated-carbon-based EDLCs in a water-based electrolyte. Liu's research [93] indicated that highly ordered TiO_2 nanotube arrays exhibited much higher capacity in SCs (911 mF cm^{-2}) than nanocrystalline TiO_2 powder. Their relatively high specific capacity was attributed to the open-ended nanotube structure, which offered rigid tubular channel paths and an extremely large surface area accessible by solvated ions [93–95]. However, their specific capacitance was still restricted by the intrinsically poor electrochemical activity and poor electrical conductivity of TiO_2 [96].

It has been reported that the semiconducting behavior of TiO_2 nanotube arrays (NTAs) can be significantly altered by a thermal treatment that involves the introduction of oxygen vacancies (Ti^{3+} sites) or surface disorders to achieve improved capacitive behavior [97, 98]. As shown in figures 3.10(a)–(c), Tong *et al* prepared hydrogenated TiO_2 by the calcination of anodized TiO_2 nanotube arrays in a hydrogen atmosphere. The hydrogenated TiO_2 yielded a high specific capacitance of 3.24 mF cm^{-2} at a scan rate of 100 mV s^{-1}, which was 40 times higher than that of air-annealed TiO_2 nanotube arrays (figures 3.10(d) and (f)). It also yielded a remarkable rate capability of 68% when the scan rate increased from 10 to

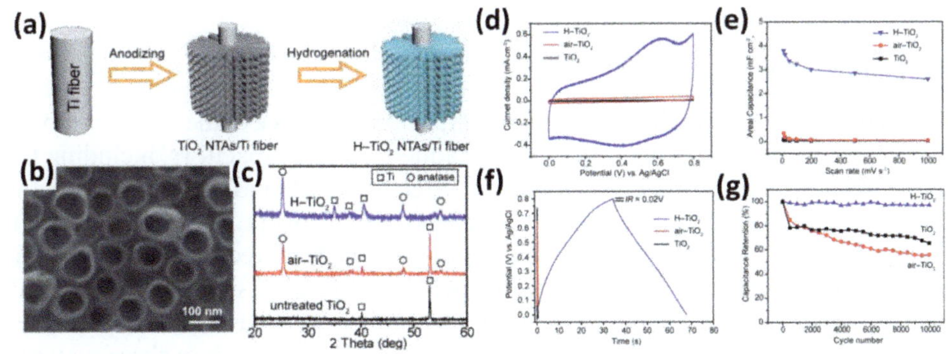

Figure 3.10. (a) A schematic diagram showing the fabrication of H–TiO_2 NTAs. (b) An SEM image of the H–TiO_2 nanotube arrays. (c) X-ray diffraction (XRD) spectra collected from untreated TiO_2, air–TiO_2, and H–TiO_2, respectively. The Ti signals originate from the Ti fiber. (d) CV curves of the untreated TiO_2, air–TiO_2, and H–TiO_2 electrodes obtained at a scan rate of 100 mV s^{-1}. (e) The areal capacitances of TiO_2 samples measured as a function of scan rate. (f) Galvanostatic charge/discharge curves of TiO_2 samples collected at a current density of 100 μA cm^{-2}. (g) The cyclic performance of TiO_2 samples measured at a scan rate of 100 mV s^{-1} for 10 000 cycles [99], reproduced with permission © American Chemical Society.

1000 mV s^{-1} (figure 3.10(e)) and long-term cyclic stability; only 3.1% of its capacity was lost after 10 000 cycles (figure 3.10(g)) [99]. The enhancement in capacitance was attributed to the combined contributions of the improved donor density and the increased density of the surface hydroxyl groups. Zhang *et al* prepared self-doped TiO_2 nanotube arrays by a simple cathodic polarization treatment on pristine TiO_2 to improved the conductivity and capacitive properties of TiO_2. The TiO_2 thus prepared exhibited a high specific capacitance of 1.84 mF cm^{-2} at a scan rate of 5 mV s^{-1}, which was 39 times higher than that of pristine TiO_2. It also exhibited excellent cyclic stability, retaining 93.1% of its capacitance after 2000 cycles. The enhancement of the self-doped TiO_2 nanotube arrays was attributed to a significant decrease in charge transport resistance through the solid material, which was accomplished by introducing oxygen vacancy sites [100]. Tang *et al* fabricated a binder-free multilayered hydrogenated TiO_2-II phase nanowire array electrodes, which possessed small TiO_2 grains (4.4 nm), a high oxygen vacancy (Ti^{3+}) concentration, and functionalized hydrophilic surfaces; the electrodes realized a significant improvement in the electric conductivity (a low bandgap of 2.39 eV), a more positive flat-band potential (−0.35 V), and high ion diffusion coefficiency. The broad bump-like peaks at −0.4 to −1.4 V (versus saturated calomel electrode) present in the CV curves were attributed to the intercalation/de-intercalation of Na^+ or H^+ ions in TiO_2 crystals, corresponding to faradic ion intercalation/de-intercalation in the prepared TiO_2-II electrode during the charge and discharge processes. The electrode exhibited a high areal specific capacitance (710.7 mF cm^{-2}). Moreover, TiO_2 nanotube arrays have been fabricated on highly conductive carbon fabric to form flexible electrodes for wearable and portable electronic devices [101].

$$Ti^{IV}O_2 + e^- + Na^+ \leftrightarrow Ti^{III}O(ONa) \tag{3.1}$$

$$Ti^{IV}O_2 + e^- + H^+ \leftrightarrow Ti^{III}O(OH) \tag{3.2}$$

3.5 Perspective

This chapter has comprehensively highlighted and discussed the most important advances in the field of anode materials for aqueous supercapacitors, including their fundamental charge storage mechanisms, effective strategies for enhancing their electrochemical performance, and the construction of tailored supercapacitor devices. Significant achievements have been attained in recent years, including achievements in the field of mechanism research and the design of flexible and/or smart devices based on metal-oxide anodes for aqueous supercapacitors. However, many research gaps still need to be filled for the wide application of these materials.

References

[1] Choi C, Ashby D S, Butts D M, DeBlock R H, Wei Q, Lau J and Dunn B 2019 *Nat. Rev. Mater.* **5** 5–19

[2] Augustyn V, Simon P and Dunn B 2014 *Energy Environ. Sci.* **7** 1597–614

[3] Nguyen T and Montemor M D 2019 *Adv. Sci.* **6** 1801797

[4] Li Y and Hu N 2020 *Instrum. Sci. Technol.* **48** 637–66
[5] Ma J, Guo X, Yan Y, Xue H and Pang H 2018 *Adv. Sci.* **5** 1700986
[6] Chen X, Chen K, Wang H and Xue D 2015 *Mater. Focus* **4** 78–80
[7] Zeng Y, Yu M, Meng Y, Fang P, Lu X and Tong Y 2016 *Adv. Energy Mater.* **6** 1601053
[8] Qu Q, Yang S and Feng X 2011 *Adv. Mater.* **23** 5574–80
[9] Kang J, Zhang S and Zhang Z 2017 *Adv. Mater.* **29** 1700515
[10] Gao Y, Wu D, Wang T, Jia D, Xia W, Lv Y, Cao Y, Tan Y and Liu P 2016 *Electrochim. Acta* **191** 275–83
[11] Chen J, Xu J, Zhou S, Zhao N and Wong C 2015 *Nano Energy* **15** 719–28
[12] Chaudhari S, Bhattacharjya D and Yu J 2013 *RSC Adv.* **3** 25120–8
[13] Wu N, Wang S, Han C, Wu D and Shiue L R 2003 *J. Power Sources* **113** 173–8
[14] Wang S, Ho G, Kuo S and Wu N 2006 *J. Electrochem. Soc.* **153** A75–80
[15] Kim J, Li W, Philips B L and Grey C P 2011 *Energy Environ. Sci.* **4** 4298–305
[16] Xia Q, Xu M, Xia H and Xie J 2016 *ChemNanoMat* **2** 588–600
[17] Chen Y, Lin Y, Hsu Y, Yen S, Chen K and Chen L 2014 *Small* **10** 3803–10
[18] Li J, Chen D and Wu Q 2019 *Eur. J. Inorg. Chem.* **10** 1301–12
[19] Pei Z, Meng Q, Wei L, Fan J, Chen Y and Zhi C 2020 *Energy Storage Mater.* **28** 55–63
[20] Zhu J, Childress A S, Karakaya M, Dandeliya S, Srivastava A, Lin Y, Rao A M and Podila R 2016 *Adv. Mater.* **28** 7185–92
[21] Zhai T, Sun S, Liu X, Liang C, Wang G and Xia H 2018 *Adv. Mater.* **30** 1706640
[22] Lu X, Zeng Y, Yu M, Zhai T, Liang C, Xie S, Balogun M S and Tong Y 2014 *Adv. Mater.* **26** 3148–55
[23] Fan X, Yu C, Yang J, Ling Z, Hu C, Zhang M and Qiu J 2015 *Adv. Energy Mater.* **5** 1401761
[24] Chen J, Xu J, Zhou S, Zhao N and Wong C 2016 *Nano Energy* **21** 145–53
[25] Owusu K A *et al Nat. Commun.* 2017**8** 14264
[26] Sun S, Zhai T, Liang C, Savilov S V and Xia H 2018 *Nano Energy* **45** 390–7
[27] Chen L, Yu Z, Wang J, Li Q, Tan Z, Zhu Y and Yu S 2015 *Nano Energy* **11** 119–28
[28] Zeng Y, Han Y, Zhao Y, Zeng Y, Yu M, Liu Y, Tang H, Tong Y and Lu X 2015 *Adv. Energy Mater.* **5** 1402176
[29] Binitha G, Soumya M S, Madhavan A A, Praveen P, Balakrishnan A, Subramanian K R V, Reddy M V, Nair S V, Nair A S and Sivakumar N 2013 *J. Mater. Chem.* A **1** 11698–704
[30] Lorkit P, Panapoy M and Ksapabutr B 2014 *Energy Procedia* **56** 466–73
[31] Zheng X, Yan X, Sun Y, Yu Y, Zhang G, Shen Y, Liang Q, Liao Q and Zhang Y 2015 *J. Colloid Interface Sci.* **466** 291–6
[32] Li J, Zhang W, Zan G and Wu Q 2016 *Dalton Trans.* **45** 12790–9
[33] Liu T, Ling Y, Yang Y, Finn L, Collazo E, Zhai T, Tong Y and Li Y 2015 *Nano Energy* **12** 169–77
[34] Li Y, Li Q, Cao L, Cui X, Yang Y, Xiao P and Zhang Y 2015 *Electrochim. Acta* **178** 171–8
[35] Shivakumara S, Penki T R and Munichandraiah N 2013 *ECS Electrochem. Lett.* **2** A60–2
[36] Wang R, Sui Y, Huang S, Pu Y and Cao P 2017 *Chem. Eng. J.* **331** 527–35
[37] Gao Y, Zhou R, Wang D, Huang Q, Cheng C and Zheng Z 2020 *Chem. Res. Chin. Univ.* **36** 97–104
[38] Shen B, Guo R, Lang J, Liu L, Liu L and Yan X 2016 *J. Mater. Chem.* A **4** 8316–27
[39] Samuel E, Joshi B, Jo H S, Kim Y I, An S, Swihart M T, Yun J M, Kim K H and Yoon S S 2017 *Chem. Eng. J.* **328** 776–84

[40] Li J, Wang Y, Xu W, Wang Y, Zhang B, Luo S, Zhou X, Zhang C, Gu X and Hu C 2019 *Nano Energy* **57** 379–87
[41] Kang L, Xie L, Li P, Liu T, Zhang X, Luo J and Liang W 2014 *Fuller. Nanotub. Carbon Nanostructures* **23** 715–20
[42] Wang H, Xu Z, Yi H, Wei H, Guo Z and Wang X 2014 *Nano Energy* **7** 86–96
[43] Zhou Z, Zhang Q, Sun J, He B, Guo J, Li Q, Li C, Xie L and Yao Y 2018 *ACS Nano* **12** 9333–41
[44] Guan C, Liu J, Wang Y, Mao L, Fan Z, Shen Z, Zhang H and Wang J 2015 *ACS Nano* **9** 5198–207
[45] Lu X, Chen X, Zhou W, Tong Y and Li G 2015 *ACS Appl. Mater. Interfaces* **7** 14843–50
[46] Wang L, Yang H, Liu X, Zeng R, Li M, Huang Y and Hu X 2017 *Angew. Chem. Int. Ed.* **56** 1105–10
[47] Wang Z, Xu Z, Huang H, Chu X, Xie Y, Xiong D, Yan C, Zhao H, Zhang H and Yang W 2020 *ACS Nano* **14** 4916–24
[48] Feng S, Wang X, Wang M, Bai C, Cao S and Kong D 2021 *Nano Lett.* **21** 7561–8
[49] Shi M, Xiao P, Lang J, Yan C and Yan X 2019 *Adv. Sci.* **7** 1901975
[50] Li H, Liu Y, Lin S, Li H, Wu Z, Zhu L, Li C, Wang X, Zhu X and Sun Y 2021 *J. Power Sources* **497** 229882
[51] Zheng Y, Zhou T, Zhang C, Mao J, Liu H and Guo Z 2016 *Angew. Chem. Int. Ed.* **55** 3408–13
[52] Tian W, Wang X, Zhi C, Zhai T, Liu D, Zhang C, Golberg D and Bando Y 2013 *Nano Energy* **2** 754–63
[53] Li Y, Xu J, Feng T, Yao Q, Xie J and Xia H 2017 *Adv. Funct. Mater.* **27** 1606728
[54] Tang X, Jia R, Zhai T and Xia H 2015 *ACS Appl. Mater. Interfaces* **7** 27518–25
[55] Li R, Ba X, Zhang H, Xu P, Li Y, Cheng C and Liu J 2018 *Adv. Funct. Mater.* **28** 1800497
[56] Dubal D P, Gomez-Romero P, Sankapal B R and Holze R 2015 *Nano Energy* **11** 377–99
[57] Szotek Z, Temmerman W M, Ködderitzsch D, Svane A, Petit L and Winter H 2006 *Phys. Rev.* B **74** 174431
[58] Selvan R K, Yuvaraj S and Lee Y S 2016 *RSC Adv.* **6** 21448–74
[59] Javed M S, Zhang C, Chen L, Xi Y and Hu C 2016 *J. Mater. Chem.* A **4** 8851–9
[60] Yu Z, Chen L and Yu S 2014 *J. Mater. Chem.* A **2** 10889–94
[61] Guan B, Guo D, Hu L, Zhang G, Fu T, Ren W, Li J and Li Q 2014 *J. Mater. Chem.* A **2** 16116–23
[62] Prasad K R and Munichandraiah N 2002 *J. Power Sources* **112** 443–51
[63] Li Y, Li Q, Wang H, Huang Y, Zhang X, Wu Q, Gao H and Yang J 2015 *Appl. Energy* **153** 78–86
[64] Wang Q, Jiao L, Du H, Wang Y and Yuan H 2014 *J. Power Sources* **245** 101–6
[65] Javed M S, Chen J, Chen L, Xi Y, Zhang C, Wan B and Hu C 2016 *J. Mater. Chem.* A **4** 667–74
[66] Vadiyar M M, Bhise S C, Ghule K, Kolekar S S, Chang J and Ghule A V 2016 *J. Mater. Chem.* A **4** 3504–12
[67] Vadiyar M M, Kolekar S S, Chang J, Ye Z and Ghule A V 2017 *ACS Appl. Mater. Interfaces* **9** 26016–28
[68] Wang Q, Zhu Y, Xue J, Zhao X, Guo Z and Wang C 2016 *ACS Appl. Mater. Interfaces* **8** 17226–32
[69] Tao T, Chen Q, Hu H and Chen Y 2012 *Mater. Lett.* **66** 102–5

[70] Rajeswari J, Kishore P S, Viswanathan B and Varadarajan T K 2009 *Electrochem. Commun.* **11** 572–5
[71] Brezesinski T, Wang J, Tolbert S H and Dunn B 2010 *Nat. Mater.* **9** 146–51
[72] Okashy S, Noked M, Zimrin T and Aurbach D 2013 *J. Electrochem. Soc.* **160** A1489–96
[73] Mai L, Hu B, Chen W, Qi Y, Lao C, Yang R, Dai Y and Wang Z 2007 *Adv. Mater.* **19** 3712–6
[74] Peng L, Wei X, Song K, Peng H, Li L, Hu J, Yang Y, Zhang H and Xiao P 2020 *Electrochim. Acta* **332** 135479
[75] Zhou L, Yang L, Yuan P, Zou J, Wu Y and Yu C 2010 *J. Phys. Chem.* C **114** 21868–72
[76] Yao B, Huang L, Zhang J, Gao X, Wu J, Cheng Y, Xiao X, Wang B, Li Y and Zhou J 2016 *Adv. Mater.* **28** 6353–8
[77] Wu Q, Zhao S, Yu L, Zheng X, Wang Y, Yu L, Nan C and Cao G 2019 *J. Mater. Chem.* A **7** 13205–14
[78] Shakir I, Nadeem M, Shahid M and Kang D J 2014 *Electrochim. Acta* **118** 138–42
[79] Ji H, Liu X, Liu Z, Yan B, Chen L, Xie Y, Liu C, Hou W and Yang G 2015 *Adv. Funct. Mater.* **25** 1886–94
[80] Zhou K *et al Nano Energy* 2015**12** 510–20
[81] Saeed G, Kumar S, Kim N H and Lee J H 2018 *Chem. Eng. J.* **352** 268–76
[82] Zhao N, Fan H, Zhang M, Ma J, Du Z, Yan B, Li H and Jiang X 2020 *Chem. Eng. J.* **390** 124477
[83] Jiang F, Li W, Zou R, Liu Q, Xu K, An L and Hu J 2014 *Nano Energy* **7** 72–9
[84] Zhang X, Zeng X, Yang M and Qi Y 2014 *ACS Appl. Mater. Interfaces* **6** 1125–30
[85] Zheng W, Halim J, Etman A S, Ghazaly A E, Rosen J and Barsoum M W 2021 *Electrochim. Acta* **370** 137665
[86] Lee T H, Pham D T, Sahoo R, Seok J, Luu T H T and Lee Y H 2017 *Energy Storage Mater.* **12** 223–31
[87] Liu Y, Zhang B, Yang Y, Chang Z, Wen Z and Wu Y 2013 *J. Mater. Chem.* A **1** 13582–7
[88] Xiao X *et al Adv. Energy Mater.* 2012**2** 1328–32
[89] Wang Q, Zhou C, Yan X, Wang J, Wang D, Yuan X and Cheng X 2018 *Energy Technol.* **6** 2367–73
[90] Xie Y, Zhou L, Huang C, Huang H and Lu J 2008 *Electrochim. Acta* **53** 3643–9
[91] Miller J R, Outlaw R A and Holloway B C 2010 *Science* **329** 1637–9
[92] Kim M S, Lee T W and Park J H 2009 *J. Electrochem. Soc.* **156** A584–8
[93] Salari M, Aboutalebi S H, Konstantinov K and Liu H 2011 *Phys. Chem. Chem. Phys.* **13** 5038–41
[94] Liu S, Pan G L, Yan N F and Gao X P 2010 *Energy Environ. Sci.* **3** 1732–5
[95] Kim J H, Zhu K, Yan Y, Perkins C L and Frank A J 2010 *Nano Lett.* **10** 4099–104
[96] Fabregat-Santiago F, Barea E M, Bisquert J, Mor G K, Shankar K and Grimes C A 2008 *J. Am. Chem. Soc.* **130** 11312–6
[97] Salari M, Konstantinov K and Liu H 2011 *J. Mater. Chem.* **21** 5128–33
[98] Wang G, Wang H, Ling Y, Tang Y, Yang X, Fitzmorris R C, Wang C, Zhang J Z and Li Y 2011 *Nano Lett.* **11** 3026–33
[99] Lu X, Wang G, Zhai T, Yu M, Gan J, Tong Y and Li Y 2012 *Nano Lett.* **12** 1690–6
[100] Zhou H and Zhang Y 2014 *J. Phys. Chem.* C **118** 5626–36
[101] Ahirrao D J, Higgins M W and Jha N 2019 *Appl. Surf. Sci.* **491** 765–78

IOP Publishing

Electrochemical Capacitors

Jianmin Ma and Junliang Yang

Chapter 4

'Rocking-chair' hybrid capacitors

Fangyuan Su and Chengmeng Chen

To combine the advantages of both supercapacitors and batteries, hybrid supercapacitors (HSCs) are constructed using a capacitive electrode and a faradic electrode simultaneously in one device. When the negative electrode is capacitive and the positive electrode is faradic, the ions in the electrolyte shuttle back and forth between the two electrodes. Hence 'rocking-chair' HSCs have been proposed. Herein, we introduce different kinds of 'rocking-chair' HSC, including Li^+, Na^+, and K^+ in organic electrolytes with different kinds of positive and negative material. First, the working mechanism of this kind of HSC is discussed. Second, the electrode materials and the corresponding electrochemical performance are systematically reviewed. Finally, we discuss the prospects of this type of capacitor.

4.1 The first proposal for 'rocking-chair' hybrid capacitors

Rechargeable batteries and electrochemical supercapacitors are two primary types of electrochemical device that are widely used for electrochemical energy storage. However, they have obvious differences in terms of energy density, power density, cyclic life, and low temperature performance due to their distinct energy storage mechanisms. For example, the energy density of commercial lithium-ion batteries (LIBs) is relatively high (200–300 Wh kg^{-1}); however, their power density (<1 kW kg^{-1}) and cyclic life (~2000 cycles) are quite low. However, in the case of supercapacitors (SCs), the situation is different. Compared with LIBs, the energy density of SCs is quite low (~5–7 Wh kg^{-1}). Nevertheless, their power density (>10 kW kg^{-1}) and cyclic life (~1 000 000) far exceed those of LIBs.

To obtain the advantages of both LIBs and SCs simultaneously, HSCs have been proposed as an advanced type of electrochemical energy storage device. Usually, HSCs are composed of an LIB-type electrode and an SC-type electrode in an aqueous or non-aqueous electrolyte. According to the different structures and energy storage mechanisms of the positive and negative electrodes, HSCs can be classified into two types: (i) physical adsorption negative electrode//electrochemical

doi:10.1088/978-0-7503-5042-6ch4

© IOP Publishing Ltd 2023

intercalative positive electrode, (ii) electrochemical intercalative (pre-lithiated or not) or pseudocapacitive negative electrode//physical adsorption positive electrode. The detailed descriptions of these HSCs have been thoroughly reviewed by many research groups [1–6]; their internal configurations and corresponding energy storage mechanisms are shown in figure 4.1 [6].

When type-(ii) HSCs are charged, the anions in the electrolyte (PF_6^-, for example) are adsorbed on the surface or into the pores of the positive electrode, while the Li ions pass through the separator to the negative electrode and take part in intercalative reactions. The electrolyte is consumed and decreases in quantity during the charge process and recovers to its initial quantity when the subsequent discharge process completes. Therefore, the concentration of the electrolyte varies during the charging/discharging process, which is called an 'accordion' mechanism. However, the depletion of the electrolyte can limit its ionic conductivity and even lead to the deterioration of the overall energy density of this type of device.

In the case of type-(i) HSCs, the situation is different. The Li ions are extracted from the positive electrode through an electrochemical reaction and then diffuse toward the negative electrode for physical adsorption during charging. During discharge, the Li ions come back to the positive electrode following desorption from the negative electrode. Therefore, the concentration of the electrolyte remains constant during the whole charge/discharge process. This kind of HSC uses 'rocking-chair' mechanism, which was first proposed for LIBs.

The first 'rocking-chair' hybrid capacitor was proposed by Xia's group in 2005 [7]. In this HSC, activated carbon (AC) is used as the negative electrode and spinel $LiMn_2O_4$ forms the positive electrode. Li ions are adsorbed on the surface of the AC electrode while they are simultaneously de-intercalated from the $LiMn_2O_4$ positive electrode. Consequently, the electrolyte depletion phenomenon is successfully suppressed.

Due to its larger specific surface area, controllable pore structure and good electronic conductivity, AC is widely used as a negative electrode for HSCs,

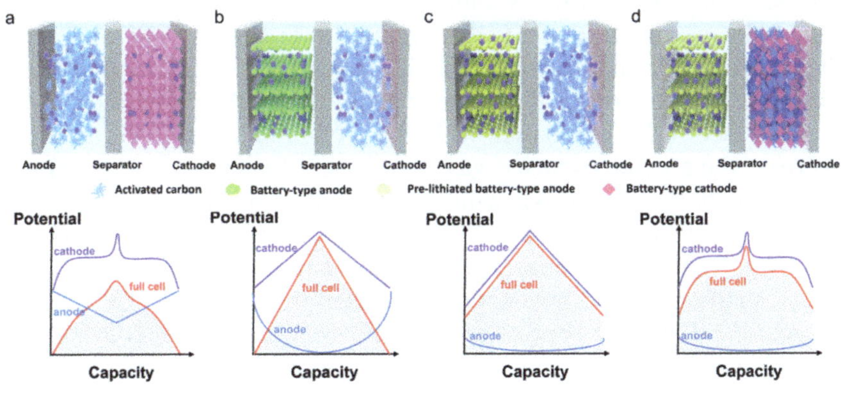

Figure 4.1. The configurations and corresponding energy storage mechanisms of HSCs: (a) 'rocking-chair'-type HSCs (battery-type positive electrode and capacitive negative electrode). (b), (c), and (d) 'Accordion'-type HSCs (battery-type negative electrode) [6], reproduced with permission © Royal Society of Chemistry.

although many kinds of porous carbon-based material, such as graphene-based materials [1, 8] and carbon nanotubes (CNTs) have also been introduced due to their unique advantages. For the cathode, the positive materials used in LIBs [2, 3, 9] and other metal-ion batteries [10, 11], such as Na and K ions, have also been widely investigated for use in constructing different kinds of HSC with excellent electrochemical performance.

4.2 Li-ion 'rocking-chair' HSCs

The concept of 'rocking-chair' HSCs was proposed for Li-ion HSCs; to date, these have been the most investigated type of HSC. In combination with AC or other kinds of porous carbon negative electrode, almost all kinds of positive LIB electrode have been used to construct 'rocking-chair'-type Li-ion HSCs.

4.2.1 Transition-metal-oxide positive materials

4.2.1.1 AC//$LiMn_2O_4$

Spinel-type $LiMn_2O_4$ (LMO) has many advantages in comparison with $LiNiO_2$ and $LiCoO_2$, such as safety, low toxicity, low cost, environmental friendliness, and the ample reserves of Mn [12]. Its theoretical reversible specific capacity is 148 mAh g^{-1} and its discharge profile has a plateau at a potential of approximately 4.0 V vs Li^+/Li [12, 13].

In AC//LMO HSCs, the potential of the LMO remains stable during the charge/discharge process, while the AC electrode's potential changes linearly. Li *et al* investigated the effect of the mass ratio of LMO versus AC on the performance of an AC//LMO HSC at different current densities and potential ranges [14]. They found that when the mass ratio of the two electrodes was larger than the corresponding capacity ratio, the voltage range was able to reach its upper limit. Hence, when the mass ratio was high, the AC electrode's potential could reach 0.8 V vs Li^+/Li, but at lower potentials (see figure 4.2(a)), the cyclic life became worse. They suggested that the potential of the AC electrode should be limited to 1.8 V vs Li^+/Li. Cericola *et al* investigated HSCs composed of AC, LMO, and $Li_4Ti_5O_{12}$ (LTO) [15]. The shape of the voltage profile of the AC//LMO HSC was very similar to that of the AC electrode. However, its potential position was shifted by the potential of the plateau of the LMO electrode, compared with the potential of the AC electrode (see figure 4.2(b)). The specific capacity was 30 mAh g^{-1} for the AC//LMO HSC (based on the mass of both electrodes), while the specific capacitance of the negatively polarized AC electrode was 120 F g^{-1}. The specific energy of their AC//LMO HSC was 45 Wh kg^{-1} and its maximum specific power reached 1.2 kW kg^{-1}.

LMO is usually prepared by a solid-state reaction between Li_2CO_3 and Mn_3O_4, which makes it difficult to control the particle size and morphology of the LMO material obtained. Therefore, its electrochemical performance is always unsatisfactory. To solve this problem, several studies involving particle morphology modification have been conducted [16–18], some typical results of which are shown in figure 4.3.

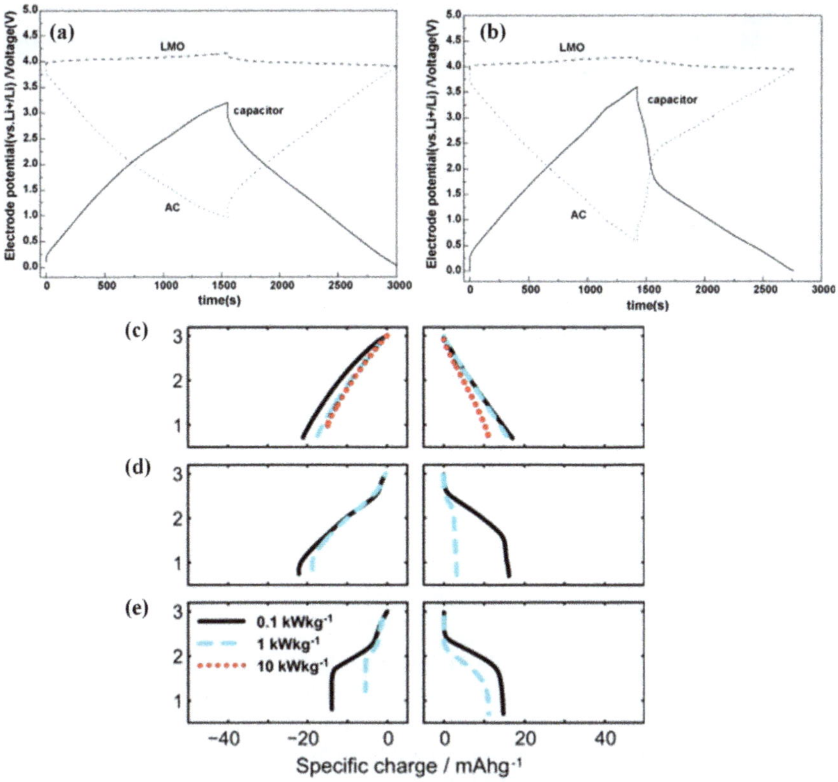

Figure 4.2. (a) and (b) The charge/discharge profiles of the AC//LMO HSC when the AC electrode potentials were ⩾1.0 V (vs Li^+/Li) and ⩾0.6 V (vs Li^+/Li), respectively [14], reproduced with permission © Elsevier. (c), (d), and (e) Profiles of the charge (left)/discharge (right) of AC//AC, (e) AC//LMO, and (f) LTO//AC HSCs at different power densities [15], reproduced with permission © Elsevier.

LMO nanorods were fabricated by a hydrothermal synthesis process using β-MnO2 nanorods as the precursor [16]. A Li-ion HSC was then constructed using the obtained LMO nanorods and AC as the positive and negative electrodes, respectively. As shown in figure 4.3(a), the diameter of the LMO nanorods was about 360 nm and their length was 2.3 μm. Figures 4.3(b) and (c) show the rate and cyclic performance of the resulting HSC. The specific capacity of the AC//LMO nanorods was twice that of an AC//AC symmetric supercapacitor when the current rate was 30 C. After 1000 cycles, the specific capacity of the AC//LMO nanorods was still 98% of its initial value.

An *et al* fabricated open-pore spherical LMO (OPS-LMO) consisting of nanosized primary particles using a spray-drying process [17]. Due to the rapid hot-air drying process, nanosized LMO particles with a porous spherical structure were obtained (figure 4.3(d)). A Li-ion HSC produced using this kind of LMO and AC as the positive electrode (anode) yielded high specific capacitances of 33.8 F cm^{-3} and 25.6 F cm^{-3} at currents of 1 A and 16 A, respectively (see figure 4.3(e)). In addition to its excellent rate ability, its capacitance retention was 83% after 5400 cycles

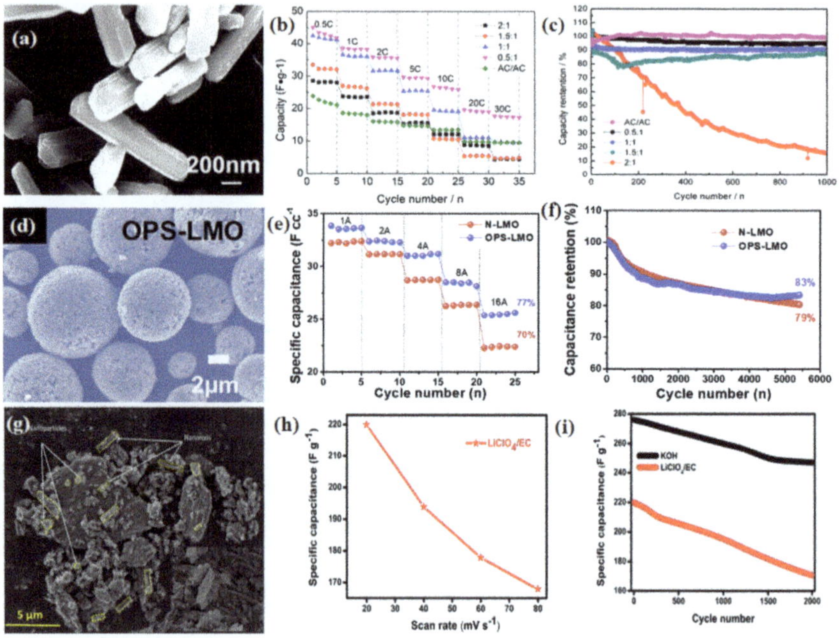

Figure 4.3. (a) Scanning electron microscopy (SEM) images of the LMO nanorods. (b) The rate performances of the AC//LMO nanorod HSC at current densities from 0.5 C to 30 C. (c) The cyclic performance of the AC//LMO nanorod HSCs at 2 C [16], reproduced with permission © ESG. (d) SEM images of open-pore spherical LMO (OPS-LMO). (e) The rate performances of an AC//OPS-LMO HSC at currents in the range of 1–16 A. (f) The cyclic performance of an AC//OPS-LMO HSC at a current of 2 A [17], reproduced with permission © Elsevier. (g) SEM images of LMO nanorods synthesized by microwave irradiation; (h) and (i) their specific capacitances at various scan rates and their cyclic performances, respectively [18], reproduced with permission © Elsevier.

(figure 4.3(f)). The authors attributed the performance enhancement to the efficient ionic diffusion resulting from the open-pore spherical structure of the LMO material.

Jose *et al* synthesized spinel LMO using a glucose-mediated microwave irradiation reaction between LiOH and MnO_2 [18]. As shown in figure 4.3(g), the LMO particles thus obtained took the shape of nanorods and had sizes from 0.8 μm to 2.4 μm. The large exposed surface area of this kind of LMO provided more sites for Li-ion diffusion. In a $LiClO_4$/EC electrolyte, the specific capacitances of AC//LMO were 220 F g^{-1} and 168 F g^{-1} at scan rates of 20 mV s^{-1} and 80 mV s^{-1}, respectively, while the capacitance retention was 78% of its initial value after 2000 cycles (figures 4.3(h) and (i)).

Low electronic conductivity is another challenge that needs to be overcome for LMO. In order to enhance its conductivity, Li *et al* synthesized an LMO/graphene composite material, which is shown in figure 4.4 [19]. In this composite material, LMO exhibited a sphere-like shape which was surrounded by many nanoparticles. Many graphene nanosheets were found coat the surfaces of the spherical LMO particles. The LMO material thus obtained demonstrated a specific capacitance of

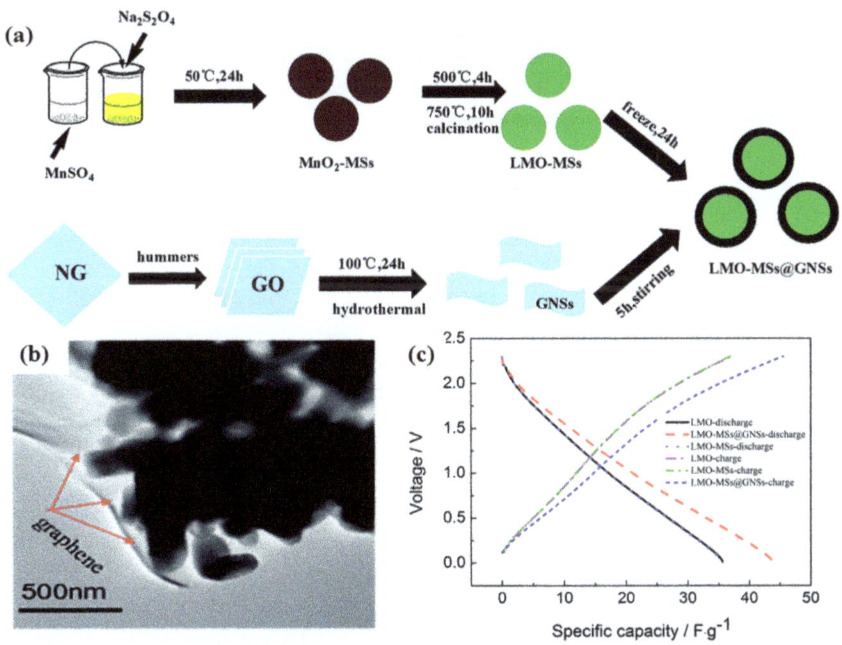

Figure 4.4. (a) A schematic illustration of the synthesis of LMO/graphene composites. (b) Transmission electron microscopy (TEM) images of the LMO/graphene composites obtained. (c) The charge/discharge profiles of Li-ion HSCs made using different kinds of LMO positive material at 2 C [19], reproduced with permission © Royal Society of Chemistry.

53.2 F g^{-1} at 0.5 C and excellent rate performance (35.6 F g^{-1}@8 C). A Li-ion HSC produced using this kind of LMO positive material displayed an energy density of 38.8 Wh kg^{-1} in an organic electrolyte.

It is well known that the Mn ions of LMO are easily dissolved in the electrolyte during the charge/discharge process. This creates another obstacle for the practical application of LMO-based materials. The Mn ion dissolution problem is usually attributed to the instability of the crystallographic structure of the LMO material and the formation of HF molecules in the electrolyte during the cyclic test. Zhang *et al* proposed lithium difluoro(oxalato)borate (LiDFOB) as an electrolyte additive for LIBs to enhance their cyclic performance [20]. Inspired by this work, Lee *et al* also introduced LiDFOB into an AC//LMO HSC, and the results are shown in figure 4.5 [21]. After the introduction of LiDFOB, the AC//LMO HSC showed a specific capacitance of 44 F g^{-1} (0.65 A g^{-1}), which increased by 3.9 F g^{-1} after 700 cycles. In addition, the polarization of the HSC was also effectively reduced. They attributed this enhancement to the fact that the LiDFOB could effectively suppress the Mn ion dissolution and the phased transition of LMO during the charging/discharging process.

The use of higher output voltages is a good way to further enhance the energy density of Li-ion HSCs. Therefore, researchers have introduced ionic liquids (ILs) that have wide potential windows as electrolytes for Li-ion HSCs [22, 23]. In

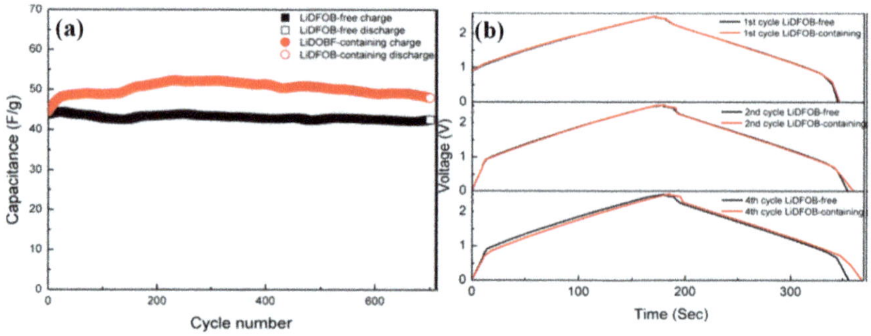

Figure 4.5. (a) The cyclic performances of AC//LMO HSCs with and without the LiDFOB additive. (b) The charge/discharge curves of Li-ion HSCs with and without the LiDFOB additive at different numbers of cycles [21], reproduced with permission © Springer Nature.

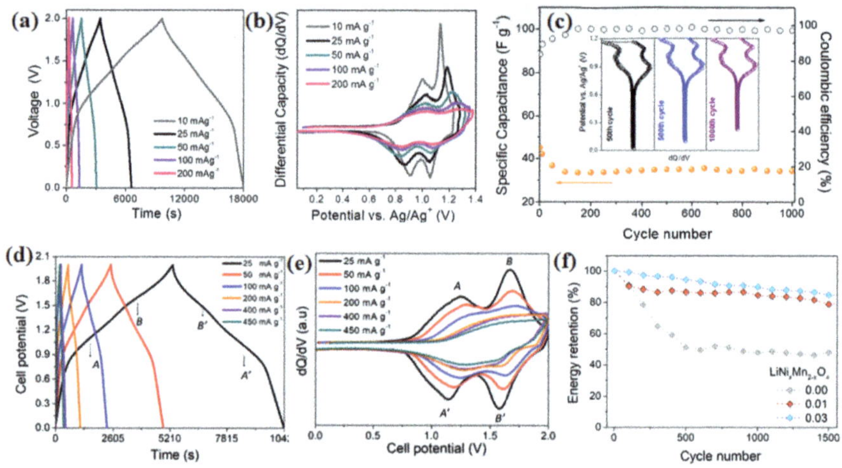

Figure 4.6. (a), (b), and (c) The charge/discharge profiles, differential capacity vs potential (d*Q*/d*V*) curves, and cyclic performance of the AC//LMO HSC with lithium bis(trifluoromethanesulfonyl)imide (LiTFSI)/1-ethyl-3-methylimidazolium bis(trifluoromethanesulfonyl)imide ([EMIm][TFSI]) as the electrolyte [24], reproduced with permission © Elsevier. (d), (e) and (f) The charge/discharge profiles, d*Q*/d*V* curves, and cyclic performance of the AC//Ni-doped LMO HSC based on LiTFSI (1 M) with 1-ethyl-3-methylimidazolium bis (trifluoromethylsulfonyl)imide (EMITFSI) as the electrolyte [27], reproduced with permission © American Chemical Society.

addition to their higher charged voltages, IL electrolytes also have the advantages of low vapor pressure, nonflammability, and higher thermal stability.

Lavall *et al* introduced lithium bis(trifluoromethanesulfonyl)imide (LiTFSI)/1-ethyl-3-methylimidazolium bis(trifluoromethanesulfonyl)imide ([EMIm][TFSI]) as an electrolyte for an HSC, using LMO as the cathode [24]. As shown in figures 4.6(a)–(c), the charge/discharge profiles of this HSC are slightly distorted. This was explained by the polarization caused by the viscosity of the IL electrolyte. A differential capacity vs potential (d*Q*/d*V*) analysis showed that two-step Li insertion reactions were taking place in the LMO cathode. The cell capacitance reached 50.7 F g^{-1} and the energy density was in the range of 17.9–28.0 Wh kg^{-1} when the power density was about 12.2–217.5 W kg^{-1}. After 1000 cycles, the specific capacitance decreased from 44.7 F g^{-1} to 33.9 F g^{-1}, i.e. a retention of 75.8%.

In order to retard the Jahn–Teller (J–T) distortion of the $[MnO_6]$ octahedral unit in LMO, metal cation doping has been a widely used methodology for the partial replacement of Mn^{3+} [25]. Among all metal cation substituents, Ni has been successfully applied [26]. In their subsequent work, Lavall *et al* systemically investigated the effect of IL-based electrolytes on the performance of Ni-doped $LiNi_xMn_{2-x}O_4$ ($0.01 \leqslant x \leqslant 0.10$) (figures 4.6(d)–(f)) [27]. In their experiment, a mixture of 1 M LiTFSI and 1-ethyl-3-methylimidazolium bis(trifluoromethylsulfonyl)imide (EMITFSI) as an IL was used as the electrolyte. They found that among all the HSCs studied, the one that used $LiNi_{0.01}Mn_{1.99}O_4$ delivered the highest energy density values of around 21.7–40.9 Wh kg^{-1}. At the same time, the Ni content had a strong effect on the cyclic life of the HSCs with the IL-based electrolyte. When the Ni content was low ($x \leqslant 0.03$), the retained capacity fell within the range of 79%–85% after 1500 cycles.

4.2.1.2 *AC//$Li_4Mn_5O_{12}$*

$Li_4Mn_5O_{12}$, which has been discovered to enhance the cyclic life of LMO, has been widely applied in both LIBs and HSCs [28–30]. Theoretically, the specific capacity of $Li_4Mn_5O_{12}$ is 163 mAh g^{-1}; this capacity is mainly available in the voltage region of 3 V vs Li^+/Li. However, nonstoichiometric $Li_4Mn_5O_{12}$ has been found to have a capacity of 60 mAh g^{-1} in the voltage range of 3.5 V–4.5 V vs Li^+/Li due to the presence of Mn^{3+} [31, 32]. Therefore, Kim *et al* synthesized $Li_4Mn_5O_{12}$ by the solution combustion method and attempted to use it in an HSC at a higher voltage than 4 V in a non-aqueous electrolyte. The discharge curves of an AC/$Li_4Mn_5O_{12}$ HSC at 1–2.5 V showed the reaction of the $Li_4Mn_5O_{12}$ electrode to be non-faradic (see figure 4.7(a)). The first specific discharge capacitances of AC//$Li_4Mn_5O_{12}$ were 54.8 F g^{-1} and 30.0 F g^{-1} at current densities of 0.1 mA cm^{-2} and 10 mA cm^{-2}, respectively. As shown in figure 4.7(b), after 100 cycles, the capacity retention of the AC//$Li_4Mn_5O_{12}$ HSC was 99.6% [33].

In another study, nonstoichiometric nanocrystalline $Li_4Mn_5O_{12}$ was synthesized using the sol–gel method at different pH values [34]. According to the results shown in figures 4.7(c) and (d), the $Li_4Mn_5O_{12}$ material fabricated at pH=9 showed the best performance. The specific capacity of $Li_4Mn_5O_{12}$ was 76 mAh g^{-1} in an organic electrolyte. The energy density of the final HSC reached 100 Wh kg^{-1}; however, its cyclic performance was unsatisfactory. After 100 cycles, the specific capacitance had decreased from 33 F g^{-1} to less than 15 F g^{-1}.

4.2.1.3 *AC//$LiNi_{0.5}Mn_{1.5}O_4$*

As mentioned above, Ni doping is an effective way of inhibiting the J–T distortion of LMO. At the same time, the introduction of Ni into crystalline LMO can also effectively improve its work voltage. $LiNi_{0.5}Mn_{1.5}O_4$, whose interaction potential is 4.7 V vs Li/Li^+, is believed to be a promising 5 V positive material for LIBs [35, 36]. The first Li-ion HSC based on $LiNi_{0.5}Mn_{1.5}O_4$ as a positive electrode in combination

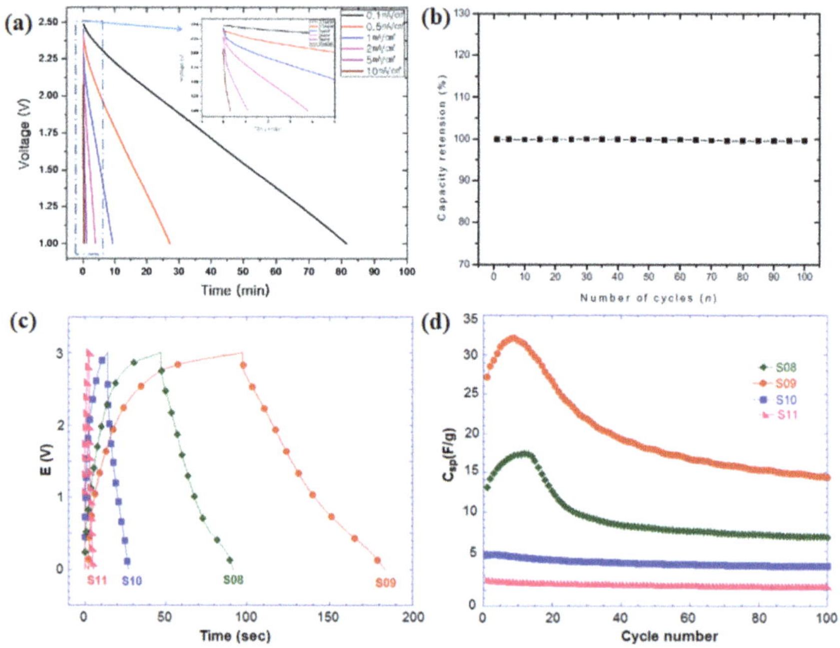

Figure 4.7. (a) The discharge profiles of the AC//$Li_4Mn_5O_{12}$ HSC at various current densities; (b) its cyclic performance at 5 mA cm^{-2} [33], reproduced with permission © IOP Publishing. (c) The discharge curve and (d) cyclic performance of hybrid capacitors based on AC//$Li_4Mn_5O_{12}$ synthesized by the sol–gel method [34], reproduced with permission © Institute of Physics.

with an AC negative electrode was exploited by Xia and co-workers [37]. In this kind of HSC, a sloping voltage profile appeared in the voltage range of 2.8–1.5 V. The specific capacity of all the active materials was 25 mAh g^{-1}, and their energy density was 55 Wh kg^{-1}. After 1000 cycles, the capacity retention was 80%.

Wu *et al* prepared $LiNi_{0.5}Mn_{1.5}O_4$ through a solid-state reaction at low temperature using metal acetate as the starting material [38]. An HSC constructed using this kind of $LiNi_{0.5}Mn_{1.5}O_4$ as the positive material delivered specific energies of 68 Wh kg^{-1} and 36 Wh kg^{-1} when the current density was in the range of 1–20 mA cm^{-2}. Figure 4.6(c) shows the cyclic performance at a current density of 20 mA cm^{-2}. It can be seen that this Li-ion HSC delivered very good cyclic performance. Even after 1000 cycles, its capacity retention was still greater than 95%.

Winter and co-workers synthesized $LiNi_{0.5}Mn_{1.5}O_4$ through a sol–gel process with a thermal treatment at 1000°C, and the results are shown in figures 4.8(d)–(f) [39]. Combined with an AC negative electrode, an HSC made using this material delivered excellent electrochemical performance over a cell voltage range of 3.3 V. The cyclic performance was outstanding. After 4000 cycles at a current density of 600 mA g^{-1}, the capacity retention was greater than 89%, and the average specific energy and power density were ~50 Wh kg^{-1} and ~1100 W kg^{-1}, respectively, during all of the cyclic tests [39].

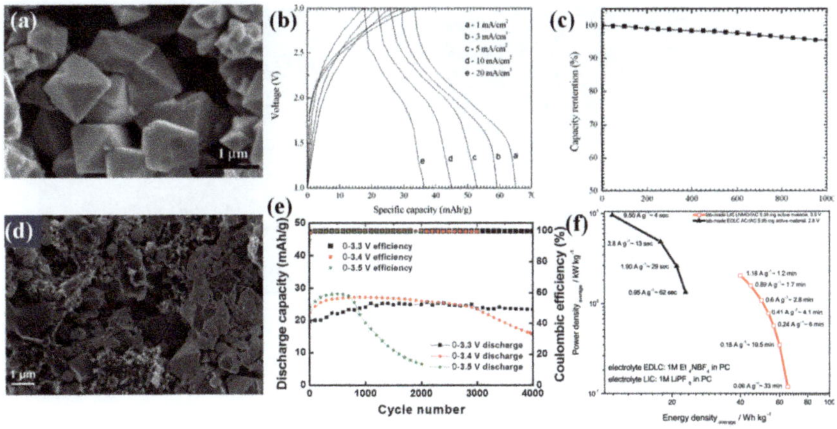

Figure 4.8. (a) An SEM image of $LiNi_{0.5}Mn_{1.5}O_4$ fabricated via a solid-state reaction [38], reproduced with permission © Elsevier. (b) and (c) The charge/discharge profiles of the AC//$LiNi_{0.5}Mn_{1.5}O_4$ HSC at different discharge current densities and its cyclic performance, respectively [38], reproduced with permission © Elsevier. (d) An SEM image of a cycled $LiNi_{0.5}Mn_{1.5}O_4$ electrode, the $LiNi_{0.5}Mn_{1.5}O_4$ of which was fabricated by the sol–gel method [39], reproduced with permission © IOP Publishing. (e) and (f) The cyclic performance and a Ragone plot of the AC//$LiNi_{0.5}Mn_{1.5}O_4$ HSC [39], reproduced with permission © IOP Publishing.

In addition, they also investigated the effect of the cell cutoff voltage (3.3 V, 3.4 V, and 3.5 V) on the performance of the AC//$LiNi_{0.5}Mn_{1.5}O_4$ Li-ion HSC. For increased cutoff voltages, the discharge capacity increased. However, the coulombic efficiency decreased from 99.4% to 94.2%. More seriously, the cutoff voltage also affected the cyclic life. The capacity retention decreased to only 23% after 2000 cycles when the cutoff voltage was 3.5 V. Finally, they claimed that a cutoff voltage of 3.3 V resulted in the best performance of the AC//$LiNi_{0.5}Mn_{1.5}O_4$ Li-ion HSC [39].

One-dimensional $LiNi_{0.5}Mn_{1.5}O_4$ nanofibers were fabricated via the electrospinning technique to enhance Li-ion diffusion kinetics [40]. The resulting material (shown in figure 4.9) delivered an energy density of ~19 Wh kg^{-1} with a good cyclic ability. The capacity retention was ~81% after 3000 cycles.

4.2.1.4 *AC//$LiNi_{1/3}Co_{1/3}Mn_{1/3}O_4$*

$LiNi_{1/3}Co_{1/3}Mn_{1/3}O_4$ (NMC) is another kind of well-known positive LIB material. NMC is also being widely studied as a positive material for Li-ion HSCs due to its large reversible capacity (270 mAh g^{-1}) and good thermal stability [41].

The morphology and particle size distribution of NMC have been adjusted in order to increase the energy density and power density of a Li-ion HSC. For example, Yoon *et al* synthesized NMC materials via carbonate and hydroxide co-precipitation [42]. As shown in figures 4.10(a) and (b), the NMC particles prepared via carbonate co-precipitation were spherical, and many large pores that were hundreds of nm in size were usually found on their surface. However, the morphology of NMC particles obtained by hydroxide co-precipitation was very different. They were secondary particles composed of rod-like primary particles,

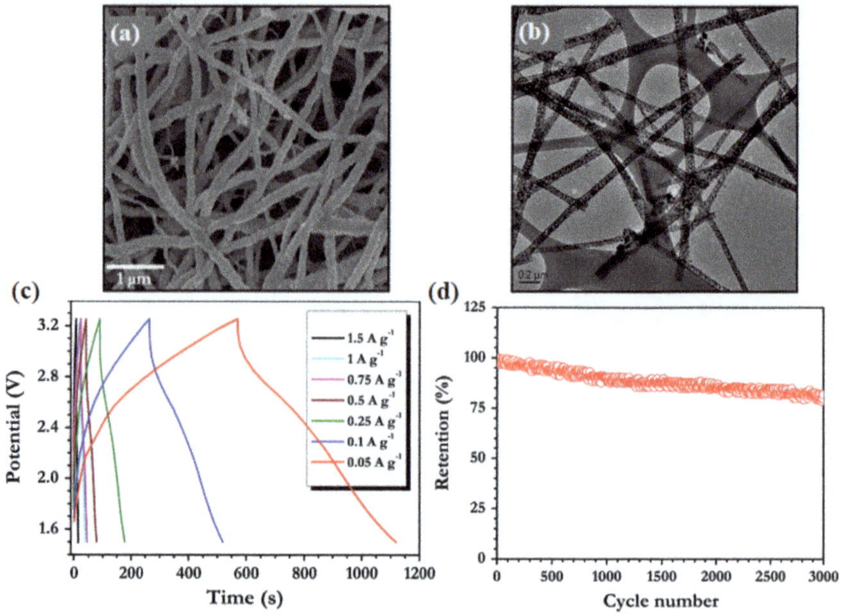

Figure 4.9. (a) and (b) SEM and TEM images of 1D $LiNi_{0.5}Mn_{1.5}O_4$ nanofibers fabricated via the electrospinning process [40]. (c) and (d) Charge/discharge profiles and cyclic performance of the AC//$LiNi_{0.5}Mn_{1.5}O_4$ nanofibers between 1.5 and 3.25 V [40], reproduced with permission © Elsevier.

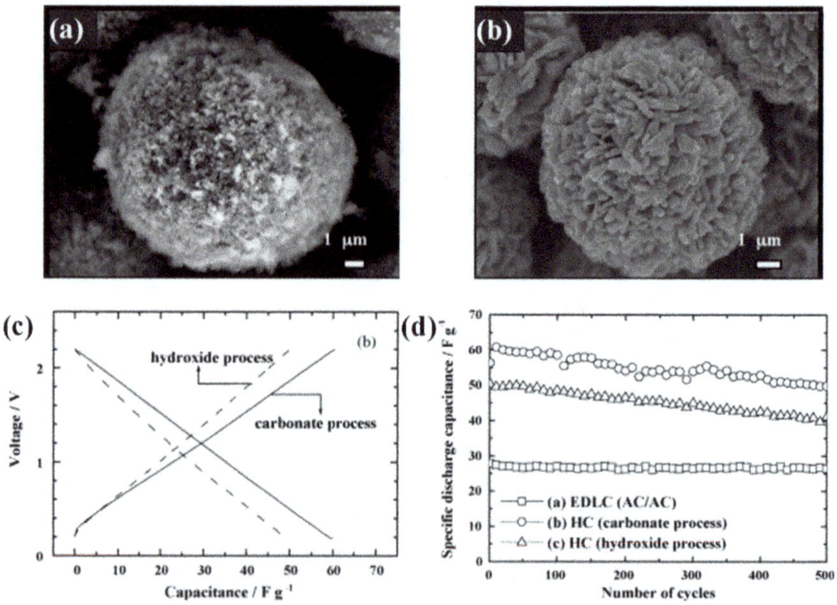

Figure 4.10. (a) and (b) SEM images of NMC prepared via carbonate co-precipitation and hydroxide co-precipitation, respectively [42], reproduced with permission © Elsevier. (c) and (d) Charge/discharge profiles at 15 C and the cyclic performance at 15 C, respectively [42], reproduced with permission © Elsevier.

which were 500 nm in width and several μm in length. The specific surface area of the former kind of particle was 5 m^2 g^{-1}, while that of the latter was 0.9 m^2 g^{-1}.

According to the charge/discharge profiles of the AC//NMC HSCs shown in figures 4.10(c) and (d), linear behaviors were observed for both kinds of positive NMC material. Based on the mass of both NMC and AC, an HSC that used NMC produced via the carbonate co-precipitation method showed a specific capacitance of 60 F g^{-1}, while an HSC that used NMC produced via hydroxide co-precipitation showed a specific capacitance of 50 F g^{-1}. Based on the cyclic performance at 15 C, the capacity retentions of the Li-ion HSCs using the two kinds of NMC were 88.4% and 81% after 500 cycles, respectively. However, both kinds of HSC showed better leakage current and self-discharge performance than those of an AC//AC supercapacitor.

To further improve the energy density of the HSC using NMC as the cathode, capacitive materials have also been introduced into the cathode together with NMC [43, 44]. Lee and co-workers synthesized layered $Li(Mn_{1/3}Ni_{1/3}Fe_{1/3})O_2$ (MNFe) instead of NMC for use as a cathode by replacing toxic Co with Fe [43, 45]. However, its electronic conductivity decreased. In order to improve its electronic conductivity, they used conductive polymers (polyaniline (PANI) and polypyrrole (PPy)) as effective coatings for the MNFe material. In this way, they constructed organic–inorganic composite positive electrodes for Li-ion HSCs.

PANI or PPy were able to bind to the MNFe particles, thus increasing the surface area and improving the electrochemical reaction sites, as well as improving the conductivity (see figures 4.11(a) and (b)). In addition, the coated polymer network was also able to adsorb more electrolyte, enhancing Li-ion diffusion during the charge/discharge process. More importantly, those polymers were even believed to contribute pseudocapacitance to the HSC.

As shown in figures 4.11(b) and (c), the electrochemical results suggested that the specific capacitance was 29 F g^{-1} for the HSC without the conductive polymer. However, when PPy and PANI were used to coat the NMFe, the specific capacitances reached 57 and even 140 F g^{-1} at 0.72 A g^{-1}, respectively. The HSC with the NMFe-PANI positive electrode delivered a capacitance retention of >90% after 1000 cycles, while the capacitance retentions of the HSCs without and with PPy were 83% and 87%, respectively, after 1000 cycles. The HSCs with conductive polymers also showed excellent rate performance. When the current rate was increased to 2.15 A g^{-1}, the specific capacitances of the HSCs with PPy and PANI were 27 F g^{-1} and 56 F g^{-1}, respectively.

AC has also been used as a capacitive electrode material in the positive electrodes of Li-ion capacitors [15, 46, 47]. Lee *et al* introduced AC into an NMC electrode; it was used both as a capacitive electrode and a conductive additive (figures 4.11(e) and (f)) [44]. Due to the contribution of AC to the composite cathode, the capacitance retention of the Li-ion HSC was 91.6% when the current rate increased to 5 A g^{-1}. After 9000 cycles, the retention was still 98.9%. When the power density was 8 kW kg^{-1}, the energy density was nearly 9 Wh kg^{-1}. The authors attributed this excellent cyclic performance to the structural stability of NMC that resulted from the introduction of AC into the cathode.

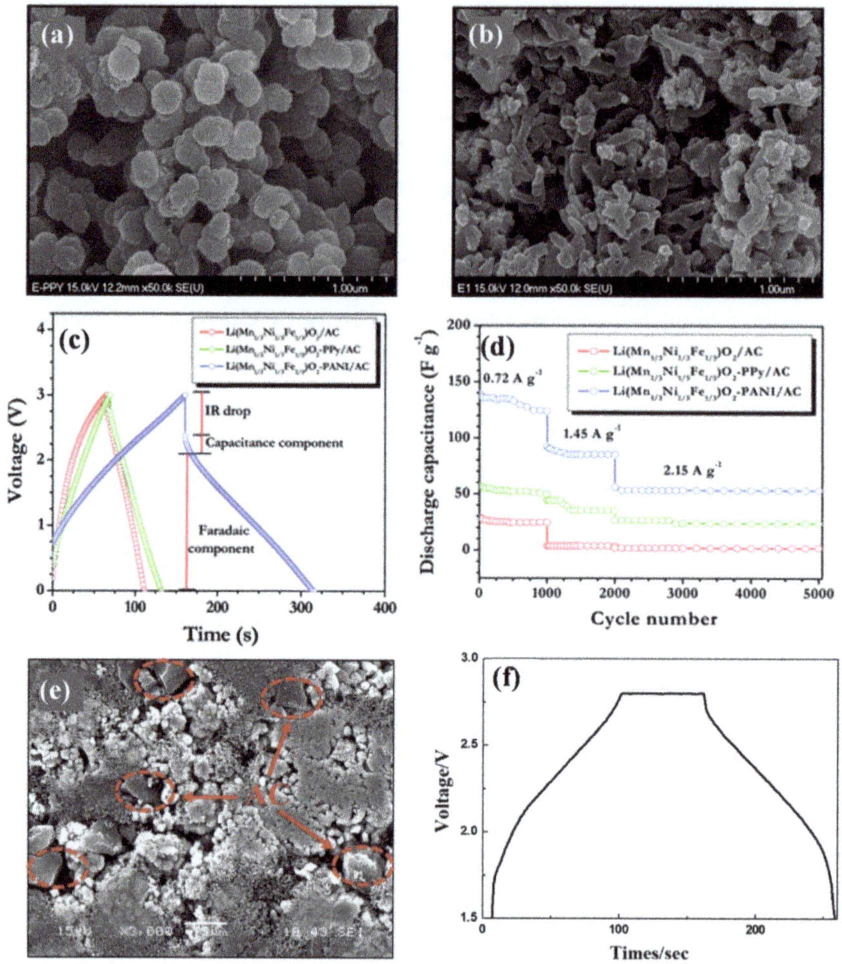

Figure 4.11. (a) and (b) SEM images of $Li(Mn_{1/3}Ni_{1/3}Fe_{1/3})O_2$–PPy and $Li(Mn_{1/3}Ni_{1/3}Fe_{1/3})O_2$–PANI composites synthesized via chemical polymerization [43], reproduced with permission © Royal Society of Chemistry. (c) and (d) Typical charge/discharge curves at 0.72 A g^{-1} and the cyclic performance of the AC//(Li $(Mn_{1/3}Ni_{1/3}Fe_{1/3})O_2$+conductive polymer) Li-ion HSCs [43], reproduced with permission © Royal Society of Chemistry. (e) and (f) SEM images of a composite electrode made from NMC+AC and its typical charge/discharge curve [44], reproduced with permission © Elsevier.

4.2.1.5 AC//$LiCoO_2$

Kim *et al* systematically compared the performances of $LiCoO_2$ (LCO) and LMO positive electrodes for Li-ion HSCs with AC as the negative electrode. In their study, LCO and LMO were also both mixed with AC to fabricate composite cathodes [48]. As shown in figure 4.12(a), they found that as the AC content decreased, the specific capacitance of LCO and LMO first increased and then decreased. The mass ratios of AC versus LMO and LCO that resulted in the highest capacitances were 0.14 and 0.16, respectively. However, the specific capacitance of LCO was lower than that of LMO for all content ratios.

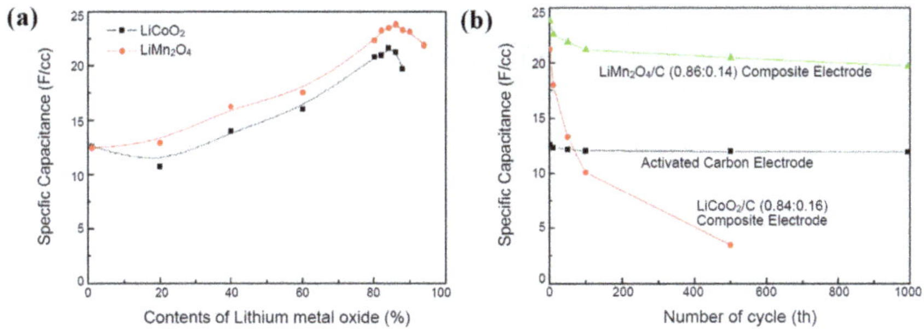

Figure 4.12. (a) and (b) Specific capacitances and the cyclic performance of Li-ion HSCs with LCO and LMO positive electrodes [48], reproduced with permission © Korean Chemical Society.

In regard to cyclic tests, as shown in figure 4.12(b), the performance of the composite positive electrode composed of LCO+AC degraded sharply as the number of cycles increased. However, the cyclic performance of the LMO+AC composite positive electrode was much better. Even after 1000 cycles, its capacitance retention was nearly 80%. This result was explained by the structural difference between LCO and LMO. The LCO had a layered structure which was easily broken by the frequent charge/discharge process, whereas the spinel structure of LMO was quite stable, promising good cyclic performance. In addition, poor LCO cyclic performance was also observed in another study [49]. Given the further consideration of its lower specific capacitance, therefore, LCO has rarely been investigated for use as the positive material of Li-ion HSCs with non-aqueous electrolytes.

4.2.2 Polyanionic compound positive materials

Polyanionic positive materials have rather stable crystal structures; therefore, their cyclic performance and safety properties are more satisfactory. This type of positive material, examples of which include $LiFePO_4$ (LFP), $Li_3V_2(PO4)_3$ (LVP), $LiMnPO_4$ (LMP), and $LiCoPO_4$ (LCP), have also been introduced into HSCs and widely investigated for use as positive materials.

4.2.2.1 AC//LiFePO4

Feng *et al* investigated the effect of the mass ratio of AC to LFP on the electrochemical performance of an AC//LFP Li-ion HSC [50]. They found that when the mass ratio increased from 0.33 to 3.19, the knee frequency decreased from 11.8 Hz to 0.34 Hz during electrochemical impedance spectroscopy (EIS) characterization, and the specific capacitance increased from 21 F g^{-1} to 166.4 F g^{-1}, indicating that the capacitive property was dominant in the HSC. With regard to the rate performance, the capacity retentions of HSCs with mass ratios of 1.93 and 3.19 were 90% and 80%, respectively, and the coulombic efficiencies of both HSCs were nearly 99%. After the electrochemical tests were complete, they found that the LFP positive electrode of the HSC with a mass ratio of 1.93 had not undergone significant structural change, and an solid electrolyte interphase (SEI) film was also not detected at the surface of the AC electrode. Therefore, the mass ratio of 1.93 was claimed to be the best value, and at this mass ratio, the AC//LFP Li-ion HSC delivered a high specific energy of 35 Wh kg^{-1}.

In order to further improve the performance of LFP-based HSCs, many kinds of LFP with different particle morphologies have been fabricated. Naoi *et al* fabricated core–shell LFP nanoparticles via an ultracentrifugation-triggered sol–gel reaction [51]. As shown in figure 4.13(a), in the innermost parts of the particles, the LFP was crystalline and had a diameter of 12–15 nm. However, the outer parts of the LFP particles (2–5 nm thick) were amorphous because of defects. The whole particle was

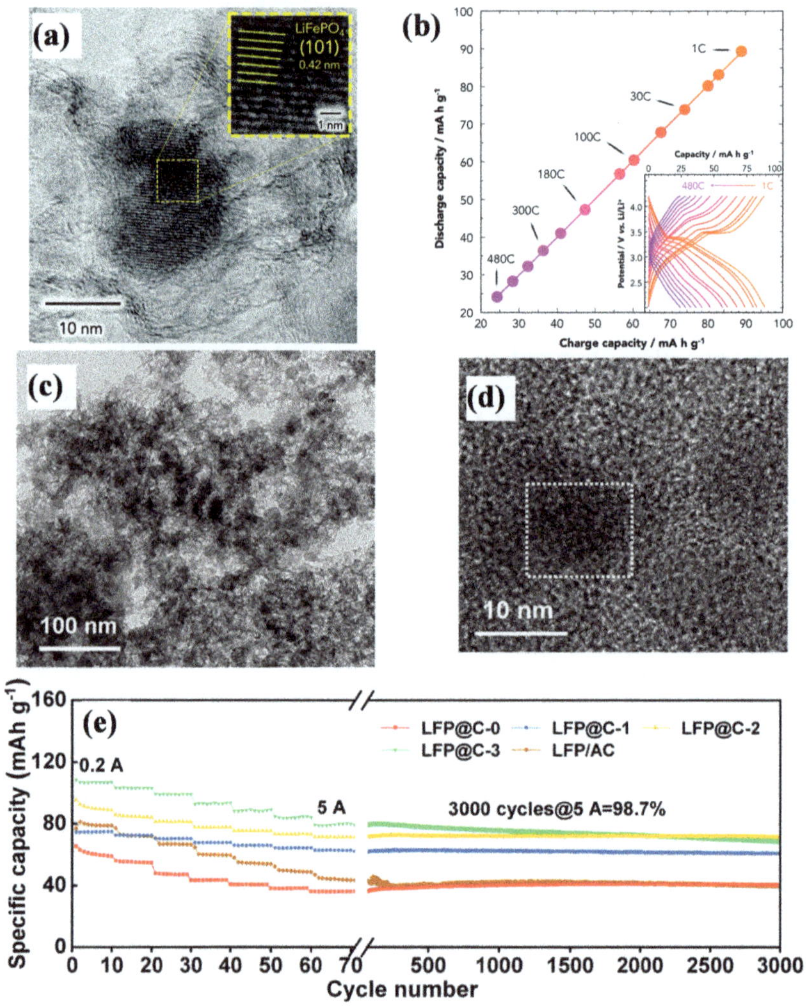

Figure 4.13. (a) A high-resolution TEM (HRTEM) image of the LFP/graphitic carbon composite and (b) its corresponding specific capacities at different rates (1–480 C) [51], reproduced with permission © Royal Society of Chemistry. (c), (d) TEM images of the nanosized LFP encapsulated in mesoporous carbon and (e) its cyclic performance [52], reproduced with permission © American Chemical Society.

then surface coated by a graphitic carbon layer to a thickness of 5–10 nm. Due to its unique structure, the charge/discharge profile showed three distinct regions. As in the case of other LFP materials, a plateau was observed at 3.4 V. However, two sloping profiles also occurred, one above and the other below the 3.4 V voltage region, which were attributed to the one-phase reaction of the amorphous defective outer LFP. The obtained LFP material delivered outstanding rate performance (figure 4.13(b)). Its specific capacities were 60 mAh g^{-1}, 36 mAh g^{-1}, and 24 mAh g^{-1} at discharge rates of 200 C, 300 C, and 480 C, respectively. Even when it was charged at high rates such as 100 C and 300 C, its specific capacity reached 60 mAh g^{-1} and 36 mAh g^{-1}, respectively.

Nanosized (~10 nm) LFP particles coated by interconnected mesoporous carbon were fabricated by Li *et al* [52]. Due to the capillary force produced by the mesopores in the carbon, the LFP particles were evenly distributed (figures 4.13(c) and (d)). SEM and N_2 adsorption characterization both indicated that all the LFP particles were inside the mesoporous carbon. The LFP/mesoporous carbon exhibited excellent rate performance. Its specific capacity was 78 mAh g^{-1} and its capacity retention at 5 A g^{-1} was 85%. A galvanostatic intermittent titration technique (GITT) test suggested that this nanostructure increased the Li-ion diffusion coefficient by two orders of magnitude. The energy density reached 131 Wh kg^{-1} when the power density increased to 25 kW kg^{-1}. As shown in figure 4.13, after 3000 cycles, the capacity had only decayed by 8%.

Lavall *et al* introduced a LiTFSI/ionic liquid as an electrolyte for AC//LFP HSCs [53]. Its energy density reached 43.4 Wh kg^{-1} at 10 mA g^{-1}, and after 1000 cycles, its capacity retention was 78%. However, due to the high viscosity of the electrolyte, its rate performance and output power density were unsatisfactory, needing further optimization.

4.2.2.2 AC//$Li_3V_2(PO4)_3$

LVP has three Li ions per molecule and hence its theoretical capacity can reach 197 mAh g^{-1} if all three Li ions are extracted [54]. According to previous studies, the extraction of Li ions in LVP occurs at 3.65–3.75 V, 4.15 V vs Li/Li^+ (1 Li), and 4.65 V vs Li/Li^+. However, LVP can also be used as a negative material, because at less than 3.0 V, LVP can also react with Li ions through a series of two-phase and single-phase processes [55]. Therefore, due to its unique electrochemical character, LVP can be used as a positive material or a negative material. It can also deliver high rates and powers. Hence, LVP is an attractive electrode material candidate for LIBs and HSCs [56].

Satish *et al* fabricated carbon-coated LVP via a fast sol–gel method and investigated its potential utilization as a positive and negative material (in combination with an AC electrode) for lithium-ion HSCs [57]. When used as a positive material (3.0 V–4.5 V), LVP underwent the following phase changes during charging: $Li_3V_2(PO_4)_3 \rightarrow Li_{2.5}V_2(PO_4)_3 \rightarrow Li_2V_2(PO_4)_3 \rightarrow Li_1V_2(PO_4)_3$. Its specific capacity in this voltage range was 125 mAh g^{-1} at a current rate of 100 mA g^{-1}. When the LVP was used as an anode, its voltage range was 1.0 V–3.0 V. Its specific capacity was lower than that of the positive electrode LVP, namely ~92 mAh g^{-1}.

The charge/discharge profiles of the AC//LVP and LVP//AC HSCs are shown in figure 4.14. The two figures show similar shapes. When LVP was used as a positive material, as shown in figure 4.14(a), the Li ions were extracted from LVP and then went to the AC negative material when the charging process started. At this time, the energy density was 27 Wh kg^{-1}. When the LVP was used as a negative material, PF_6^- anions in the electrolyte were adsorbed onto the surface of the AC, and Li ions were inserted into the LVP when the HSC was charged. The energy density was ~22 Wh kg^{-1}. Both kinds of HSC exhibited a good cyclic life. After 1000 cycles, the retained capacity was 66%.

As in the case of other kinds of positive LIB material, the electronic conductivity of LVP is poor, inhibiting its rate performance and hence the power density of the devices made from it. Carbon coating is usually used to enhance its electronic conductivity [58–60]. Metal-ion doping is also a good method. For example, Ni doping was adopted to improve the conductivity of LVP [61]. After it was doped by Ni, the $Li_3V_{1.95}Ni_{0.05}(PO_4)_3/C$ thus obtained delivered a specific capacity of 93 mAh g^{-1} at 100 C and retained 97% of its capacity after 1000 cycles. The energy density of a Li-ion HSC containing this kind of LVP positive electrode was 18.7 Wh l^{-1} even when its power density reached 2.8 kW l^{-1} [62].

4.2.2.3 AC//LiMnPO$_4$

LMP is another kind of phosphor-olivine compound which can be used as a positive material for LIBs. The theoretical capacity of LMP is 170 mAh g^{-1} and its redox potential is also comparatively high [63–65]. One study [66] synthesized nanosized LMP single crystallites coated by carbon, and their electrochemical performance as the positive material in a Li-ion HSC was investigated using different kinds of electrolyte. The charge/discharge profiles in all electrolytes, i.e. aqueous and organic, were battery-like (figure 4.15) when the current rate was low (for example, 0.6 A g^{-1}). However, when the current increased (>3 A g^{-1}), sloping voltage profiles appeared.

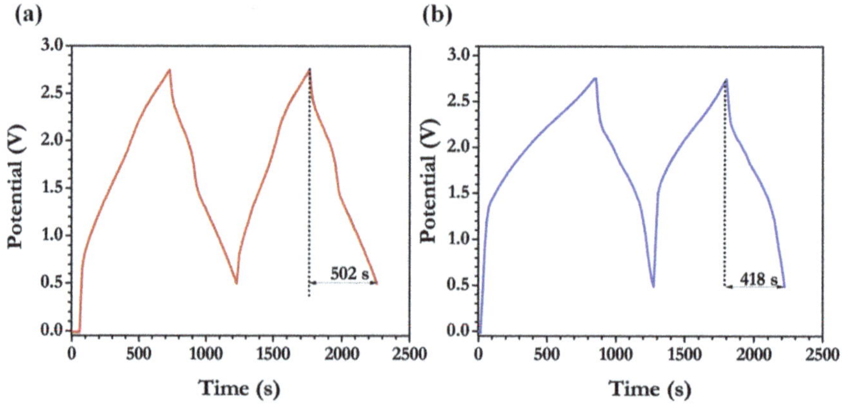

Figure 4.14. Charge/discharge profiles of Li-ion HSCs in which LVP was used as (a) the positive electrode or (b) the negative electrode at a current density of 100 mA g^{-1} [57], reproduced with permission © Elsevier.

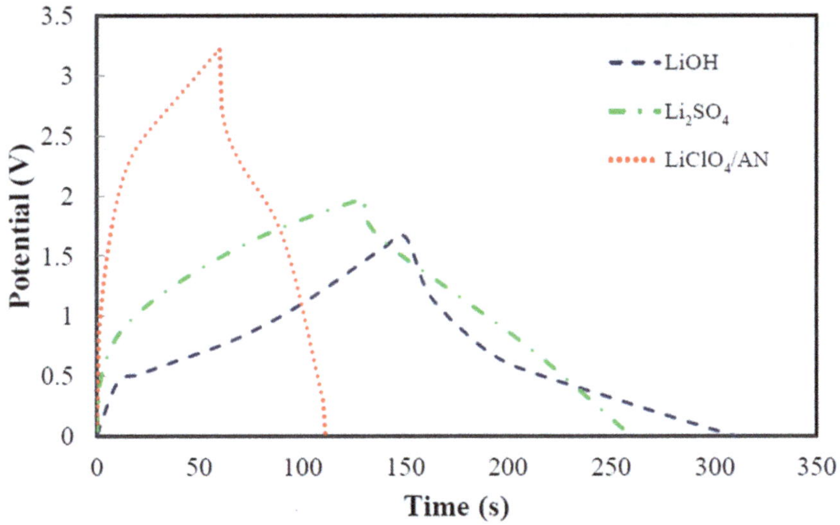

Figure 4.15. Typical charge/discharge curves of AC//LMP Li-ion HSCs at 0.6 A g^{-1} with different kinds of electrolyte [66], reproduced with permission © Elsevier.

4.2.2.4 AC//LiCoPO$_4$

LCP has also been investigated for use as the positive material in HSCs [67]. A porous nanostructured LCP/C composite material was fabricated via a mechano-chemical process by Kosova *et al* [68]. The total porosity of the material was in the range of 0.3 cm^3 g^{-1}–0.34 cm^3 g^{-1}, and it had submicron-sized primary particles. As shown in figure 4.16, its specific capacity decreased with increasing current density, and the fabricated LCP/C material showed pure capacitive behavior when the current density was 6.2 mA cm^{-2}. As a result of its pore structure, the power density of the AC//LCP Li-ion HSC reached 2.3 kW kg^{-1}. In order to improve its theoretical capacity and redox potential, F was introduced into LCP to form fluoro-oxyanion compounds [69]. The presence of the F ions also improved the stability of the crystal structure due to its high electronegativity. The resulting LCPF showed two negative peaks at 4.4 V and 4.8 V. The discharge capacity of LCPF was 70 mAh g^{-1} at a current density of 400 mA g^{-1} (see figure 4.16(b)). The specific capacitances of an AC//LCPF Li-ion HSC were 42 F g^{-1}, 34 F g^{-1}, 30 F g^{-1}, 24 F g^{-1}, and 21 F g^{-1} when the current rates were 1.05 C, 1.5 C, 2.5 C, 5 C, and 8 C, respectively. Based on the total weight of the materials used in the two electrodes, the energy density reached 47 Wh kg^{-1} when the power density was 215 W kg^{-1}. After 30 000 cycles at 1.1 A g^{-1}, the capacity retention was still greater than 92%.

4.2.3 Graphene-based porous negative electrodes

AC was typically used as the non-faradic negative electrode in the HSCs discussed above. However, due to its complex pore structure, its rate performance can also be limited by its accessible surface area. To solve this problem, graphene nanosheets have been used in capacitive electrodes [70–72].

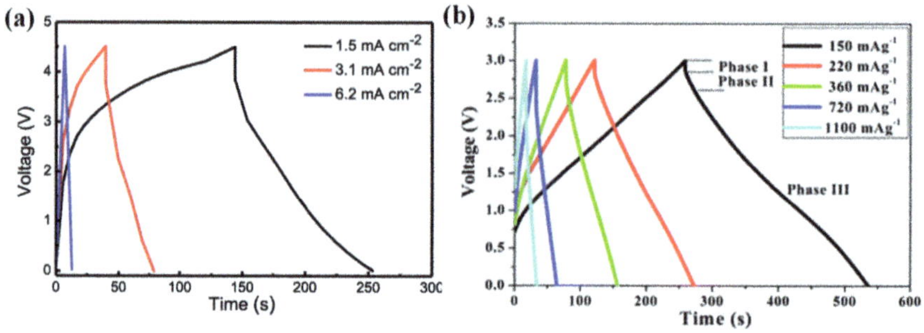

Figure 4.16. Charge/discharge profiles at different current densities: (a) the porous LCP/C composite [68], reproduced with permission © Springer Nature; (b) LCPF [69], reproduced with permission © Royal Society of Chemistry.

Dai *et al* adopted a freestanding reduced graphene oxide (rGO)/CNT membrane as the negative material for an HSC with an LFP cathode [73]. They found that this HSC could exhibit an energy density of more than 180 Wh kg^{-1} when the power density was 218 W kg^{-1}. Due to the irreversible electrochemical reaction, the coulombic efficiency of the rGO/CNT negative film during the first cycle was only 50%. However, it reached 92.7% in the second cycle. The specific capacity of this film was 570 mAh g^{-1} at a low current rate, and it still retained a specific capacity of 230 mAh g^{-1} when the current rate increased to 1 A g^{-1}. The highly porous structure created by the linked CNTs and rGO provided enough room for Li-ion storage and quick diffusion (figure 4.17(a)–(c)).

N-doped rGO foam was fabricated by Huang *et al* via a hydrothermal process and used as the negative material together with a $Ni_{0.25}V_2O_5$ (NVO) cathode [74]. The pore size of the foam was in the range of 5–75 nm and its surface area was 150 m^2 g^{-1}. Therefore, it provided many active sites for Li-ion adsorption during the charge/discharge process. This foam showed excellent rate performance in the voltage range of 0.01–2 V, and even after 500 cycles, its specific capacity still reached 160 mAh g^{-1} at a current rate of 4.5 A g^{-1}. A full HSC cell delivered very good electrochemical performance. Its energy densities were 63 Wh kg^{-1} and 32 Wh kg^{-1} for current densities of 5 A g^{-1} and 10 A g^{-1}, respectively. After over 5000 cycles, its energy retention was 87% (figure 4.17(d), (e)).

Many kinds of oxygen-containing functional group can be found on the surface of graphene oxide. Inspired by the fact that carbonyl groups (>C=O) in polymeric electrodes can react with Li ions to form redox pairs, Jang *et al* attempted to use graphene oxide as both the positive and negative electrode material to store Li ions [75]. The energy storage mechanism and its performance are shown in figure 4.18. During the first discharge process, the Li ions from the Li powder placed on the negative side penetrated the separator and entered the surface of the rGO positive electrode (figure 4.18(a)). In the subsequent charge and discharge process, Li ions shuttled between the positive electrode and the negative electrode. The energy storage and release of the system relied on the electrochemical reaction and

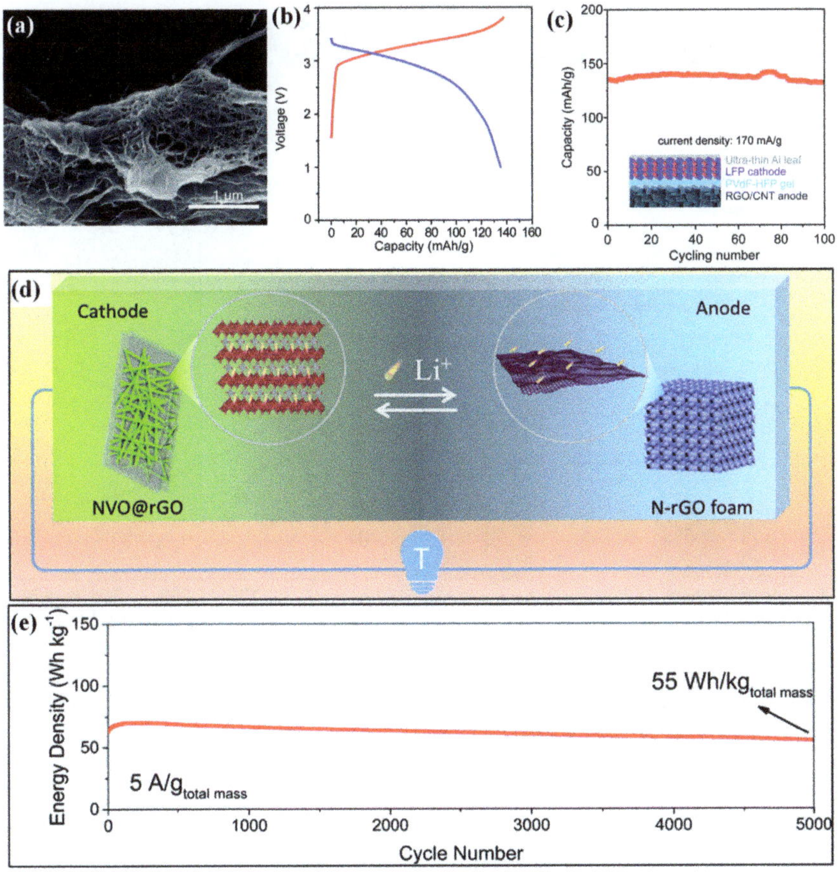

Figure 4.17. (a), (b), and (c) SEM images of the porous network of the rGO/CNT film, its charge/discharge profile, and its cyclic performance, respectively [73], reproduced with permission © Royal Society of Chemistry. (d) and (e) A schematic of the HSC combining the NVO@rGO and N-rGO foam and its cyclic performance, respectively [74], reproduced with permission © Royal Society of Chemistry.

adsorption of Li ions on the surfaces of the positive and negative rGO-based materials. Due to the rich pore structure and high specific surface area of the positive and negative rGO materials, Li ions were quickly released from the surfaces of the electrode materials, promising outstanding rate performance (figures 4.18(b) and (c)). At a power density of 100 W kg^{-1}, the highest energy density reached 160 Wh kg^{-1} (based on the overall mass), which was equivalent to that of LIBs. The highest power density was 100 kW kg^{-1}, ten times that of supercapacitors.

Therefore, graphene-based active materials can achieve an energy density close to that of LIBs through the oxidation/reduction reactions of Li ions and surface functional groups. At the same time, the high specific surface area of graphene-based electrodes can adsorb the electrolyte liquid well, shortening the transportation distance of Li ions in the electrolyte. Such electrodes have both the capacitance characteristics of the electric double layer and redox battery characteristics.

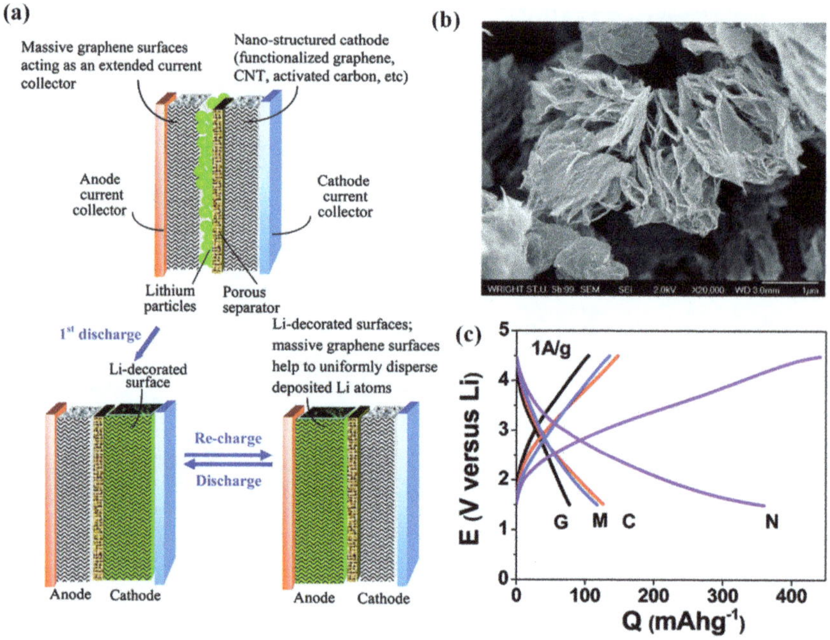

Figure 4.18. (a) The Li-ion storage mechanism of the HSC that used rGO as both the positive electrode and the negative electrode. (b) and (c) The morphology of the rGO and the charge/discharge profiles of the rGO Li-ion HSC, respectively, [75], reproduced with permission © American Chemical Society.

Subsequently, 35.4% TiN was added to the graphene electrode to improve the conductivity of the composite electrode and prevent the agglomeration of graphene sheets by Han *et al* [76]. The energy density of a device constructed using this material reached 162 Wh kg^{-1} when the power density was 150 W kg^{-1} and the potential range was 0.005 V–3.0 V (figures 4.19(a)–(c)). The same author later prepared a rGO/MoO_2 composite electrode containing 10 wt% graphene. The specific capacitance of the rGO/MoO_2 electrode was 624 F g^{-1}, much higher than that of rGO without MoO_2. Even at a current density of more than 1 A g^{-1}, its specific capacitance was still very high (~173.2 F g^{-1}). Therefore, a specific energy density of 142.6 Wh kg^{-1} was obtained at a power density of 150 W kg^{-1} (figures 4.19(d)–(f)) [77].

4.3 Na- and K-ion 'rocking-chair' HSCs

4.3.1 Na-ion HSCs

Nowadays, Na-ion batteries (NIBs) are being widely investigated as an alternative to LIBs, because Na sources are abundant and low cost, while Li is very scarce on the Earth [78–81]. The basic electrochemical process of NIBs is similar to that of LIBs. During the charge/discharge process, the Na ions migrate across the separator toward the negative material and the cathode [82].

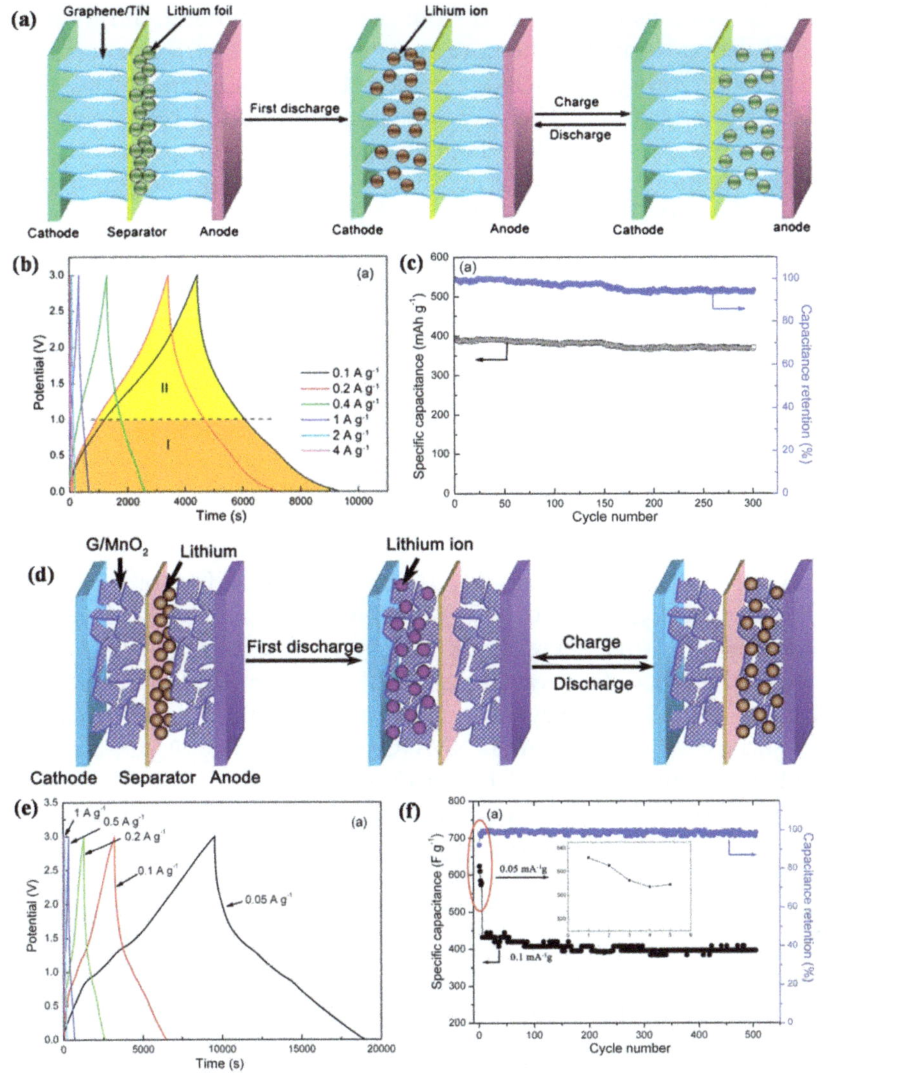

Figure 4.19. (a) The working mechanism of the rGO Li-ion HSC with TiN; (b), (c) its charge/discharge profiles and cyclic performance [76], reproduced with permission © Royal Society of Chemistry. (d) The working mechanism of the rGO Li-ion HSC with MOO_2; (e), (f) its charge/discharge profiles and cyclic performance [77], reproduced with permission © Royal Society of Chemistry.

Na-ion capacitors (NICs) have also been proposed, in which both capacitive and battery-type electrode materials are used [83–86]. Although the research related to NICs is in its early stages, many kinds of configuration have been investigated. Among them, 'rocking-chair'-type Na-ion HSCs, which are based on the energy storage mechanism of Li-ion HSCs, have been proposed [87]. In these Na-ion HSCs, various types of Na-containing positive electrode material from NIBs are used for the positive electrode, while porous carbon materials are used for the negative electrode. The Na ions are extracted from the positive material during the charging process and then adsorbed in the pores or on the surface of the capacitive negative carbon material. During the discharge process, the Na ions return to the positive side from the negative electrode. Therefore, the concentration of Na ions in the electrolyte remains constant during the whole charge/discharge process.

Layered transition-metal oxide compounds are possibly the most investigated class of positive materials for NIBs and NICs. These kinds of material can be classified into O3, P2, or P3 types, depending on the stacking sequence of O layers in their crystal structures [88]. O3-structured electrodes, in which the O layers are stacked in an ABCABC structure, usually experience a complex phase change when they are charged to high voltages. Their cyclic life is always unsatisfactory. However, due to their wider diffusion channel, P2 structures (ABBA stacking) are more stable, even when they work within a wider potential window [89]. To further enhance the electrochemical performance of P2-type layered cathodes, metal-ion doping [90] and coating are adopted.

4.3.1.1 P2-type cathodes

Chen *et al* introduced an ultrathin Al_2O_3 film onto the surface of a P2-type positive material, $Na_{0.66}Mn_{0.54}Ni_{0.13}Co_{0.13}O_2$ (NMNC) via the atomic layer deposition (ALD) method to further increase its Na-ion storage ability at high voltage [91]. Using a sol–gel method, they synthesized NMNC material with an average particle size of 1.5 μm. After 84 ALD cycles, a uniform 9.4 nm thick Al_2O_3 layer was coated onto the surface of the prepared NMNC material. The Al_2O_3-coated NMNC showed a high specific capacity of 125 mAh g^{-1} within the range of 2 V–4.5 V when it worked as a positive material vs a Na electrode; this specific capacity was much higher than that of NMNC without Al_2O_3. Commercial AC was used as the counter electrode for the Al_2O_3-coated NMNC cathode. Through EIS characterization, it was found that the presence of the Al_2O_3 layer could effectively reduce the interfacial resistance. The specific capacitances of Na-ion HSCs based on NMNC with and without an Al_2O_3 layer were 66 F g^{-1} and 59 F g^{-1} at 0.35 A g^{-1} in the voltage range of 0–3 V, respectively. The energy density of the Na-ion HSC based on the Al_2O_3-coated NMNC positive electrode was 63 Wh kg^{-1} when the power density was 6.6 kWh kg^{-1}. After 5000 cycles, the HSC that employed the Al_2O_3-coated NMNC showed a capacitance retention of 98%, while the HSC without the Al_2O_3 layer only retained 78%. The presence of Al_2O_3 formed a film on the surface of the NMNC material that protected it from harmful electrolyte attack and also reduced electrode dissolution (figure 4.20).

In another study, a kind of porous carbon, ZIF-8-derived carbon (ZDC), was derived from zeolitic imidazolate framework-8 (ZIF-8). The ZDC was used as the negative electrode together with a P2-Na0.67Co0.5Mn0.5O2 (P2-NCM) positive material to form Na-ion HSCs [92]. The pore diameter of the ZDC-type negative material was ~150 nm and it had graphitic pore walls. Its specific surface area was 1259 m^2 g^{-1}, higher than that of commercial AC. The high surface area–pore ratio indicated that the ZDC thus obtained had a high exterior surface area. After 2000 cycles, its capacity still remained greater than 92%. The energy density of HSC

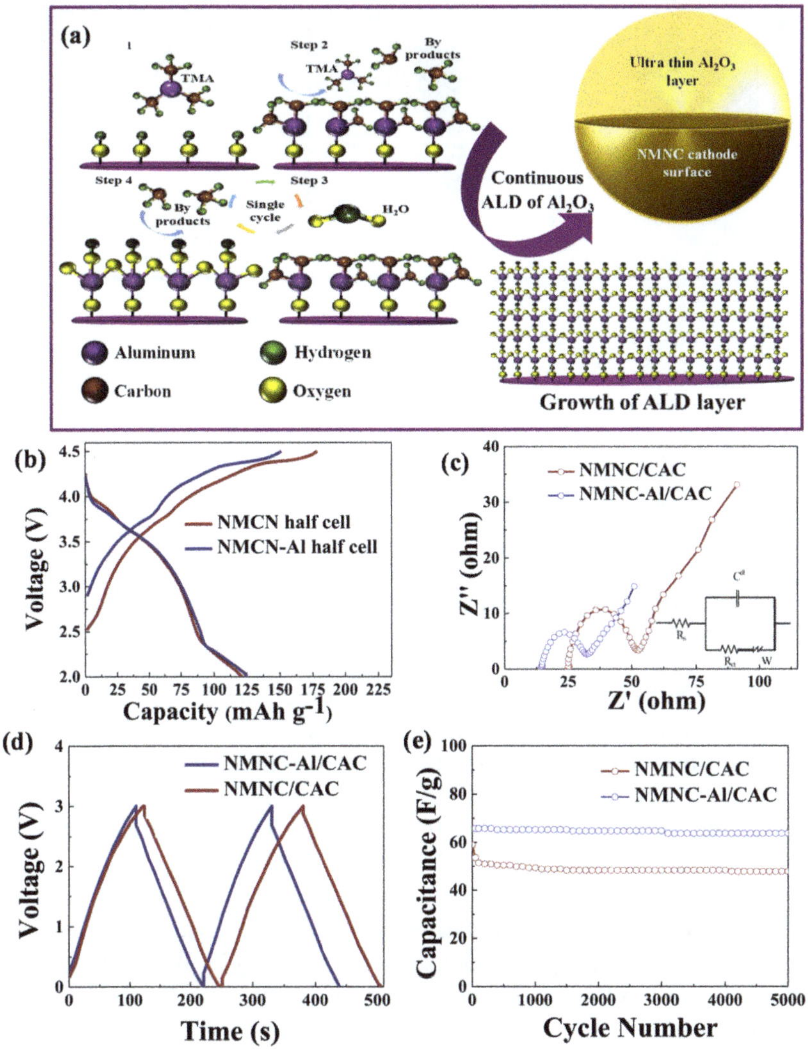

Figure 4.20. (a) A schematic representation of atomic layer deposition (ALD) on NMNC. (b) and (d) The charge/discharge curves of the NMNC positive electrode. (c) EIS test results. (e) The cyclic performances of the Na-ion HSCs [91], reproduced with permission © Elsevier.

constructed from the ZDC negative electrode and the P2-NCM positive electrode reached 18.8 Wh kg^{-1} at 0–3 V, even when the power density reached 12.75 kW kg^{-1}. After 2000 cycles, the capacity retention was still near 80%, and the coulombic efficiency remained near 100% for the whole period of cyclic testing (figure 4.21).

4.3.1.2 $Na_3V_2O_2(PO_4)_2F$-type cathodes

In addition to layered cathodes, polyanionic compounds have also drawn a lot of attention in the NIB research field. The most obvious advantages of this kind of positive material are its structural stability and diversity. Polyanionic positive

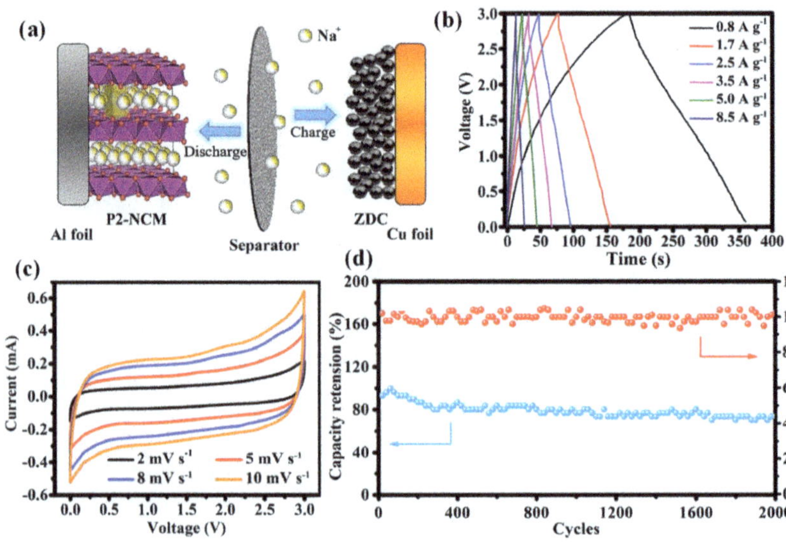

Figure 4.21. (a) A schematic of the Na-ion HSC based on ZDC-carbon//P2-NCM. (b) Charge/discharge profiles. (c) and (d) cyclic voltammetry (CV) and cyclic performance, respectively [92], reproduced with permission © Elsevier.

materials have also been introduced into 'rocking-chair' Na-ion HSCs in combination with AC negative electrodes. Among these kinds of material, $Na_3V_2O_2(PO_4)_2F$ (NVOPF) has much higher working potentials of ~3.6 and ~4.0 V vs Na^+/Na. Wu *et al* fabricated a conductive polymer from poly(3,4-ethylenedioxythiophene) (PEDOT)-coated $Na_3V_2O_2(PO_4)_2F$ nanorods [93]. The discharge specific capacity of this material was 120 mAh g^{-1} at 0.1 C. After 8000 cycles at 5 C, its retention was higher than 83%, indicating an excellent cyclic performance. It was then assembled into a Na-ion HSC with an AC negative material, and it was found that its energy density was 158 h kg^{-1} at a power density of 230 W kg^{-1}. When the power density was increased to 7 kW kg^{-1}, the energy density still reached 24 Wh kg^{-1} (figure 4.22).

4.3.2 K-ion hybrid superconductors

In addition to NIBs, K-ion batteries (KIBs) have also been investigated for use as next-generation electrochemical energy storage devices because K sources are affordable and abundant (about 2.09 wt% of the Earth's crust) [94, 95].

As in the cases of LICs and NICs, the development of high-capacity positive materials is necessary to enhance the energy density of K-ion capacitors (KICs). Layered transition-metal oxides have been intensively studied for application in KIBs and KICs. For example, Sada *et al* found that $Na_{0.84}CoO_2$ had a capacity of over 80 mAh g^{-1} when it was used as a K^+ intercalation material [96]. Meanwhile, other materials, such as $K_{0.5}MnO_2$ and $K_{0.6}CoO_2$ have also been discussed [97, 98]. The doping of positive materials with metal ions in order to further improve their cyclic performance has also been reported [99].

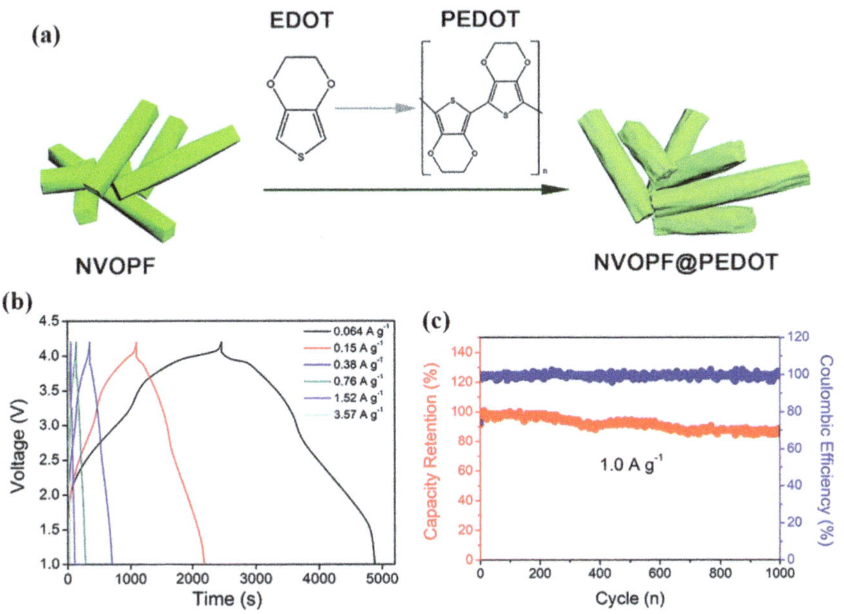

Figure 4.22. (a) The synthesis process used to obtain PEDOT-coated $Na_3V_2O_2(PO_4)_2F$ nanorods. (b) and (c) The charge/discharge profiles and cyclic performance of the final Na-ion HSC, respectively [93], reproduced with permission © Royal Society of Chemistry.

Lee and co-workers demonstrated a 'rocking-chair'-type K-ion HSC using novel P3-type $K_{0.45}Mn_{0.5}Co_{0.5}O_2$ and AC as the positive and negative electrode, respectively [100]. The $K_{0.45}Mn_{0.5}Co_{0.5}O_2$ material, which was fabricated via the sol–gel method, delivered a capacity of 140 mAh g^{-1} in the voltage range of 1.2–3.9 V vs K^+/K. The presence of Mn and Co metals in this material made the charge/discharge profiles smooth. This kind of K-ion HSC delivered a specific capacitance of 34.69 F g^{-1} when the current density was 0.1 A g^{-1}. When the power density was 150 W kg^{-1}, its energy density was more than 43 Wh kg^{-1}, and its energy density remained at 5 Wh kg^{-1} when the power density was more than 30 kW kg^{-1}. After 30 000 cycles, its energy density retention was 87.8%. Therefore, the K-ion HSC is a very promising kind of novel electrochemical energy storage device (figure 4.23).

4.4 Prospects

At the time of writing, 'rocking-chair'-type HSCs have been investigated for more than 15 years. Although many papers are still published on this topic every year, research in this area is reducing, compared with research into HSCs with capacitive positive electrodes (AC, for example). Among the most limiting factors are their low output voltage and poor cyclic stability. Since the specific surface area of negative AC electrodes is very high, a severe irreversible reaction occurs with organic electrolytes when the potential of the negative electrode exceeds the voltage limit. Therefore, the negative electrode should keep at a relatively high electrochemical potential to inhibit the reduction of the electrolyte. For Li-ion HSCs, this value may

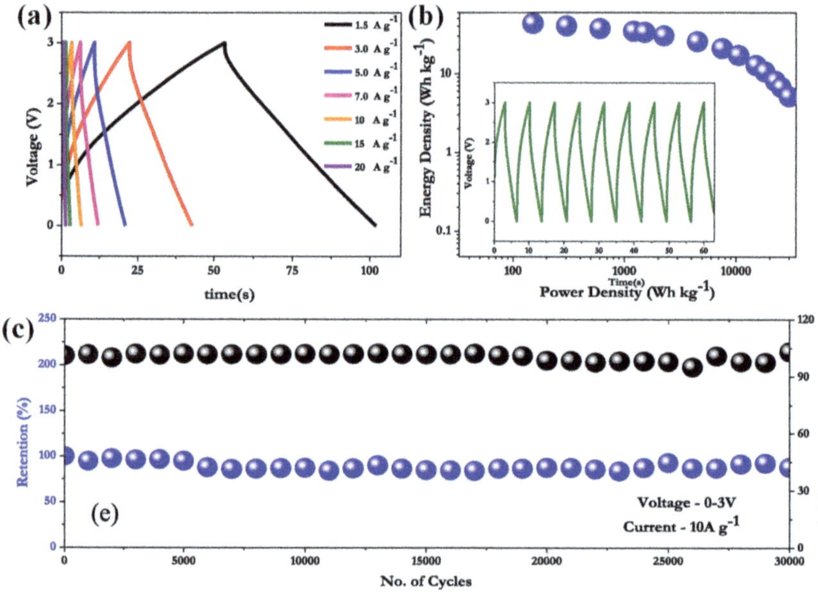

Figure 4.23. (a) The charge/discharge curves of a K-ion HSC at varying current densities. (b) and (c) A Ragone plot and cyclic performance at 10 A g^{-1} [100], reproduced with permission © Elsevier.

be >2.0 V vs Li^+/Li or even higher. Therefore, the output voltage of the whole Li-ion HSC, including the electrolyte, becomes very low (usually less than 3.0 V). Hence, the energy density and cyclic performance are rather limited in real applications.

Considering this fact, 'rocking-chair'-type Li-ion HSCs using aqueous electrolytes might be a good choice for further study [101–104]. According to the reported results, even though the electrochemical window of aqueous electrolytes (Li_2SO_4, $LiNO_3$, LiCl, etc.) is narrower than those of organic electrolytes, Li-ion HSCs with aqueous electrolytes can still deliver satisfactory energy density and cyclic life. Considering the aspects of performance cost and environmental impact, this kind of 'rocking-chair'-type Li-ion HSC has unique advantages and will also be used for future energy storage.

In addition to the 'rocking-chair' Li-ion HSCs, Na-ion and K-ion HSCs are novel electrochemical energy storage devices which deserve further exploration. To date, the research into these devices is still in its early stages. Novel positive materials should be developed to further enhance their performance, such as their specific capacity and working voltage. At the same time, the pore structures of AC electrodes should also be optimized according to the different ion sizes. Novel types of capacitive negative material, such as graphene and CNT materials, are also good choices.

References

[1] Ma Y, Chang H, Zhang M and Chen Y 2015 *Adv. Mater.* **27** 5296–308
[2] Zuo W, Li R, Zhou C, Li Y, Xia J and Liu J 2017 *Adv. Sci. (Weinh.)* **4** 1600539
[3] Han P, Xu G, Han X, Zhao J, Zhou X and Cui G 2018 *Adv. Energy Mater.* **8** 1801243

[4] Muzaffar A, Ahamed M B, Deshmukh K and Thirumalai J 2019 *Renew. Sustain. Energy Rev.* **101** 123–45
[5] Tie D, Huang S, Wang J, Ma J, Zhang J and Zhao Y 2019 *Energy Storage Mater.* **21** 22–40
[6] Jin L, Shen C, Shellikeri A, Wu Q, Zheng J, Andrei P, Zhang J-G and Zheng J P 2020 *Energy Environ. Sci.* **13** 2341–62
[7] Wang Y-g and Xia Y-y 2005 *Electrochem. Commun.* **7** 1138–42
[8] Zheng S, Wu Z-S, Wang S, Xiao H, Zhou F, Sun C, Bao X and Cheng H-M 2017 *Energy Storage Mater.* **6** 70–97
[9] Li B *et al Adv. Mater.* 2018**30** 1705670
[10] Ding J, Hu W, Paek E and Mitlin D 2018 *Chem. Rev.* **118** 6457–98
[11] Chen J, Yang B, Liu B, Lang J and Yan X 2019 *Curr. Opin. Electrochem.* **18** 1–8
[12] Wu M, Zhang Q, Lu H and Chen A 2004 *Solid State Ion.* **169** 47–50
[13] Zhan D, Zhang Q, Hu X, Zhu G and Peng T 2013 *Solid State Ion.* **239** 8–14
[14] Li J and Gao F 2009 *J. Power Sources* **194** 1184–93
[15] Cericola D, Novák P, Wokaun A and Kötz R 2011 *J. Power Sources* **196** 10305–13
[16] Li J, Guo J Q, Zhang X, Huang Y and Guo L 2017 *Int. J. Electrochem. Sci.* **12** 1157–66
[17] Lee B-G, Shin S-I, Ha M-W and An G H 2020 *Curr. Appl. Phys.* **20** 419–24
[18] Prasankumar T, Vigneshwaran J, Bagavathi M and Jose S 2020 *J. Alloys Compd.* **834** 155060
[19] Li J, Zhang X, Peng R, Huang Y, Guo L and Qi Y 2016 *RSC Adv.* **6** 54866–73
[20] Shui Zhang S 2006 *Electrochem. Commun.* **8** 1423–8
[21] Kwon Y K, Choi W, Choi H-S and Lee J K 2013 *Electron. Mater. Lett.* **9** 751–4
[22] Qin B, Zhang S, Hu Z, Liu Z, Zhang J, Zhao J, Xiong J and Cui G 2017 *Ionics* **23** 1399–406
[23] Rock S E, Wu L, Crain D J, Krishnan S and Roy D 2013 *ACS Appl. Mater. Interfaces* **5** 2075–84
[24] dos Santos G A Jr, Fortunato V D S, Silva G G, Ortega P F R and Lavall R L 2019 *Electrochim. Acta* **325** 134900
[25] Tang X, Zhou J, Bai M, Wu W, Li S and Ma Y 2019 *J. Mater. Chem.* A **7** 13364–71
[26] Wang S, Xiang M, Lu Y, Guo J, Su C, Bai H and Liu X 2020 *J. Mater. Sci., Mater. Electron.* **31** 6036–44
[27] dos Santos Junior G A, Fortunato V D S, Bastos G A A, Silva G G, Ortega P F R and Lavall R L 2020 *ACS Appl. Energy Mater.* **3** 9028–39
[28] Ivanova S, Zhecheva E, Nihtianova D and Stoyanova R 2011 *J. Mater. Sci.* **46** 7098–105
[29] Ivanova S, Zhecheva E, Nihtianova D, Mladenov M and Stoyanova R 2013 *J. Alloys Compd.* **561** 252–61
[30] Jiang Y P, Xie J, Cao G S and Zhao X B 2010 *Electrochim. Acta* **56** 412–7
[31] Takada T, Hayakawa H, Akiba E, Izumi F and Chakoumakos B C 1997 *J. Power Sources* **68** 613–7
[32] Shin Y and Manthiram A 2003 *Electrochim. Acta* **48** 3583–92
[33] Kim H, Sun Y-K, Shin K H and Jin C S 2010 *Phys. Scr.* **T139** 014053
[34] Tran V M, Huynh L T N, Ha C T D, Van Nguyen T M and Le M L P 2016 *Adv. Nat. Sci.-Nanosci. Nanotechnol.* **7** 015012
[35] Arrebola J C, Caballero A, Cruz M, Hernán L, Morales J and Castellón E R 2006 *Adv. Funct. Mater.* **16** 1904–12
[36] Talyosef Y, Markovsky B, Lavi R, Salitra G, Aurbach D, Kovacheva D, Gorova M, Zhecheva E and Stoyanova R 2007 *J. Electrochem. Soc.* **154** A682

[37] Li H, Cheng L and Xia Y 2005 *Electrochem. Solid-State Lett.* **8** A433
[38] Wu H, Rao C V and Rambabu B 2009 *Mater. Chem. Phys.* **116** 532–5
[39] Brandt A, Balducci A, Rodehorst U, Menne S, Winter M and Bhaskar A 2014 *J. Electrochem. Soc.* **161** A1139–43
[40] Arun N, Jain A, Aravindan V, Jayaraman S, Chui Ling W, Srinivasan M P and Madhavi S 2015 *Nano Energy* **12** 69–75
[41] Wu Z-S, Zhou G, Yin L-C, Ren W, Li F and Cheng H-M 2012 *Nano Energy* **1** 107–31
[42] Yoon J H, Bang H J, Prakash J and Sun Y K 2008 *Mater. Chem. Phys.* **110** 222–7
[43] Karthikeyan K, Amaresh S, Aravindan V, Kim H, Kang K S and Lee Y S 2013 *J. Mater. Chem.* A **1** 707–14
[44] Lee S-H and Lee T-H 2018 *Int. J. Hydrog. Energy* **43** 15365–9
[45] Karthikeyan K, Kim S H, Kim K J, Lee S N and Lee Y S 2013 *Electrochim. Acta* **109** 595–601
[46] Ahn S, Haniu Y, Nara H, Momma T, Sugimoto W and Osaka T 2020 *J. Electrochem. Soc.* **167** 040519
[47] Yang B, Chen J, Liu B, Ding Y, Tang Y and Yan X 2021 *J. Mater. Chem.* A **9** 6352–60
[48] Kim K-H, Kim M-S and Yeu T-W 2010 *Bull. Korean Chem. Soc.* **31** 3183–9
[49] Zhang S L, Ma L H, Li X G, Song Y H and Li W 2011 *Adv. Mater. Res.* **287–90** 1565–8
[50] Feng J, Chernova N A, Omenya F, Tong L, Rastogi A C and Stanley Whittingham M 2018 *J. Solid State Electrochem.* **22** 1063–78
[51] Naoi K *et al Energy Environ. Sci.* 2016**9** 2143–51
[52] Yan D, Lu A-H, Chen Z-Y, He L and Li W-C 2021 *ACS Appl. Energy Mater.* **4** 1824–32
[53] Ortega P F R, Dos Santos G A Jr, Montoro L A, Silva G G, Blanco C, Santamaría R and Lavall R L 2018 *J. Phys. Chem.* C **122** 1456–65
[54] Saïdi M Y, Barker J, Huang H, Swoyer J L and Adamson G 2003 *J. Power Sources* **119–21** 266–72
[55] Rui X H, Yesibolati N and Chen C H 2011 *J. Power Sources* **196** 2279–82
[56] Khomenko V, Raymundo-Piñero E and Béguin F 2008 *J. Power Sources* **177** 643–51
[57] Satish R, Aravindan V, Ling W C and Madhavi S 2015 *J. Power Sources* **281** 310–7
[58] Liao Y, Li C, Lou X, Hu X, Ning Y, Yuan F, Chen B, Shen M and Hu B 2018 *Electrochim. Acta* **271** 608–16
[59] Deng Z, Jiang H, Hu Y, Liu Y, Zhang L, Liu H and Li C 2017 *Adv. Mater.* **29** 1603020
[60] Böckenfeld N and Balducci A 2013 *J. Power Sources* **235** 265–73
[61] Wu W-l, Liang J, Yan J and Mao W-f 2013 *J. Solid State Electrochem.* **17** 2027–33
[62] Secchiaroli M, Giuli G, Fuchs B, Marassi R, Wohlfahrt-Mehrens M and Dsoke S 2015 *J. Mater. Chem.* A **3** 11807–16
[63] Zhou F, Cococcioni M, Kang K and Ceder G 2004 *Electrochem. Commun.* **6** 1144–8
[64] Padhi A K, Nanjundaswamy K S and Goodenough J B 1997 *J. Electrochem. Soc.* **144** 1188–94
[65] Nie Z X, Ouyang C Y, Chen J Z, Zhong Z Y, Du Y L, Liu D S, Shi S Q and Lei M S 2010 *Solid State Commun.* **150** 40–4
[66] Prabaharan S R S, Star A, Kulkarni R, Michael A R and Michael M S 2015 *Curr. Appl. Phys.* **15** 1624–33
[67] Vasanthi R, Kalpana D and Renganathan N G 2007 *J. Solid State Electrochem.* **12** 961–9
[68] Kosova N V, Kulova T L, Nikolskaya N F, Podgornova O A, Rychagov A Y, Sosenkin V E and Volfkovich Y M 2019 *J. Solid State Electrochem.* **23** 1981–90

[69] Karthikeyan K, Amaresh S, Kim K J, Kim S H, Chung K Y, Cho B W and Lee Y S 2013 *Nanoscale* **5** 5958–64
[70] Peng L, Zhu Y, Li H and Yu G 2016 *Small* **12** 6183–99
[71] Wang H, Mitlin D, Ding J, Li Z and Cui K 2016 *J. Mater. Chem.* A **4** 5149–58
[72] Lee J H, Shin W H, Ryou M-H, Jin J K, Kim J and Choi J W 2012 *Chem. Sus. Chem.* **5** 2328–33
[73] Dai J, Fu K, Palanisamy R, Gong A, Pastel G, Kornfeld R, Zhu H, Sanghadasa M, Bekyarova E and Hu L 2017 *J. Mater. Chem.* A **5** 15266–72
[74] Huang H, Pan L, Chen X, Tervoort E, Sologubenko A and Niederberger M 2019 *J. Mater. Chem.* A **7** 22444–52
[75] Jang B Z, Liu C, Neff D, Yu Z, Wang M C, Xiong W and Zhamu A 2011 *Nano Lett.* **11** 3785–91
[76] Han P, Yue Y, Wang X, Ma W, Dong S, Zhang K, Zhang C and Cui G 2012 *J. Mater. Chem.* **22** 24918
[77] Han P, Ma W, Pang S, Kong Q, Yao J, Bi C and Cui G 2013 *J. Mater. Chem.* A **1** 5949
[78] Grosjean C, Miranda P H, Perrin M and Poggi P 2012 *Renew. Sustain. Energy Rev.* **16** 1735–44
[79] Li W-J, Han C, Wang W, Gebert F, Chou S-L, Liu H-K, Zhang X and Dou S-X 2017 *Adv. Energy Mater.* **7** 1700274
[80] Liu M, Chang L, Le Z, Jiang J, Li J, Wang H, Zhao C, Xu T, Nie P and Wang L 2020 *Chem. Sus. Chem.* **13** 5837–62
[81] Zhang Y D *et al Chem. Sus. Chem.* 2020**13** 2522–39
[82] Pan H, Hu Y-S and Chen L 2013 *Energy Environ. Sci.* **6** 2338–60
[83] Zhu Y E, Yang L, Sheng J, Chen Y, Gu H, Wei J and Zhou Z 2017 *Adv. Energy Mater.* **7** 1701222
[84] Chen Z, Augustyn V, Jia X, Xiao Q, Dunn B and Lu Y 2012 *ACS Nano* **6** 4319–27
[85] Thangavel R, Kaliyappan K, Kang K, Sun X and Lee Y S 2016 *Adv. Energy Mater.* **6** 1502199
[86] Zhang X, Zhang Z and Zhou Z 2018 *J. Energy Chem.* **27** 73–85
[87] Wasiński K, Półrolniczak P and Walkowiak M 2018 *Electrochim. Acta* **259** 850–4
[88] Kim H, Kim H, Ding Z, Lee M H, Lim K, Yoon G and Kang K 2016 *Adv. Energy Mater.* **6** 1600943
[89] Carlier D, Cheng J H, Berthelot R, Guignard M, Yoncheva M, Stoyanova R, Hwang B J and Delmas C 2011 *J. Chem. Soc., Dalton Trans.* **40** 9306–12
[90] Kim H J, Ramasamy H V, Jeong G H, Aravindan V and Lee Y S 2020 *ACS Appl. Mater. Interfaces* **12** 10268–79
[91] Kaliyappan K and Chen Z 2018 *Nano Energy* **48** 107–16
[92] Gu H, Kong L, Cui H, Zhou X, Xie Z and Zhou Z 2019 *J. Energy Chem.* **28** 79–84
[93] Wu L, Dong S, Pang G, Li H, Xu C, Zhang Y, Dou H and Zhang X 2019 *J. Mater. Chem.* A **7** 1030–7
[94] Xu J, Dou S, Cui X, Liu W, Zhang Z, Deng Y, Hu W and Chen Y 2021 *Energy Storage Mater.* **34** 85–106
[95] Liu X, Sun Y, Tong Y, Wang X, Zheng J, Wu Y, Li H, Niu L and Hou Y 2021 *Nano Energy* **86** 106070
[96] Sada K, Senthilkumar B and Barpanda P 2017 *Chem. Commun.* **53** 8588–91

[97] Kim H, Seo D-H, Kim J C, Bo S-H, Liu L, Shi T and Ceder G 2017 *Adv. Mater.* **29** 1702480
[98] Kim H, Kim J C, Bo S-H, Shi T, Kwon D-H and Ceder G 2017 *Adv. Energy Mater.* **7** 1700098
[99] Liu C, Luo S, Huang H, Wang Z, Hao A, Zhai Y and Wang Z 2017 *Electrochem. Commun.* **82** 150–4
[100] Ramasamy H V, Senthilkumar B, Barpanda P and Lee Y-S 2019 *Chem. Eng. J.* **368** 235–43
[101] Wang Y-g and Xia Y-y 2006 *J. Electrochem. Soc.* **153** A450
[102] Hao Y-J, Wang L and Lai Q-Y 2011 *J. Solid State Electrochem.* **15** 1901–7
[103] Hao Y J, Lai Q Y, Wang L, Xu X-Y and Chu H Y 2010 *Synth. Met.* **160** 669–74
[104] Tang W, Hou Y, Wang F, Liu L, Wu Y and Zhu K 2013 *Nano Lett.* **13** 2036–40

Chapter 5

Nonaqueous electrochemical capacitors based on battery anodes and porous carbon cathodes

Zhanheng Yan, Junliang Yang, Jianmin Ma and Xinhua Liu

The development of high power electrical equipment (electric vehicles, large-scale grid storage) requires energy storage equipment to have higher energy densities, higher power densities, and longer cyclic life. Hybrid metal-ion capacitors are a new type of energy storage device that is in between batteries and supercapacitors. These devices are prepared using a battery anode and a supercapacitor cathode; therefore, they have the advantages of both batteries and capacitors, showing high energy densities, high power densities, and long lifespans. Lithium-ion capacitors (LICs), sodium-ion capacitors (SICs), potassium-ion capacitors (PICs), and zinc-ion capacitors (ZICs) are several common hybrid ion capacitors. At present, the development of hybrid ion capacitors is at an early stage; it still suffers from a kinetics mismatch between the cathode and the anode and a lack of appropriate electrode materials. In this chapter, we mainly introduce the principle, structure, and electrode material design of hybrid ion capacitors in nonaqueous electrolyte systems.

5.1 Lithium-ion capacitors

With the development of electronic devices and electric vehicles, there is an increasing demand for energy storage equipment with high energy density and power density. Due to their high energy density, excellent rate performance, and long cyclic life, lithium-ion batteries (LIBs) have become the most widely used secondary batteries. LIBs possess high energy densities (150–200 Wh kg^{-1}) but low power densities (<1 kW kg^{-1}). Electrochemical capacitors have higher power densities (>10 kW kg^{-1}) and ultralong cyclic lives (10^4–10^5 cycles) [1, 2]. Given the characteristics of these two types of energy storage device, researchers decided to combine their advantages to synthesize energy storage devices with high energy density and high power density. The resulting devices are constructed using an LIB anode, a capacitor cathode, and a nonaqueous Li^+ electrolyte and are called LICs.

doi:10.1088/978-0-7503-5042-6ch5

© IOP Publishing Ltd 2023

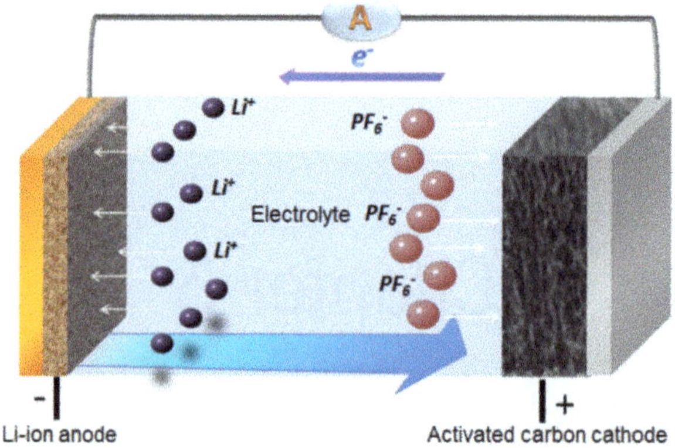

Figure 5.1. A schematic diagram of a lithium-ion capacitor [3], reproduced with permission © John Wiley and Sons.

A schematic diagram of a lithium-ion capacitor is shown in figure 5.1. In the charge–discharge process, LICs realize the storage and conversion of energy through ion adsorption–desorption on the surface of the cathode carbon material and Li^+ insertion–removal in the anode.

At present, the biggest obstacle to the development of LICs is the mismatch between the electrode kinetics of the cathode and those of the anode. Thus, the realization of high-performance LICs will require appropriate cathode and anode materials. Activated carbon is the usual cathode material for LICs. For anode materials, conversion-type materials (Fe_2O_3, MnO), insertion-type materials (Nb_2O_5, $Li_4Ti_5O_{12}$), and alloy-type materials (Sn, Sb, Si) are commonly used.

Conversion-type anodes: as they possess several obvious advantages (high specific capacity, plentiful raw materials, and eco-friendliness), conversion-type anodes have attracted much attention. In various anode materials that undergo conversion reactions, MnO has attracted great attention due to its high specific capacity (756 mAh g^{-1}) and low redox potential. In practical applications, however, MnO always exhibits lower conductivity and severe volume changes during charging and discharging. At present, reducing the sizes of nanomaterials and compositing MnO with carbon materials are considered to be two effective ways of addressing these problems [4–6]. Yan's group adopted a simple method of fabricating graphene nanoscroll-wrapped MnO nanoparticles (GNS@MnO-600) (figure 5.2). Their unique structure not only maintained the structural integrity of the MnO but also effectively improved the conductivity of the material. When used as an LIB anode, GNS@MnO-600 exhibited high specific capacity (766 mAh g^{-1} at 100 mA g^{-1}), excellent rate performance (437 mAh g^{-1} at 5 A g^{-1}), and ultralong cyclic life (555 mAh g^{-1} at 2 A g^{-1} after 800 cycles). In addition, GNS@MnO-600 was selected as the anode and 3D framework activated carbon (3DFAC) was selected as the cathode to fabricate LICs (GNS@MnO-600//3DFAC).

As shown in figure 5.3, the LICs composed of GNS@MnO-600 (anode), 3DFAC (cathode), the separator, and the electrolyte contained Li^+(figure 5.3(a)). Figures 5.3(b) and (e) show the cyclic voltammetry (CV) and galvanostatic charge/

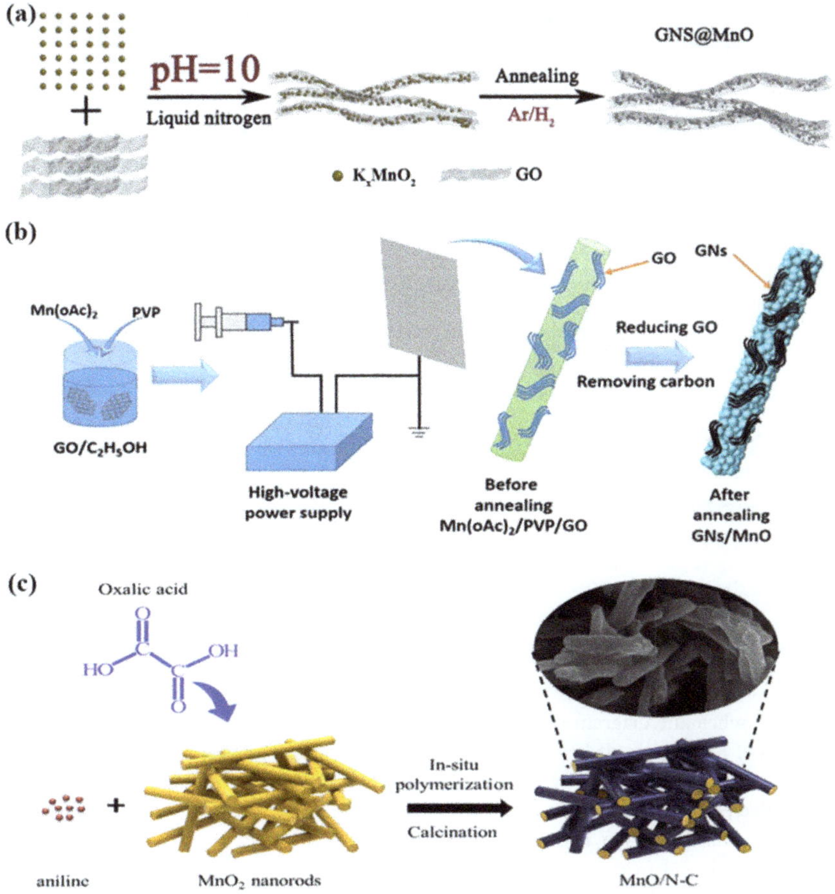

Figure 5.2. (a) The synthesis route of the GNS@MnO-600 composite [4], reproduced with permission © Royal Society of Chemistry. (b) The synthesis route of the GNs/MnO [5], reproduced with permission © American Chemical Society. (c) The synthesis route of the MnO/N-C [6], reproduced with permission © Royal Society of Chemistry.

discharge (GCD) curves of GNS@MnO-600//3DFAC at different current densities at voltages ranging from 0.5 to 4.2 V, respectively. All the curves exhibit obvious capacitive behavior. Although GNS@MnO-600//3DFAC possesses smaller polarization during the charge/discharge process, the value of IR only drops to 362 mV at 10 A g^{-1} (figure 5.3(d)). Specifically, GNS@MnO-600//3DFAC has a high energy density and power density (197 Wh kg^{-1} at 235 W kg^{-1}) (figure 5.3(e)). More importantly, the energy–power performance of GNS@MnO-600//3DFAC is superior to those of most previously reported materials (figure 5.3(f)). Among LICs with different mass ratios of the cathode and the anode, the best performance was achieved when the mass ratio of the cathode and to the anode was 1:1. After 3000 cycles, the GNS@MnO-600//3DFAC (1:1) retained 84.8% of its capacity (figure 5.3(g)). In addition to MnO, the anode materials of LIBs which undergo conversion reactions (Fe_3O_4, Mn_3O_4) [7, 8] have shown excellent electrochemical performance in LICs.

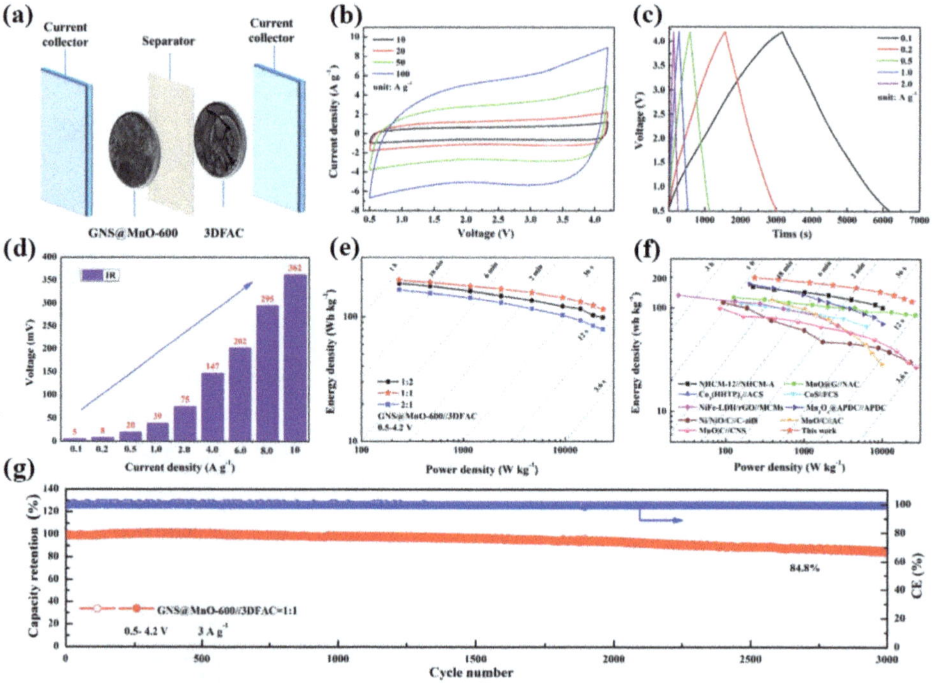

Figure 5.3. (a) A schematic diagram of an LIC. (b) CV curves at different scan rates (10–100 mV^{-1}), and (c) GCD profiles of LICs (1:1). (d) IR drops at different current densities when the mass ratio of the cathode and the anode is 1:1. (e) Energy–power densities of different mass ratios of LICs. (f) A comparison with previously reported performances. (g) Long-term cycling of the GNS@MnO-600//3DFAC (1:1) at 3 A g^{-1} [4], reproduced with permission © Royal Society of Chemistry.

Insertion-type anodes: in insertion-type materials, lithium ions undergo relatively fast insertion and transport; in addition, they have higher power outputs. Due to these advantages, insertion-type materials have attracted the attention of many researchers [9]. Ti-based materials, Nb-based materials, and V-based materials are three common insertion-type anode materials for LIBs and LICs. For example, $Li_4Ti_5O_{12}$ possesses a 'zero-strain' property and a lower voltage plateau (1.5 V vs Li^+/Li), which give $Li_4Ti_5O_{12}$ excellent safety and stability [10]. $Li_4Ti_5O_{12}$ stores lithium ions through a reversible reaction that occurs at the interface between the disordered spinel phase ($Li_4Ti_5O_{12}$) and the rock-salt phase ($Li_7Ti_5O_{12}$). Further research revealed that Li^+ migration in $Li_4Ti_5O_{12}$ occurred at the two-phase boundaries [11]. At the moment, the main problem facing $Li_4Ti_5O_{12}$ is its lower Li^+ migration efficiency. To improve its electrical conductivity, researchers have tried several things. For example, the confinement of $Li_4Ti_5O_{12}$ nanoparticles in porous carbon or carbon materials can improve ion/electron transport efficiency. LICs with $Li_4Ti_5O_{12}$ and activated carbon as the anode and cathodes made from activated carbon have excellent electrochemical performance (105 mAh g^{-1} at 61 A g^{-1}) [12]. In addition, modification through doping is another effective method. Dong's group adopted simple Ti^{3+} self-doping [13], and Wang adopted a novel molecular design [14]. Nb-based materials have attracted a lot of attention due to their chemical stability, unique redox couples (Nb^{4+}/Nb^{3+} and Nb^{5+}/Nb^{4+}), and suitable voltage windows. Among a variety of Nb-based materials, Nb_2O_5 is considered to be a promising anode material for LIBs. Nb_2O_5 possesses a new type of intercalative pseudocapacitive reaction mechanism that results in higher energy density. Such unique reaction mechanisms, including the pseudocapacitive reaction and intercalative/deintercalative reactions, occur on the electrode surface. Based on the above advantages, Nb_2O_5 can easily obtain rapid reaction kinetics and achieve higher specific capacity. However, Nb_2O_5 still faces some obstacles, such as its inherently poor electron conductivity in practical applications, which limits its rate performance and charge transport efficiency.

Unique nanostructures can effectively promote ion/electron transport efficiency, enhancing energy and power outputs. Wang's group adopted a sequential template strategy to synthesize triple-shelled Nb_2O_5 hollow spheres (3S-Nb_2O_5) [16]. LICs with 3S-Nb_2O_5 as the anode and activated carbon (AC) as the cathode possessed ultralong cyclic life; their capacity retention rate was 89.9% after 10 000 cycles. Carbon coating is another effective way of improving electron conductivity. For example, Kong's group adopted graphene as a basis with which to synthesize a layer-by-layer graphene/T-Nb_2O_5 nanocomposite; its graphene structure effectively inhibited structural collapse and improved the overall electrical conductivity of the composite material [11]. Doping can apparently improve the intrinsic conductivity of materials. Zhang's group adopted the antisolvent precipitation method to synthesize urchin-like Nb_2O_5 nanospheres composed of many ultrathin nanowires. Through a nitrogenization treatment, they obtained nitrogenated urchin-like Nb_2O_5 (NUNO), whose synthetic path is shown in figure 5.4 [15]. As shown in figure 5.5(a), lithiated NUNO was selected as the anode and AC derived from peanut shells (peanut shell carbon, PSC) served as the cathode in the production of LICs (NUNO//PSC). The CV curves of NUNO//PSC at different sweep rates (1–100 mV s^{-1}) took the form

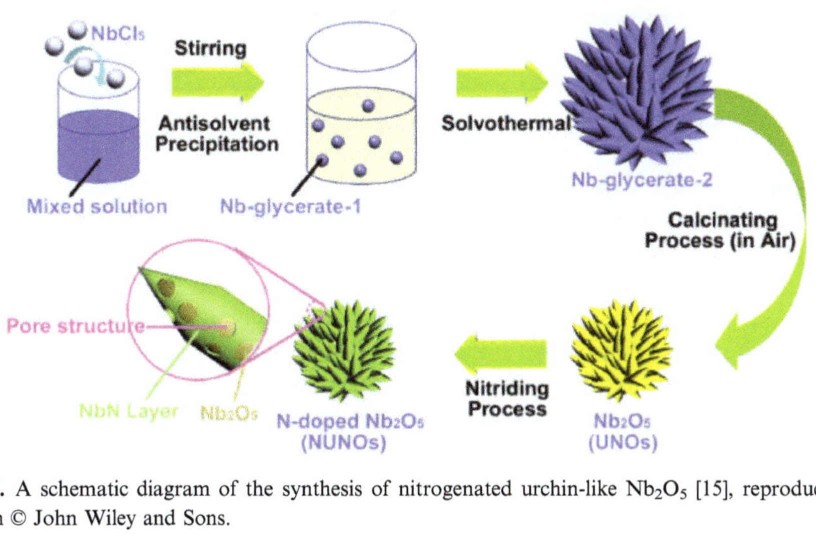

Figure 5.4. A schematic diagram of the synthesis of nitrogenated urchin-like Nb_2O_5 [15], reproduced with permission © John Wiley and Sons.

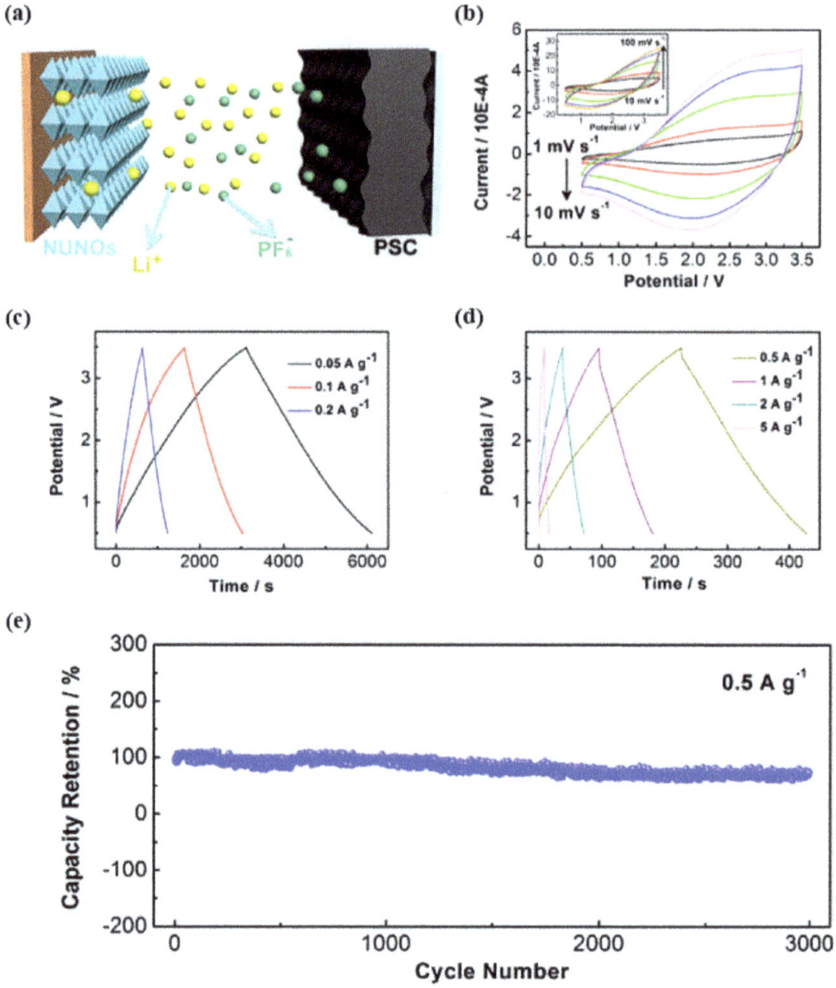

Figure 5.5. (a) A schematic diagram of NUNO//PSC, (b) its CV curves at various scan rates, (c, d) its galvanostatic charge–discharge curves at different current densities. (e) Cyclic stability test at 0.5 A g^{-1} [15], reproduced with permission © John Wiley and Sons.

of irregular rectangles (figure 5.5(b)), suggesting that Li^+ intercalation at the anode occurred simultaneously with anion transportation at the cathode. Figures 5.5(c) and (d) show the charge–discharge curves of NUNO//PSC at different current densities (0.05–5 A g^{-1}). The results of analysis indicated that the energy density achieved 83.5 Wh kg^{-1} at 0.05 A g^{-1} and the energy density remained at 19 Wh kg^{-1} at 10 KW kg^{-1}. In addition, the NUNO//PSC possessed excellent cyclic stability; its capacity retention rate was 80% at 0.5 A g^{-1} after 3000 cycles (figure 5.5(e)). The results indicated that nitrogenated urchin-like Nb_2O_5 microspheres have excellent practical application prospects. In addition to Nb_2O_5, other oxygen and niobium compounds, for example, Ti–Nb–O compounds ($TiNb_2O_7$, $Ti_2Nb_{10}O_{29}$) [17] and Nb–W–O compounds (Nb_2O_5–WO_3, $Nb_{12}WO_{33}$, $Nb_{14}W_3O_{44}$, $Nb_{26}W_4O_{77}$, etc.), have significant application prospects in LICs [18]. In contrast to Ti-based and Nb-based anode materials, VN has many inherent advantages, including lower redox potential, high specific capacity (>1000 mAh g^{-1}), and excellent electron conductivity (10^6 S m^{-1}). Based on the above advantages, VN is considered to be a promising anode material for LIBs and LICs. However, in practical applications, VN still faces the problem of electrochemical instability. A common way of improving the performance of VN is to form a composite material with a carbon material. For example, Yan's group adopted a hydrothermal method to synthesize a porous VN nanowire/reduced graphene oxide (VN-RGO) composite. In this composite, the graphene kept the VN structure stable and improved the electron conductivity. In addition, the porous structure of the VN can effectively improve the ion/electron transfer efficiency [19]. The synergistic effects of structure and composition endow VN with excellent lithium-ion storage performance.

Alloy-type anodes: anode materials that participate in alloying reactions (Si-based, Sn-based, Sb-based, Bi-based) have higher specific capacities. Metallic Sn has attracted great attention from many researchers because of its high specific capacity (990 mAh g^{-1}) and lower operational potential (0.25 V vs Li/Li^+) [20]. However, Sn-based anode materials are confronted with serious volume expansion during the charge/discharge process. The volume expansion associated with the alloy reaction results in the pulverization of the electrode material, which falls off the collector. This series of negative effects leads to the irreversible loss of electrode capacity and short cyclic life. One effective way to reduce the volumetric expansion effect is to reduce the size of the active material and combine it with carbon materials. Kang's groups encapsulated Sn in N-rich nanotubes (Sn@NRT) [21]. These unique structures prevented the agglomeration of the active material, improving the capacity and cyclic life of the electrode. In addition, the one-dimensional nanotube structure offered a pathway for ion/electron transport and provided sufficient active sites to store lithium ions. As a result of the above advantages, Sn@NRT had excellent electrochemical performance, namely a capacity of more than 200 mAh g^{-1} at a high current 12 A g^{-1}. Its capacity remained at 450 mAh g^{-1} at 2 A g^{-1} after 3000 cycles. Moreover, LICs constructed using an Sn@NRT anode and an NRT cathode (Sn@NRT@NRT) exhibited high energy densities of 274 and 127 Wh kg^{-1} at power densities of 153 and 22 800 W kg^{-1}, respectively. The capacitance retention rate was more than 85% after 2000 cycles at 2 A g^{-1}. The excellent performance of Sn@NRT@NRT was attributed to its open mesoporous structure, ultrafine Sn size, and heteroatomic doping.

Another effective way to solve metal agglomeration and volume expansion during charging and discharging is electrospinning. Yang's group adopted the electrospinning technique to incorporate Sn into N/P co-doped carbon nanofibers (Sn-NP-CNFs) [23]. The synthetic route used for the Sn-NP-CNFs is shown in figure 5.6(a). Melamine polyphosphate (MP) was used as the source of the N and P and polyacrylonitrile (PAN) was the source of C and N. Embedding Sn into carbon nanofibers can effectively inhibit its volume expansion, improving the cyclic stability of Sn-NP-CNFs. Moreover, high levels of N and P dopants increased the defects and expanded the carbon interlayer distance, enhancing the pseudocapacitance and rate performance of the Sn-NP-CNFs. In addition, the cyclic stability can be significantly

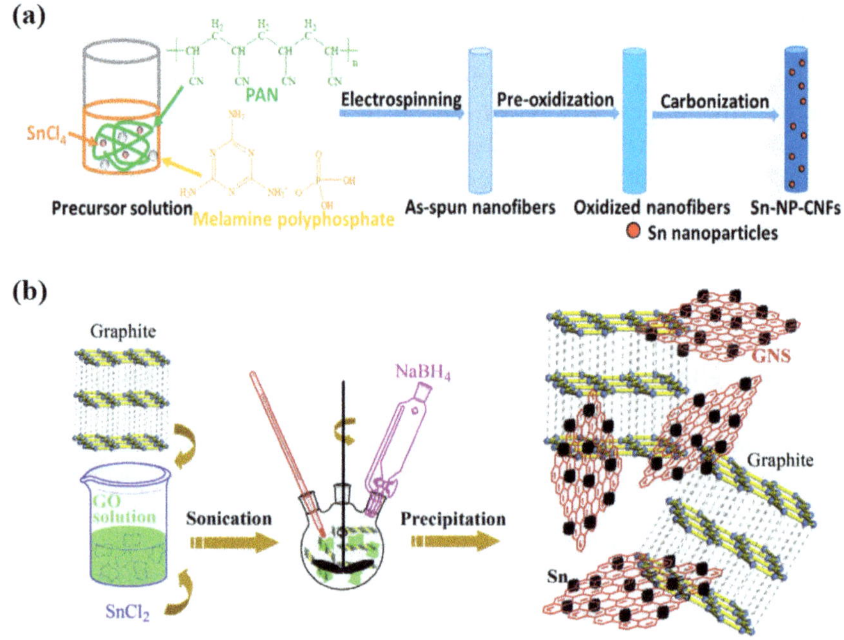

Figure 5.6. (a) A schematic diagram of the Sn-NP-CNFs [23], reproduced with permission © Elsevier. (b) A schematic diagram of the Sn/graphite/graphene [22], reproduced with permission © Elsevier.

improved by combining Sn with graphene (figure 5.6(b)) [22]. LICs composed of a Sn-NP-CNF anode and AC as the cathode exhibited excellent electrochemical performance. Figure 5.7(a) shows the CV curves of AC/Sn-NP-CNF-4 (i.e. the mass ratio of MP to PAN was 1:4). With an increase in sweep rate, the rectangular CV curves were gradually distorted, reflecting the fact that the energy storage mechanism of the LICs was a battery-type anode and a capacitor-type cathode. The GCD curves of AC/Sn-NP-CNF-4 confirm the presence of a typical pseudocapacitive storage mechanism (figure 5.7(b)). Figure 5.6(c) shows that the energy density of AC/Sn-NP-CNFs-4 reached 186.0 Wh kg^{-1}. At a high power density of 20.0 kW kg^{-1}, the energy density remained at 22.2 Wh kg^{-1}. The performance of the AC/Sn-NP-CNF-4 LIC was superior to the performances of previously reported LICs. Moreover, one coin-type lithium-ion capacitor could light up to five small bulbs for a few minutes (figure 5.7(c)). AC/Sn-NP-CNF-4 also had excellent cyclic stability; its capacitance retention was 83.7% after 10 000 cycles (figure 5.7(d)). The results confirm that reducing the size of the metal crystals and combining them with carbon materials can significantly improve their properties. In addition to Sn, Si-based materials have broad application prospects in lithium-ion capacitors. Wang's groups synthesized a boron-doped Si-based anode (B–Si/SiO_2/C) [3]. Boron doping apparently improved the rate performance of the B–Si/SiO_2/C. The capacity of the B–Si/SiO_2/C was 2.4 times that of the undoped Si-based material (Si/SiO_2/C) at the same current density (6.4 A g^{-1}). A lithium-ion capacitor composed of B–Si/SiO_2/C as the anode and porous spherical carbon (PSC) as the cathode exhibited excellent

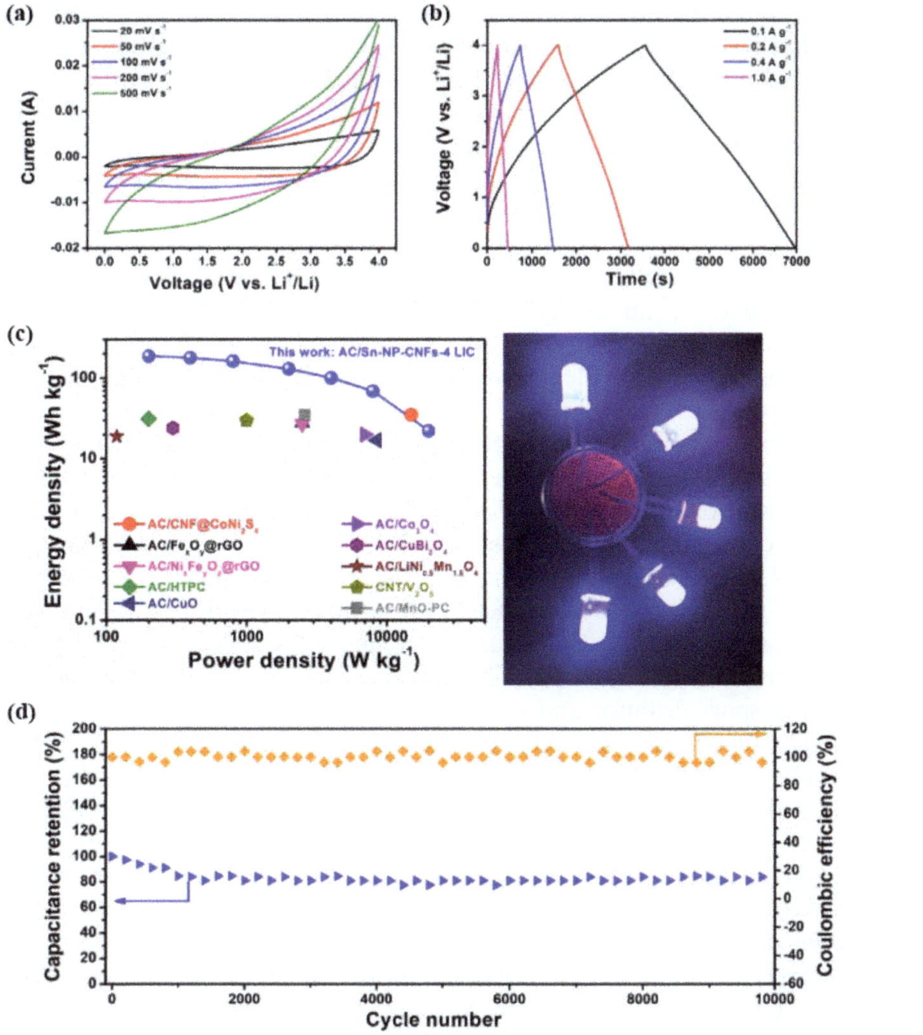

Figure 5.7. (a) CV and (b) GCD curvesof AC/Sn-NP-CNF-4. (c) A Ragone plot for AC/Sn-NP-CNF-4 versus those of other reported LICs; light-emitting diodes powered by a coin-type AC/Sn-NP-CNF-4 LIC. (d) The long-term cyclic performance of the AC/Sn-NP-CNF-4 LIC.

electrochemical performance. Its energy density reached 128 Wh kg^{-1} at a power density of 1229 W kg^{-1}. When the power density increased to 9704 W kg^{-1}, the energy density still reached 89 Wh kg^{-1}. In addition, the LIC possessed excellent cyclic stability (the capacity retention rate reached 70% after 6000 cycles) and a low self-discharge rate (the voltage remained at 82% of the original level for 50 h).

5.2 Sodium-ion capacitors

SICs are similar to LICs: they have high energy density, high power density, and ultralong cyclic life. Moreover, sodium sources are abundant, and sodium has a reasonable redox potential (Na/Na^+= −2.7 V) [24]. SICs are new energy storage devices that have the potential to replace LICs. When SICs are charged, Na^+ is embedded into the anode material to form sodium-containing compounds, and the anions are absorbed by cathode carbon materials. Conversely, during the discharge process, Na^+ ions are released from the anode material back into the electrolyte and the anions dissociate from the porous carbon surface. The radius of Na^+ (1.02 Å) is larger than that of Li^+ (0.76 Å), which means that the Na^+ migration required for the processes of charging and discharging is difficult. It is necessary to develop suitable anode materials for SICs because the anode materials of LICs are not suitable for SICs. The anode materials of SICs need to have large lattice spacing and they must support fast ion–electron diffusion. At present, the anode materials commonly used for SICs include transition-metal oxides and compounds, transition-metal chalcogenides, and alloys.

Transition-metal oxides: transition-metal oxides such as TiO_2, Nb_2O_5, V_2O_5, and Fe_3O_4 have sizable tunnels that can improve ion/electron diffusion rates. Moreover, transition-metal oxides usually exhibit typical pseudocapacitive behavior and have high specific capacity, but they still suffer from poor conductivity and other problems. The most common approach for improving the conductivity of transition-metal oxides is to combine them with carbon materials (e.g. graphene, carbon cloth, carbon nanotubes). For example, Liang's group grew TiO_2 nanorods on carbon fiber cloth (TiO_2/CFC). The carbon cloth effectively shortened the ion/electron transport distance and enhanced conductivity. The synthesis path of the TiO_2/CFC is shown in figure 5.8(a) [25]; it is a facile hydrothermal method that successfully grows TiO_2 nanorods on carbon fiber cloth. This unique structure retains its structural stability and offers sufficient interspaces for Na+ diffusion. Liang's group adopted TiO_2/CFC as the anode and electro-spun carbon fibers (CFs) as the cathode and assembled an SIC (TiO_2/CFC/CF). Figure 5.8(b) shows the GCD curves of the TiO_2/CFC/CF device at different current densities (0.2–5 A g^{-1}). All the curves have the form of irregular triangles, confirming that the storage mechanism is a mixture of battery-type and capacitor-type mechanisms. The energy density of the TiO_2/CFC/CF device reached 73.8 Wh kg^{-1} at a power density of 13 750 W kg^{-1}. Moreover, the flexible TiO_2/CFC/CF SIC possessed ultralong cyclic stability: its capacity retention rate was 90% after 4000 cycles (figure 5.8(c)). The SIC also possessed outstanding flexibility; there was no obvious capacity loss under different bending conditions (figure 5.8(d)). A Ragone plot of the SIC suggested that its performance was better than those of most previously reported devices; its energy density reached 73.8 Wh kg^{-1} at 0.2 A g^{-1} (figure 5.8(e)). The excellent performance of the SIC was attributed to the synergistic effect of its structure and components. The binder-free structure effectively alleviated volume expansion during the charging and discharging processes and enhanced the cyclic performance. In addition, cross-linked carbon fiber shortened the Na+ transport paths, helping to improve the rate performance. Another effective way of enhancing TiO_2 performance is to design nanostructured TiO_2 to improve the Na^+ diffusion rate. In addition, some heteroatom dopants (N, B) of TiO_2 form abundant oxygen defects which improve its electrochemical performance. For example, Liang's group adopted a facile

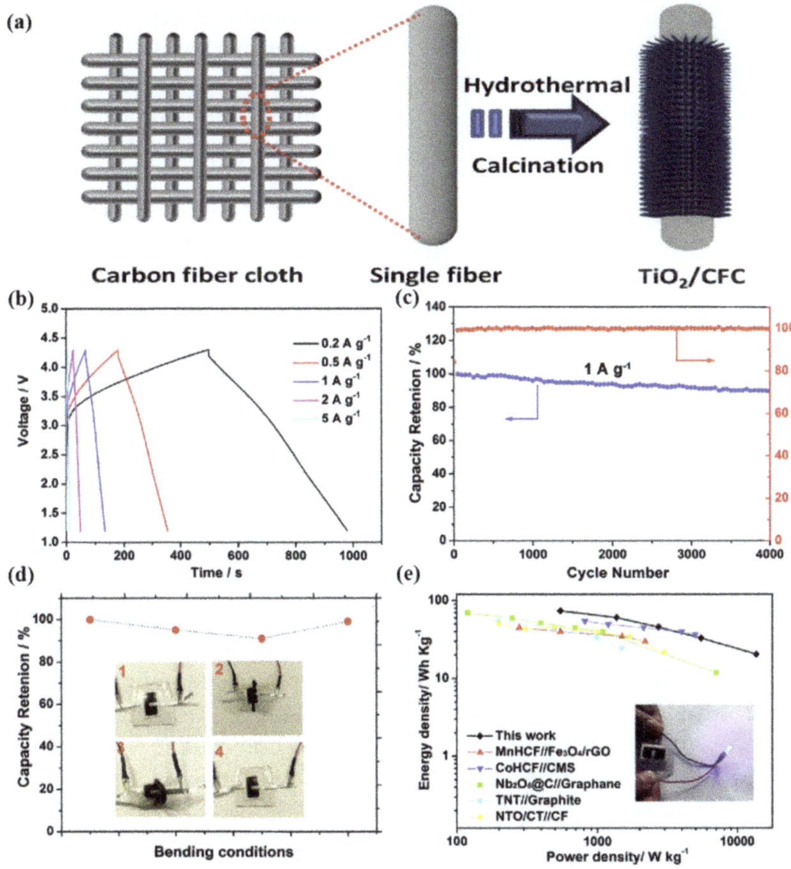

Figure 5.8. (a) The synthesis path of TiO_2/CFC. (b) Charge–discharge curves and (c) cyclic performance at 1 A g^{-1} of TiO_2/CFC//CF. (d) Electrochemical performance under different bending conditions. (e) A Ragone plot for the TiO_2/CFC//CF SIC compared with results previously reported for SICs [25], reproduced with permission © Elsevier.

hydrothermal and subsequent calcination method to synthesize nitrogen-doped TiO_2 nanospheres (N-TiO_2). The N-TiO_2 possessed excellent electrochemical performance when used as the anodes of sodium-ion batteries. Moreover, SICs composed of N-TiO_2 anodes and AC cathodes (N-TiO_2/AC) exhibited high energy density (80.3 Wh kg^{-1}) and power density (12 500 W kg^{-1}) [26]. In addition to Ti-based oxides, layer-structured V-based oxides (e.g. V_2O_5) which have a large lattice distance of 0.95 nm are ideal anode materials for sodium-ion batteries. However, in practical applications, their insufficient conductivity (10^{-6} S cm^{-1}) inhibits the large-scale application of V_2O_5. To improve the electrochemical performance of V_2O_5, Chen's group used a one-pot hydrothermal method to synthesize a V_2O_5/carbon nanotube composite for use as an anode and selected AC as the cathode to assemble an SIC [27]. The carbon nanotubes used in this study effectively enhanced the conductivity of the V_2O_5 and increased the ion/electron transfer rates. As a result of the above advantages, the SIC exhibited a maximum energy density of 38 Wh kg^{-1} at an energy density of 5000 W kg^{-1}. Nb-based materials (Nb_2O_5) also have good prospects for use in SICs. For example, Lim's group used core–shell structured Nb_2O_5@carbon nanoparticles combined with reduced graphene oxide (rGO) as an anode to form an SIC that exhibited high energy density (76 Wh kg^{-1}) and high power density (20 800 W kg^{-1}) [28].

Transition-metal chalcogenides: since they possess unique layered structures and large interplanar distances, transition-metal sulfides have been widely applied in the energy storage field. MoS_2 is a typical transition-metal sulfide. In 2017 Wang's group used an atomic interface modification technique to synthesize a monolayer MoS_2–carbon (MoS_2–C) composite material with a large lattice spacing (0.98 nm) [29]. SICs constructed using MoS_2–C as the anode and polyaniline-derived porous carbon as the cathode exhibited excellent performance; the maximum energy density and power density achieved were 111.4 W kg^{-1} and 12 000 W kg^{-1}, respectively. More importantly, the SICs showed ultralong cyclic stability; their capacitance retention rate was 77.3% after 10 000 cycles. Because of its high theoretical specific capacity (609 mAh g^{-1}), abundant natural resources, and environmental benignity, FeS has been widely applied in energy storage devices. However, in practical applications, FeS has always exhibited huge volume variation, poor conductivity, and sluggish kinetics. An effective way of solving the above problems is to synthesize nanometer-sized materials, which can apparently reduce the volume expansion and enhance Na^+ transport, thus improving cyclic stability and rate performance. A widely used way of enhancing its conductivity is to combine it with porous carbon materials. Porous carbon materials have many advantages in the field of energy storage and transport. First, porous structures can provide a pathway for ion/electron transport. Second, an open porous framework benefits electrolyte penetration, which helps to improve energy density. Finally, the carbon can alleviate volume expansion during the charging and discharging processes. Based on the above principles, Wen's group synthesized FeS quantum dots embedded in three-dimensional inverse opal-structured N-doped carbon (3D-IO FeS-QDs@NC) nanocomposites [30]. Figure 5.9(a) shows a schematic illustration of the synthesis of the 3D-IO FeS-QDs@NC. Silica nanospheres were used as the pore-foaming agent (template). Both sulfurization and carbonization were performed at high temperatures. Finally, the template was removed, leaving a 3D ordered porous architecture. Yang's group adopted a freeze-drying technique and a carbothermal reduction strategy to fabricate FeS nanocrystals coated by 3D carbon network nanocomposites (FeS@C) (figure 5.9(b)) [31]. This unique structure effectively alleviated the volume change that occurs during charging and discharging and enhanced the cyclic stability of the material. SICs were prepared using the 3D-IO FeS-QDs@NC as the anode and commercial AC as the cathode (AC//3D-IO FeS-QDs@NC) (figure 5.10(a)). The SICs thus prepared exhibited excellent electrochemical performance, including high energy density, high power density, and ultralong cyclic stability. Figure 5.10(b) shows the GCD curves of the AC//3D-IO FeS-QDs@NC SICs at different current densities, which resemble irregular triangles, showing obvious characteristics of pseudocapacitance. The GCD curve analysis results

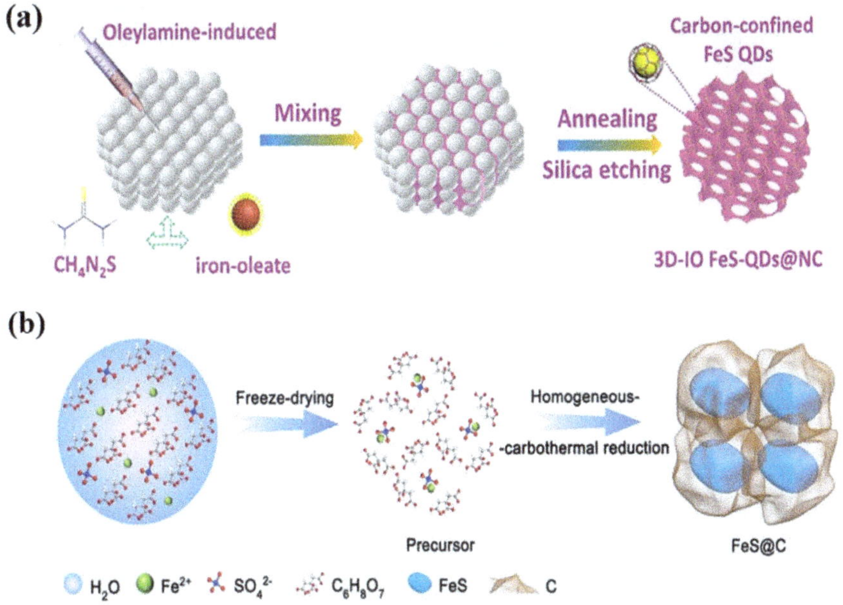

Figure 5.9. A schematic illustration of the formation processes of (a) 3D-IO FeS-QDs@NC and (b) FeS@C [30, 31], reproduced with permission © Royal Society of Chemistry.

indicate that Na^+ storage in the AC//3D-IO FeS-QDs@NC SICs involves both capacitor and battery storage behaviors. The capacities of the AC//3D-IO FeS-QDs@NC SICs were 129, 108, 97, 83, 72, and 62 F g^{-1} at 0.1, 0.4, 0.8, 1.6, 3.2, and 6.4 A g^{-1}, respectively (figure 5.10(c)). The capacity retention of the AC//3D-IO FeS-QDs@NC SICs reached 91% after 5000 cycles, confirming their excellent cyclic stability (figure 5.10(d)). More importantly, in the GCD curves, no significant changes occurred during the entire cycle. The performance of the AC//3D-IO FeS-QDs@NC SICs was better than those of most SICs previously reported (figure 5.10(e)). The SICs achieved a maximum energy density of 151.8 Wh kg^{-1} at a power density of 145 W kg^{-1}. Figure 5.10(f) shows that AC//3D-IO FeS-QDs@NC SICs have great advantages over energy storage devices (fuel cells, LIBs, and supercapacitors) in terms of energy density and power density. Other transition-metal sulfides such as manganese sulfide (MnS) and titanium sulfide (TiS_2) have also exhibited excellent electrochemical performances when used in SICs. Lee's group embedded MnS in N, S co-doped carbon (MnS@NSC) [32]. SICs prepared using MnS@NSC as the anode and N-doped porous carbon (NC) as the cathode exhibited high energy density (139.8 Wh kg^{-1}) and excellent cyclability (their capacitance retention rate was 84.5% after 3000 cycles). Wang's group used polyvinyl pyrrolidone (PVP)-modified Ti3C2Tx MXene to synthesize carbon-coated TiS_2 nanosheets (TiS_2@Cpvp), which exhibited outstanding electrochemical performance. SICs prepared using a TiS2@Cpvp anode and an AC cathode exhibited high energy density (101.7 Wh kg^{-1}) and high power density (200 W kg^{-1}) [33].

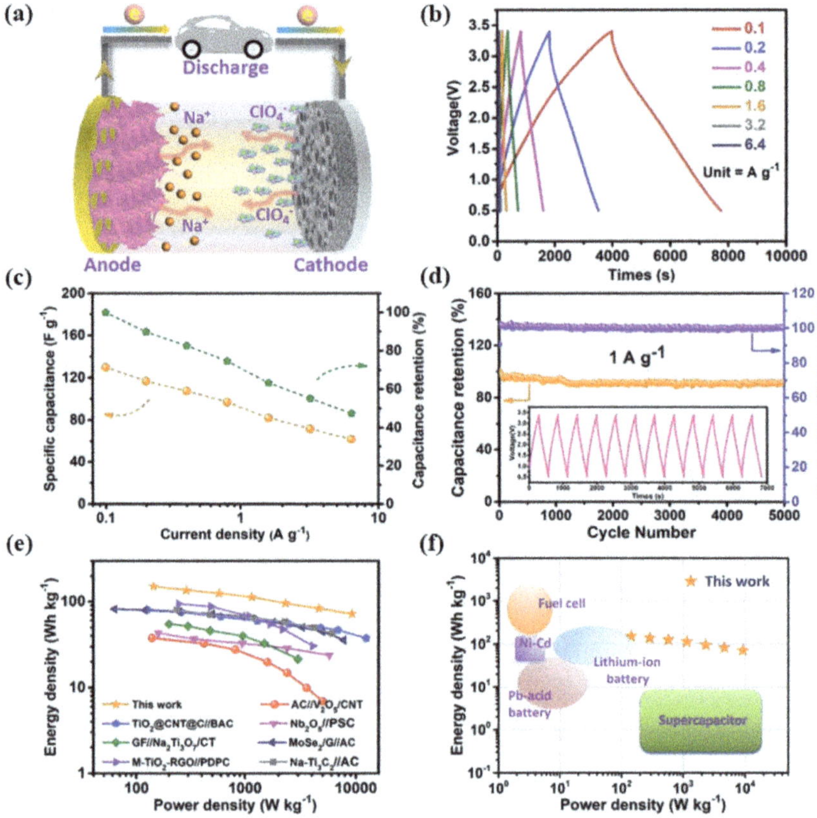

Figure 5.10. (a) A schematic illustration of the 3D-IO FeS-QDs@NC//AC SIC. (b) GCD curves of 3D-IO FeS-QDs@NC//AC SIHCs at different current densities. (c) Rate performance and capacity retention rate of the 3D-IO FeS-QDs@NC//AC SIC. (d) Long-term cyclic performance of the 3D-IO FeS-QDs@NC//AC SIC. A Ragone plot of the 3D-IO FeS-QDs@NC//AC as compared with those of (e) other studies and (f) other types of energy storage device.

Transition-metal selenides have attracted a lot of research attention for use in SICs. $MoSe_2$ is a common transition-metal selenide with a high theoretical capacity (422 mAh g^{-1}), a unique layered structure, a large interlayer distance (0.64 nm), and a narrow bandgap (1.1 eV). Cao's group used a hydrothermal method to synthesize a $MoSe_2$/graphene nanocomposite [34]. The graphene improved the conductivity of the $MoSe_2$ nanocomposite, alleviated volume expansion during the charge and discharge processes, and inhibited $MoSe_2$ aggregation. In addition, Mo–C chemical bonding enhanced the charge transfer efficiency. An SIC prepared using the $MoSe_2$/graphene anode and an AC cathode exhibited high energy (82 Wh kg^{-1}) and power density (10 752 W kg^{-1}). To prevent the agglomeration and stacking of $MoSe_2$ during the charging and discharging processes, the researchers also combined it with MoO_2. Zhao's group added MoO_2 grown on a $MoSe_2$–graphene interface to enhance the reversible conversion of $MoSe_2$ [35]. The synergistic effect of each component endowed the MoO_2/$MoSe_2$–graphene heterostructure with excellent

Na^+ storage performance. SICs composed of $MoO_2/MoSe_2$–graphene and AC exhibited an energy density of 70.3 Wh kg^{-1} at a power density of 63.7 W kg^{-1}. Moreover, the $MoO_2/MoSe_2$–graphene//AC exhibited ultralong cyclic stability; no obvious decay in capacity was apparent after 7000 cycles at 6 A g^{-1}. Similarly, Xia's group grew $CoSe_2$ on carbon cloth, which also significantly improved the electrochemical performance of $CoSe_2$ [36].

Other metallic materials: in addition to the common metal oxides and transition-metal chalcogenides, other metallic materials have also been used as SIC anode materials, such as Sn_4P_3, Sb_2X_3 (X=O, S) and NaBi. Beguin's group adopted a high energy ball-milling technique to synthesize Sn_4P_3 [37]. A novel hybrid SIC based on a Sn_4P_3 anode and an AC cathode displayed ultralong cyclic stability and a capacitance retention rate as high as 94% after 6500 cycles. Liang's group reported the use of a hydrothermal method to grow Sb_2O_3 nanosheets and Sb_2S_3 micronanospheres on carbon fiber cloth (denoted by SO/CFC and SS/CFC, respectively), which were then used as flexible electrodes for SIC devices. To prepare the SIC devices, carbon fibers (CFs) were produced by an electrospinning method for use as cathodes. The energy density of the SO/CFC/CF SICs was 124 Wh kg^{-1} at a power density of 5800 W kg^{-1}, and the SS/CFC/CF SICs delivered a high energy density of 117 Wh kg^{-1} at a power density of 580 W kg^{-1} [38]. In 2018, Chen's group used NaBi as an anode and AC as a cathode to research the performance of SICs (NaBi/AC) [39]. The NaBi/AC SICs exhibited high energy density (106.5 Wh kg^{-1}) and power density (11 100 W kg^{-1}). The capacity retention rate of the NaBi/AC SICs was 98.6% after 1000 cycles.

5.3 Potassium-ion capacitors

Potassium, lithium, and sodium have similar electrochemistries; however, potassium resources are more abundant and cheaper than lithium resources. More importantly, K^+ possesses a lower redox potential and a smaller solvated ionic radius than Na^+. Therefore, a potassium-ion battery is more suitable for large-scale storage than sodium-ion batteries or lithium-ion batteries. PICs combine the advantages of potassium-ion batteries and supercapacitors, namely high energy density, high power density, and excellent cyclic stability. In addition, due to their high working voltage and abundant resources, PICs are expected to replace LICs and SICs. At present, research into PICs is in its early stages. The large radius of K^+ makes it difficult to move during the charge and discharge processes, resulting in poor rate performance and cyclic stability. When LIC and SIC anode materials are directly used in PICs, they are often incompatible. Therefore, it is very important to develop anode materials that are suitable for PICs.

Transition-metal oxides: transition-metal oxides have been widely applied as anode materials for PIBs due to their high theoretical capacity, high safety, and environmental benignity. However, their lower ionic conductivity results in poor rate performance and unsatisfactory cyclic stability. It has been shown that the electrochemical performance of the transition-metal oxides can clearly be improved by designing them into nanostructures, such as hierarchical hybrid structures and hollow structures. For example, Zhang's group adopted a supramolecular assembly strategy to fabricate hierarchical nest-like TiO_2-N doped carbon hybrid nanostructures (TiO_2/NC-HN) [40]. The synergistic effect of their porous structure, hierarchical morphology, and TiO_2 endowed the TiO_2/NC-HN with excellent electrochemical performance. In order to assess the practical application potential of TiO_2/NC-HN, PICs were prepared using TiO_2/NC-HN anodes and AC cathodes (TiO_2/NC-HN/AC). The energy storage process is schematically illustrated in figures 5.11(a) and (b). In the charging process, K^+ is inserted into the TiO_2/NC-HN anode and FSI^- is absorbed on the face of the AC cathode. The opposite reaction occurs during discharge. The charge–discharge curves exhibit approximately linear trends at different current densities, which reflect the interfacial reactions with fast kinetics (figure 5.11(c)). The energy density of TiO_2/NC-HN/AC was 108.6 Wh kg^{-1} at a power density of 97.5 W kg^{-1}, which was superior to those of other PICs previously reported (figure 5.11(d)). More importantly, TiO_2/

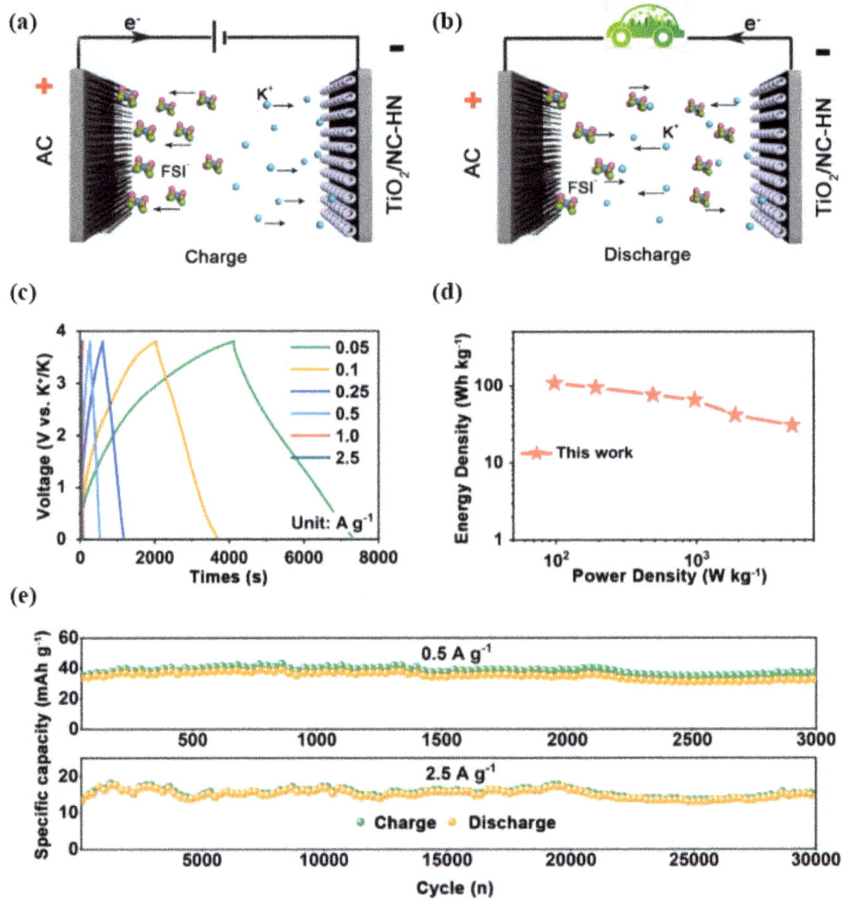

Figure 5.11. (a and b) Schematic illustrations of the TiO_2/NC-HN//AC SIC. (c) Charge–discharge curves at different rates, (d) a Ragone plot and (e) the cyclic performance of the TiO_2/NC-HN//AC SIC [40], reproduced with permission © Elsevier.

NC-HN/AC exhibited excellent cyclic stability; its capacity retention rate was almost 100% after 3000 cycles at a current density of 0.5 A g^{-1}. The capacity was not affected by obvious decay after 30 000 cycles at high current density (2.5 A g^{-1}) (figure 5.11(e)).

Transition-metal-based compounds have attracted a lot of attention due to their high theoretical capacity and excellent cyclic stability. Among various transition-metal compounds, V-based materials exhibit smaller volume changes during the charging and discharging processes due to the presence of a V–O bond. Among the V-based electrode materials, cobalt vanadate has attracted the attention of researchers due to its high activity and multiple accessible valence states. However, its intrinsic poor conductivity has inhibited its large-scale application. To solve this problem, Mai's group combined $Co_2V_2O_7$ with graphene to synthesize $Co_2V_2O_7$ nanosheets@graphene oxide composites [41]. Li's group synthesized porous CoV_2O_6 nanosphere and graphene oxide composites (P-CVO–N@GO), then researched their electrochemical performance as the anode material of PIBs and PICs [42]. P-CVO–N@GO PIBs exhibited high reversible capacity (557 mAh g^{-1} at 0.1 A g^{-1}) and excellent rate performance (97 mAh g^{-1} at 10 A g^{-1}). PICs prepared with P-CVO–N@GO as the anode and AC as the cathode delivered a high energy density of 150.8 Wh kg^{-1} at a power density of 22 500 W kg^{-1}. The capacity retention rate reached 96.8% after 3000 cycles at 1 A g^{-1}.

Transition-metal chalcogenides: metal sulfides and metal selenides have a typical layered structure, which is beneficial to K^+ intercalation and delamination. WS_2 has attracted attention due to its high reversible capacity and tunable interlayer distance. The large layer spacing of WS_2 (0.618 nm) can provide a fast channel for the insertion and migration of K^+, which can also effectively alleviate the volume expansion that occurs during cycling. In addition, the potential plateau of K^+ intercalation into WS_2 is lower, which helps to improve the energy density of PICs [43]. At the same time, the electron mobility of WS_2 (1103 cm^2 V^{-1} s^{-1}) is superior to those of MoS_2 (340 cm^2 V^{-1} s^{-1}) and $MoSe_2$ (240 cm^2 V^{-1} s^{-1}); this promotes electron transfer and improves the rate performance of the material [44]. Although WS_2 has many advantages in the energy storage field, the volume expansion phenomenon is still intractable when conversion reactions are used. Therefore, the development of a new structure for WS_2 is of great importance for its large-scale application. Li's group fabricated freestanding carbon-coated WS_2 nanosheets supported on carbon nanofibers (C-WS_2@CNFs) [45]. Figure 5.12(a) shows the synthetic routes of C-WS_2@CNF. Carbon nanofibers were prepared by electrospinning a substrate, and WS_2 nanosheets grown on the carbon nanofibers were synthesized by a hydrothermal method. Finally, the samples were polydopamine (PDA) coated and annealed to obtain C-WS_2@CNFs. The low layer count of the WS_2 enhanced the conversion reaction efficiency, and the carbon coating improved the conductivity of the WS_2 and inhibited volume expansion during cycling. Due to these advantages, PIBs prepared using C-WS_2@CNFs as the anode and AC nanofibers (ACNFs) as the cathode (C-WS_2@CNFs||ACNFs) exhibited excellent electrochemical performance.

Figure 5.12(b) shows PICs prepared with C-WS_2@CNFs as the anode, ACNFs as the cathode, and 1 M potassium bisfluorosulfonylimide (KFSI) as the electrolyte. The

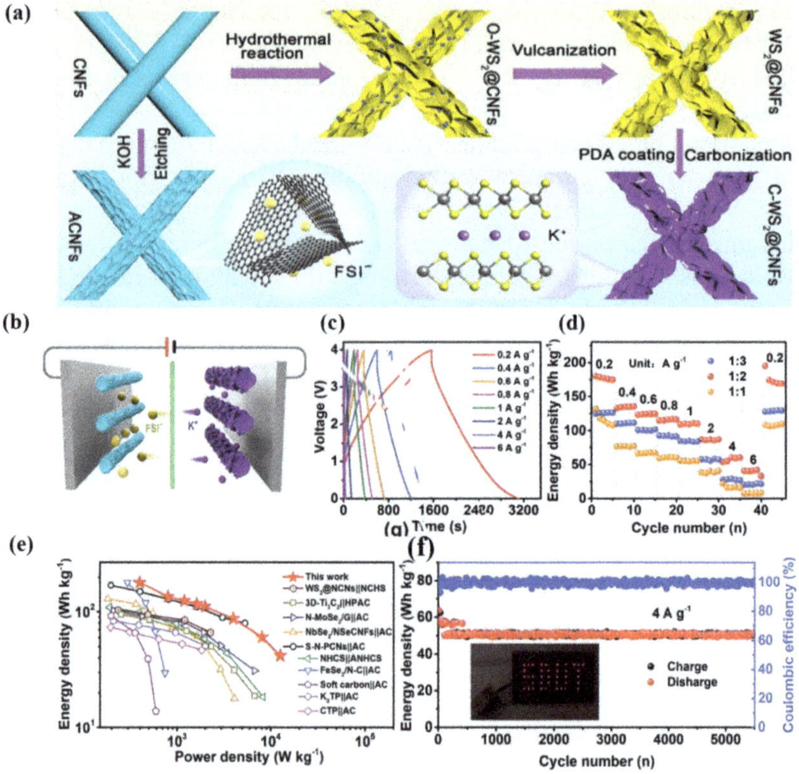

Figure 5.12. (a) The synthetic routes of C-WS_2@CNF. (b) A schematic illustration of the C-WS_2@CNFs||ACNF. (c) Charge–discharge curves of the PIC with a mass ratio of 1:2 and (d) the rate performances of the PIC at various mass ratios. (e) The electrochemical performance of the optimal PIC compared with previously reported results and (f) the cyclic stability of the optimal PIC at 4 A g^{-1}, (inset) an LED powered by the PIC [45], reproduced with permission © Royal Society of Chemistry.

K^+ was inserted into the C-WS_2@CNFs and KFS^- was absorbed on the surface of the ACNFs during the charging process. The nonlinear charge/discharge curves of the C-WS_2@CNFs||ACNFs indicate that the charge storage mechanism of C-WS_2@CNFs||ACNFs is a hybrid of the battery and capacitor mechanisms (figure 5.12(c)). To optimize the performance, PICs were prepared with anodes and cathodes at different mass ratios (1:1, 1:2, and 1:3). The results showed that the C-WS2@CNFs||ACNFs at a mass ratio of 1:2 had the best K^+ storage performance. It exhibited an energy density of 180.4 Wh kg^{-1} at a power density of 399.6 W kg^{-1} (figure 5.12(d)). Ragone plots for these PICs indicate that the performance of the optimal C-WS_2@CNFs||ACNFs was superior to the performances described in most previous reports (figure 5.12(e)). The C-WS_2@CNFs||ACNFs showed excellent cyclic stability; their capacity retention rate was 81.4% after 5500 cycles at 4.0 A g^{-1}. Moreover, LEDs were powered by the PICs, demonstrating that they have the potential for practical application (figure 5.12(f)).

Transition-metal diselenides with layered structures possess fascinating physical and chemical properties. $MoSe_2$, which has a large interlayer distance, is beneficial for K^+ insertion and diffusion [46]. Therefore, $MoSe_2$ has great potential for use as an anode material in PICs. Liu's group adopted a diatomite template method to synthesize a 3D biomorphic N-doped $MoSe_2$/graphene nanocomposite (N-$MoSe_2$/G) [47]. The synthetic routes of N-$MoSe_2$/G are depicted in figure 5.13(a). The diatomite templates, graphite oxide, sodium molybdate, and Se powder were evenly mixed and then a hydrothermal process was used to form $MoSe_2$/G@diatomite. The $MoSe_2$/G@diatomite was annealed in NH_3 to dope it with nitrogen, and the diatomite was finally removed to obtain 3D biomorphic N-$MoSe_2$/G. The biomass template endowed N-$MoSe_2$/G with an open framework, which was beneficial for ion/electron transport and mitigated the volume changes that occur during the charge and discharge processes. As a result of its ample void structures, a PIC based on a N-$MoSe_2$/G anode and a bioderived AC cathode exhibited excellent electrochemical performance. As illustrated in figure 5.13(b), during charging, PF_6^- anions were adsorbed on the surface of the AC cathode and K^+ intercalated into the N-$MoSe_2$/G anode. The reverse reaction occurred during discharge. The potential window of the PIC avoided the decomposition of the electrolyte, as determined by individual cathode and anode CV tests. The test results confirmed that the ideal potential window was 0.5–4.0 V (figure 5.13(c)). The PIC exhibited excellent rate performance; at current

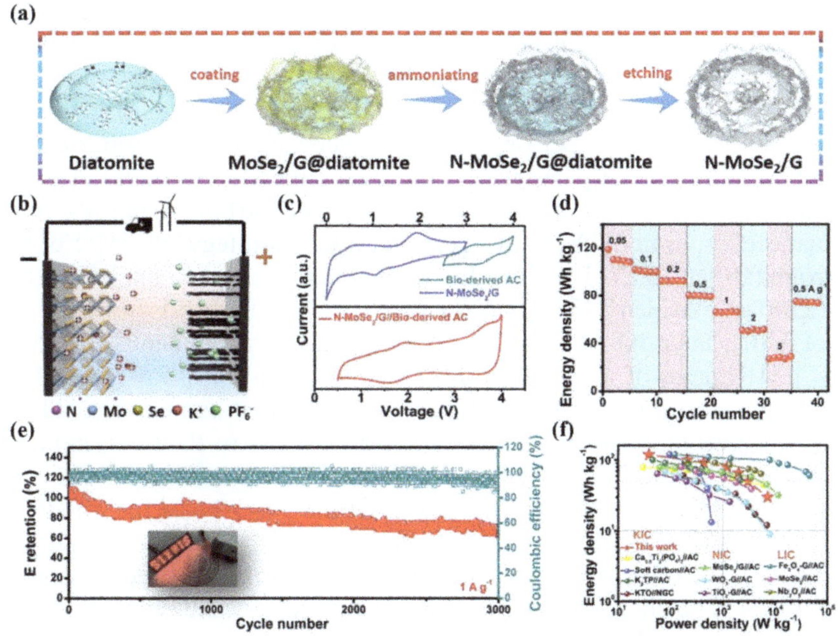

Figure 5.13. (a) The synthetic routes of N-$MoSe_2$/G. (b) A schematic illustration of the N-$MoSe_2$/G/AC PIC. (c) CV curves of N-$MoSe_2$/G and AC in half cells and a PIC. (d) The rate performance and (e) long-term cyclic performance of the PIC. (f) Ragone plots comparing the performance of the N-$MoSe_2$/G/AC PIC with results previously reported for LICs, SICs, and PICs.

densities of 0.1 A g^{-1} and 5.0 A g^{-1}, its energy densities were 101 Wh kg^{-1} and 29 Wh kg^{-1}, respectively (figure 5.13(d)). The PIC had favorable cyclic performance; its energy retention rate was 75.2% after 3000 cycles at 1.0 A g^{-1}. Moreover, the PIC had practical applications prospects, as it could easily light a 3 V LED pad (figure 5.13(e)). A Ragone plot for the PIC showed that it had advantages in terms of energy density and power density compared with previously reported PICs, SICs, and LICs (figure 5.13(f)). The excellent electrochemical performance of the PIC was attributed to its unique biomorphic nanostructure, which effectively enhanced ion/electron transport, provided abundant active reaction sites, and buffered volume expansion during cycling. In addition to $MoSe_2$, other metal selenides have also been used as the PIC anodes. For example, Lu's group used a hydrothermal method to successfully synthesize $FeSe_2$ nanoclusters and carbon composites [48]. The $FeSe_2$ coated by a N-doped carbon layer ($FeSe_2$/N-C) improved conductivity and alleviated volume changed during the charge and discharge processes. Lu's group selected AC as cathode and the $FeSe_2$/N-C as an anode to form a PIC ($FeSe_2$/N-C/AC). As a result of its unique construction, the $FeSe_2$/N-C/AC exhibited high energy density (230 Wh kg^{-1}) and power density (920 W kg^{-1}). Zhu's group adopted the electrospinning method to synthesize $NbSe_2$ nanosheets embedded in N, Se co-doped carbon nanofibers (NbSe2/NSeCNFs) [49]. A PIC prepared using a $NbSe_2$/NSeCNF anode and an AC cathode had high energy density (145 Wh kg^{-1} at 50 mA g^{-1}) and excellent cyclic stability (a capacity retention of about 83% after 10 000 cycles at 2 A g^{-1}).

Transition-metal phosphides: Due to their high conductivity and electrochemical activity, transition-metal phosphides have attracted a lot of attention for use as energy storage materials. However, in practical applications, transition-metal phosphides have suffered from serious volume changes and sluggish redox kinetics, which resulted in poor rate performance and cyclic stability. Previous efforts have been made to improve the properties of transition-metal phosphides. The dispersal of transition-metal phosphides on the surfaces of carbon materials (graphene, carbon nanotubes, or nanofibers) has been a common strategy [50, 51]. For example, Zou's group fabricated Co_2P nanorods 4 nm in diameter and 65 nm long attached to reduced graphene oxide (U-Co_2P@rGO-14) [52]. Figure 5.14(a) shows the synthetic routes of U-Co_2P@rGO-14. The low-magnification transmission electron microscopy (TEM) image of U-Co_2P@rGO-14 (figure 5.14(b)) shows that the Co_2P nanorods were embedded in the rGO framework without aggregation. This unique structure effectively improved the electrical conductivity of the Co_2P nanorods and inhibited the crushing of the electrode material in the charge and discharge processes. The PICs were prepared using U-Co_2P@rGO-14 as the anode and AC as the cathode (U-Co_2P@rGO-14/AC). The optimal potential window of the U-Co_2P@rGO-14/AC was shown by its CV curves to lie between 1.0 and 4.0 V (figure 5.14(c)). The charge–discharge curves showed excellent capacitive behavior, which was mainly attributed to the hybrid energy storage mechanism that included faradic and non-faradic reactions (figure 5.14(d)). When the energy density of the U-Co_2P@rGO-14/AC PIC was 87 Wh kg^{-1}, its power density is 12 W kg^{-1}. At a high power density of 4264.7 W kg^{-1}, the energy density was still 10 Wh kg^{-1}. A Ragone plot showed that the U-Co_2P@rGO-14//AC PIC had a high energy/power

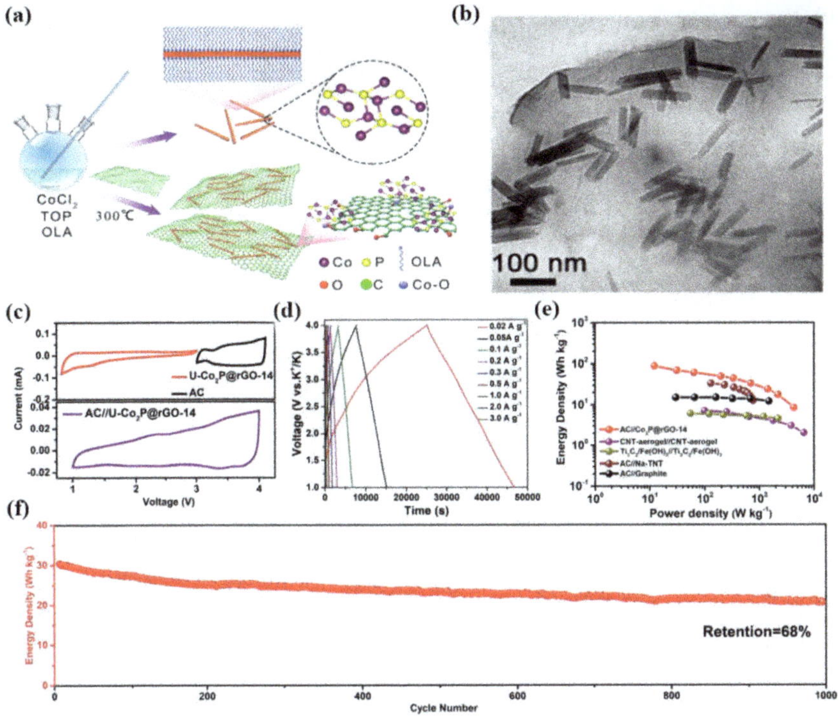

Figure 5.14. (a) A schematic illustration of the synthetic routes of U-Co_2P@rGO-14. (b) A low-magnification TEM image of the U-Co_2P@rGO-14. (c) CV curves of U-Co_2P@rGO-14 and AC. (d) Charge–discharge curves of the U-Co_2P@rGO-14//AC PIC at different current densities. (e) A Ragone plot for the U-Co_2P@rGO-14//AC PIC showing a comparison with previously reported results. (f) The cyclic stability of the PIC [52], reproduced with permission © Royal Society of Chemistry.

density that was comparable to those previously reported (figure 5.14(e)), proving that this device has confirmed practical value. In addition, the U-Co_2P@rGO-14// AC possessed excellent cyclic stability; its capacity retention was 68% after 1000 cycles at 1 A g^{-1} (figure 5.14(f)).

Traditional dispersed transition-metal phosphides on the surfaces of carbon materials only slightly relieve volume expansion during cycling. Liu's group innovatively fabricated MoP nanoparticles uniformly confined in 3D nitrogen-doped carbon nanosheets (MoP@NC) [53]. Figure 5.15(a) shows the synthetic routes of MoP@NC. The hydrogen bonds of water-soluble polyacrylamide (PAM) hydrogel were used to anchor the MoP, forming composites of MoP nanoparticles embedded in 3D porous nitrogen-doped carbon. This unique construction helped to reduce volume expansion and enhanced the ion/electron transport rate. Furthermore, the same group adopted a 3D printing technique to fabricate 3D-printed potassium-ion hybrid capacitors (3DP-PIHCs). The PIHCs were composed of a 3D-printed MoP@NC-1 anode and a 3D-printed AC cathode (figure 5.15(b)). The 3D-printed electrode had excellent mechanical stability and sufficient charge transport channels and thus excellent electrochemical performance. The CV curves resembled nearly

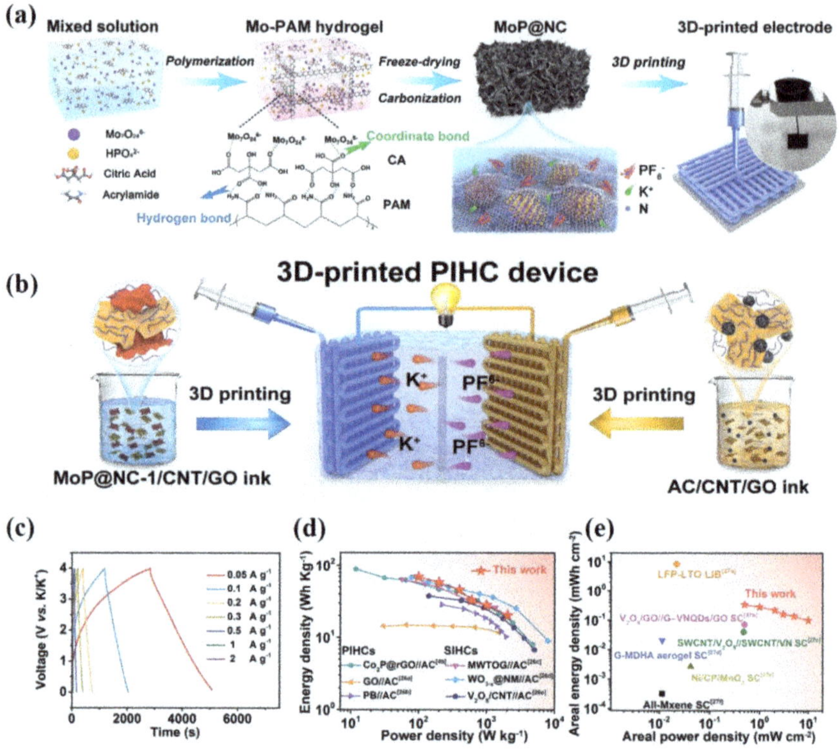

Figure 5.15. (a) The synthetic routes of MoP@NC. (b) A schematic illustration of the MoP@NC-1//AC 3DP-PIHCs. (c) The charge–discharge curves of the MoP@NC-1//AC 3DP-PIHCs at different current densities. (d), (e) Ragone plots for the MoP@NC-1//AC 3DP-PIHCs showing a comparison with previously reported devices [53], reproduced with permission © John Wiley and Sons.

symmetrical triangles, confirming that the electrochemical reactions were dominated by capacitive behavior (figure 5.15(c)). The energy density of the MoP@NC-1//AC 3DP-PIHCs was 69.7 Wh kg^{-1} at a power density of 2041.6 W kg^{-1}, which was superior to those of most previously reported SICs and PICs (figure 5.15(d)). More importantly, the areal energy density of the MoP@NC-1//AC 3DP-PIHCs was 0.34 mWh cm^{-2} at an areal power density of 9.97 mW cm^{-2}, which was superior to those of batteries and supercapacitors made by 3D printing (figure 5.15(e)).

5.4 Zinc-ion capacitors

Zinc-ion capacitors (ZICs) constitute a new type of energy storage device which combines the advantages of zinc-ion batteries (which have high energy density) and supercapacitors (which have exceptional power density and cyclic stability). Depending on the electrode material used, ZICs are classified into two types (capacitor-type cathode//battery-type anode versus battery-type cathode//capacitor-type anode). In this paper, we mainly discuss the type characterized by capacitor-type cathodes and battery-type anodes [54, 55]. As shown in figure 5.16(a), during the charging process, the anions are adsorbed on the surface

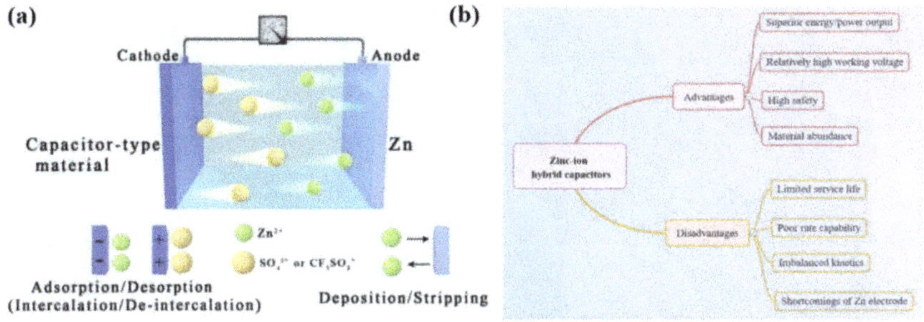

Figure 5.16. (a) A schematic illustration of a ZIC (capacitor-type cathode//battery-type anode). (b) The advantages and disadvantages of ZICs [54], reproduced with permission © John Wiley and Sons.

of the cathode, while the Zn^{2+} is deposited on the Zn anode. Compared to traditional supercapacitors, ZICs have several advantages, including high energy density, high working voltages, and outstanding safety. However, in practice, ZICs still have some problems: the formation of Zn dendrites results in poor cyclic stability, and kinetic imbalance between the battery-type and capacitor-type electrodes results in poor rate performance. The advantages and disadvantages of ZICs are detailed in figure 5.16(b). The electrolytes for ZICs are generally divided into two categories, aqueous and nonaqueous (either organic solutions or ionic liquids). Aqueous electrolytes possess higher power density because they can deliver high ionic conductivity. Due to their low working voltages, aqueous ZICs deliver lower energy densities. ZICs based on nonaqueous electrolytes can provide a wider voltage window [54] and therefore achieve higher energy densities. In this section, we focus on ZICs based on nonaqueous electrolytes.

Tang's group reported a novel zinc-ion hybrid supercapacitor (Zn-HSC) [56]. As shown in figure 5.17(a), the Zn-HSC adopted Zn foil as both anode and collector and AC as the cathode. Two kinds of nonaqueous electrolyte were studied for this system. One was 1 M $Zn(CF_3SO_3)_2$ dissolved in 1,2-dimethoxyethane (DME) and 1,3-dioxolane (DOL) (1:1), the other was 1 M $Zn(CF_3SO_3)_2$ in acetonitrile (AN). In the charging process, $(CF3SO3)^-$ moved towards and was adsorbed on the surface of the AC, while Zn^+ moved towards and was deposited on the Zn anode. Acetonitrile, which has high conductivity and excellent capacitance retention, performed better than the DME–DOL electrolyte. Figure 5.17(b) shows the CV curves of Zn-HSC in acetonitrile at different scan rates (100–1000 mV s^{-1}); they have an irregular rectangular shape. The charge–discharge curves of Zn-HSC at different current densities are triangular, and the shapes barely change with increasing current. The specific capacitances of Zn-HSC in AN were 170, 156, and 144 F g^{-1} at 0.1, 0.5, and 1 A g^{-1}, respectively (figure 5.17(c)). The specific capacitance was 120 F g^{-1} after 20 000 cycles, and the capacitance retention was 91%. Moreover, in the process of cycling, the coulombic efficiency was maintained at 100% (figure 5.17(d)). In addition to acetonitrile, researchers have also adopted ionic liquids as electrolytes for ZICs. For example, Chen's group adopted a PANI-derived carbon material as the cathode and an ionic liquid as the electrolyte in the

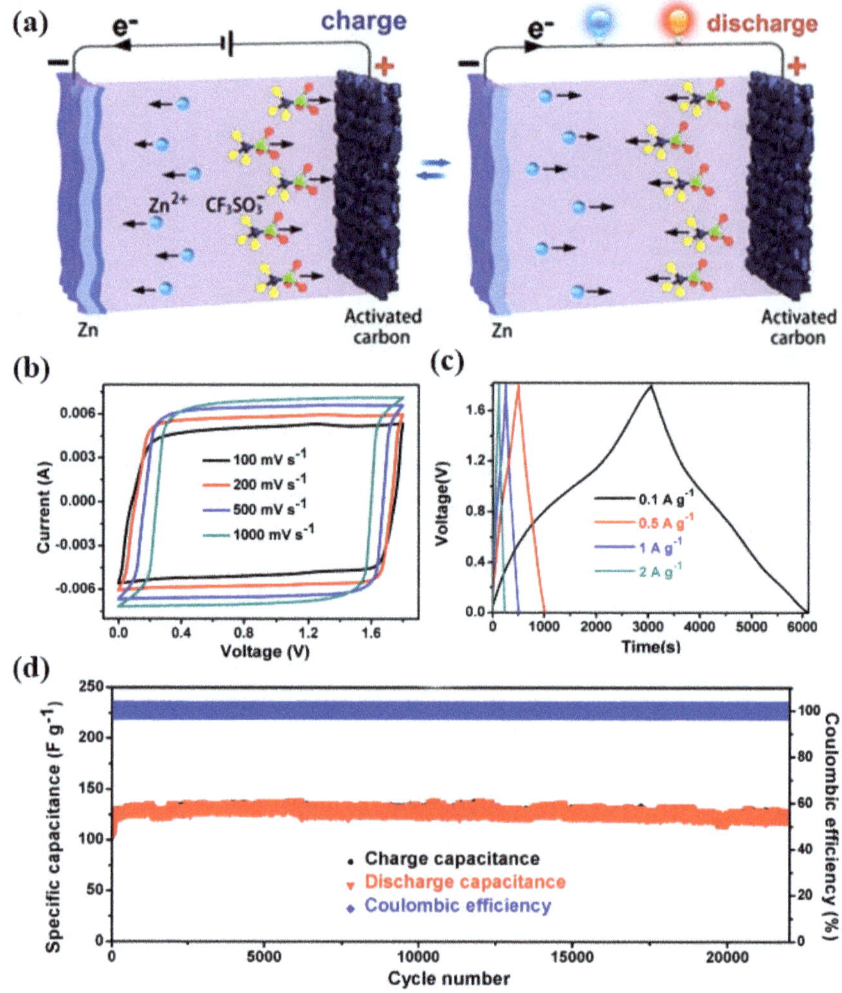

Figure 5.17. (a) A schematic illustration of the Zn-HSC. (b) CV curves of the Zn-HSC at various sweep rates (100–1000 mV s^{-1}). (c) GCD curves of the Zn-HSC at different current densities. (d) The cyclic stability of Zn-HSC at 1 A g^{-1} over 22 000 cycles [56], reproduced with permission © Elsevier.

construction of a new type of ZIC [57]. Because the ionic liquid had a relatively wide voltage window (2.4 V), the ZICs based on the ionic liquid had high energy density. The volumetric energy density of a ZIC constructed using neat ionic liquid was as high as 54 Wh l^{-1}, which approximately matches the corresponding value for lead–acid batteries. After 35 000 cycles, the capacitance retention was 82%, confirming excellent cyclic stability. Zhi's group adopted a novel organic solvent as the electrolyte to increase the voltage to 2.5 V [58]. The electrolyte was fabricated by adding 0.2 M $ZnCl_2$ into tetraethylammonium tetrafluoroborate in a propylene carbonate ($Et_{4N}BF_4$/PC) solvent. A ZIC based on the novel electrolyte not only exhibited excellent cyclic performance (105.9 F g^{-1} after 9500 cycles) but also possessed outstanding anti-self-discharge performance; after resting for 300 h, its capacitance retention was 76.16%. This kind of high voltage and low self-discharge-rate ZIC has broad application prospects.

5.5 Future prospects of hybrid ion capacitors

Although some achievements have been made, the energy densities and power densities of hybrid ion capacitors are still far from practical. In addition, research into various ion capacitors is still at the laboratory stage and commercial devices have not been assembled. There is still a long way to go to improve the energy density, power density, and industrial applications of ion capacitors. In the future, research into ion capacitors can take the following directions: (1) the selection of suitable methods (e.g. the nanoengineering, doping, and tailoring of the 3D electrode structure) to synthesize battery-type materials with high-efficiency ion/electron mixed conductors; (2) the adoption of various strategies (e.g. increasing the specific surface area, heteroatomic doping, adjustment of the void structure and crystallinity) to improve the adsorption capability of cathode materials; (3) the choice of suitable electrolytes, as suitable electrolytes can enhance the stability of the interface between the electrode material and the electrolyte and improve the cyclic performance. In addition, a suitable electrolyte can also increase the working voltage of ion capacitors and improve their energy density. Solid electrolytes make hybrid ion capacitors safer, which is also the trend of future development. (4) Flexible power supply devices have wide application prospects for future wearable devices, so the development of flexible hybrid ion capacitors is of great significance to future industrial development.

References

[1] Xia Q, Yang H, Wang M, Yang M, Guo Q, Wan L, Xia H and Yu Y 2017 *Adv. Energy Mater.* **7** 1701336

[2] Wang H, Zhu C, Chao D, Yan Q and Fan H J 2017 *Adv. Mater.* **29** 1702093

[3] Yi R, Chen S, Song J, Gordin M L, Manivannan A and Wang D 2014 *Adv. Funct. Mater.* **24** 7433

[4] Yang B, Chen J, Liu B, Ding Y, Tang Y and Yan X 2021 *J. Mater. Chem.* A **9** 6352

[5] Sun Q, Wang Z, Zhang Z, Yu Q, Qu Y, Zhang J, Yu Y and Xiang B 2016 *ACS Appl. Mater. Interfaces* **8** 6303

[6] Zhang W, Sheng J, Zhang J, He T, Hu L, Wang R, Mai L and Mu S 2016 *J. Mater. Chem.* A **4** 16936

[7] Zhang S, Li C, Zhang X, Sun X, Wang K and Ma Y 2017 *ACS Appl. Mater. Interfaces* **9** 17136

[8] Wang R, Liu P, Lang J, Zhang L and Yan X 2017 *Energy Storage Mater.* **6** 53

[9] Deng Q, Fu Y, Zhu C and Yu Y 2019 *Small* **15** 1804884

[10] Han C, He Y-B, Liu M, Li B, Yang Q-H, Wong C-P and Kang F 2017 *J. Mater. Chem.* A **5** 6368

[11] Zhang W, Seo D-H, Chen T, Wu L, Topsakal M, Zhu Y, Lu D, Ceder G and Wang F 2020 *Science* **367** 1030

[12] Zhao E, Qin C, Jung H-R, Berdichevsky G, Nese A, Marder S and Yushin G 2016 *ACS Nano* **10** 3977

[13] Dong S, Wang X, Shen L, Li H, Wang J, Nie P, Wang J and Zhang X 2015 *J. Electroanal. Chem.* **757** 1

[14] Wang C *et al Angew. Chem. Int. Ed.* 2019**58** 15863

[15] Chen Z, Li H, Lu X, Wu L, Jiang J, Jiang S, Wang J, Dou H and Zhang X 2018 *Chem. Electro. Chem.* **5** 1516

[16] Bi R, Xu N, Ren H, Yang N, Sun Y, Cao A, Yu R and Wang D 2020 *Angew. Chem. Int. Ed.* **59** 4865

[17] Han J-T, Huang Y-H and Goodenough J B 2011 *Chem. Mater.* **23** 2027

[18] Saritha D, Pralong V, Varadaraju U V and Raveau B 2010 *J. Solid State Chem.* **183** 988

[19] Wang R, Lang J, Zhang P, Lin Z and Yan X 2015 *Adv. Funct. Mater.* **25** 2270

[20] Dailly A, Willmann P and Billaud D 2002 *Electrochim. Acta* **48** 271

[21] Won J H, Jeong H M and Kang J K 2017 *Adv. Energy Mater.* **7** 1601355

[22] Zhu J and Ding X 2019 *J. Alloys Compd.* **809** 151870

[23] Yang C, Ren J, Zheng M, Zhang M, Zhong Z, Liu R, Huang J, Lan J, Yu Y and Yang X 2020 *Electrochim. Acta* **359** 136898

[24] Zhang Y *et al Chem. Sus. Chem.* 2020 **13** 2522

[25] Liu S, Luo Z, Tian G, Zhu M, Cai Z, Pan A and Liang S 2017 *J. Power Sources* **363** 284

[26] Liu S, Cai Z, Zhou J, Pan A and Liang S 2016 *J. Mater. Chem.* A **4** 18278

[27] Chen Z, Augustyn V, Jia X, Xiao Q, Dunn B and Lu Y 2012 *ACS Nano* **6** 4319

[28] Lim E *et al Adv. Funct. Mater.* 2016 **26** 3711

[29] Wang R, Wang S, Peng X, Zhang Y, Jin D, Chu P K and Zhang L 2017 *ACS Appl. Mater. Interfaces* **9** 32745

[30] Hu X, Liu Y, Chen J, Jia J, Zhan H and Wen Z 2019 *J. Mater. Chem.* A **7** 1138

[31] Yang D, Chen W, Zhang X, Mi L, Liu C, Chen L, Guan X, Cao Y and Shen C 2019 *J. Mater. Chem.* A **7** 19709

[32] Li S, Chen J, Xiong J, Gong X, Ciou J and Lee P S 2020 *Nano-Micro Lett.* **12** 34

[33] Tang J, Huang X, Lin T, Qiu T, Huang H, Zhu X, Gu Q, Luo B and Wang L 2020 *Energy Storage Mater.* **26** 550

[34] Zhao X, Cai W, Yang Y, Song X, Neale Z, Wang H-E, Sui J and Cao G 2018 *Nano Energy* **47** 224

[35] Zhao X, Wang H-E, Yang Y, Neale Z G, Massé R C, Cao J, Cai W, Sui J and Cao G 2018 *Energy Storage Mater.* **12** 241

[36] Xia Z, Sun H, He X, Sun Z, Lu C, Li J, Peng Y, Dou S, Sun J and Liu Z 2019 *Nano Energy* **60** 385

[37] Chojnacka A, Pan X, Jeżowski P and Béguin F 2019 *Energy Storage Mater.* **22** 200

[38] Liu S, Cai Z, Zhou J, Zhu M, Pan A and Liang S 2017 *J. Mater. Chem.* A **5** 9169

[39] Yuan Y, Wang C, Lei K, Li H, Li F and Chen J 2018 *ACS Cent. Sci.* **4** 1261

[40] Wang G, Li Y, Liu Y, Jiao S, Peng B, Li J, Yu L and Zhang G 2021 *Chem. Eng. J.* **417** 127977

[41] Luo Y, Xu X, Zhang Y, Chen C-Y, Zhou L, Yan M, Wei Q, Tian X and Mai L 2016 *ACS Appl. Mater. Interfaces* **8** 2812

[42] Liang H, Zhang Y, Hao S, Cao L, Li Y, Li Q, Chen D, Wang X, Guo X and Li H 2021 *Energy Storage Mater.* **40** 250

[43] Zhang R, Bao J, Pan Y and Sun C-F 2019 *Chem. Sci.* **10** 2604

[44] Zhang W, Huang Z, Zhang W and Li Y 2014 *Nano Res.* **7** 1731
[45] Geng S *et al Energy Environ. Sci.* 2021**14** 3184
[46] Huang H, Cui J, Liu G, Bi R and Zhang L 2019 *ACS Nano* **13** 3448
[47] Yi Y, Sun Z, Li C, Tian Z, Lu C, Shao Y, Li J, Sun J and Liu Z 2020 *Adv. Funct. Mater.* **30** 1903878
[48] Ge J, Wang B, Wang J, Zhang Q and Lu B 2020 *Adv. Energy Mater.* **10** 1903277
[49] Chen M *et al Adv. Funct. Mater.* 2020 **30** 2004247
[50] Yi Z, Liu Y, Li Y, Zhou L, Wang Z, Zhang J, Cheng H and Lu Z 2020 *Small* **16** 1905301
[51] Wang X, Ma J, Wang J and Li X 2020 *J. Alloys Compd.* **821** 153268
[52] Wang Y, Zhang Z, Wang G, Yang X, Sui Y, Du F and Zou B 2019 *Nanoscale Horiz* **4** 1394
[53] Zong W *et al Adv. Sci.* 2021 **8** 2004142
[54] Tang H, Yao J and Zhu Y 2021 *Adv. Energy Mater.* **11** 2003994
[55] Shao Y, Shen F and Shao Y 2021 *Chem. Electro. Chem.* **8** 484
[56] Wang H, Wang M and Tang Y 2018 *Energy Storage Mater.* **13** 1
[57] Zhou H, Liu C, Wu J-C, Liu M, Zhang D, Song H, Zhang X, Gao H, Yang J and Chen D 2019 *J. Mater. Chem.* A **7** 9708
[58] Huang Z *et al Adv. Energy Mater.* 2020**10** 2001024

IOP Publishing

Electrochemical Capacitors

Jianmin Ma and Junliang Yang

Chapter 6

Low-temperature electrochemical capacitors

Linfei Lai

It is important for electrochemical capacitors to be usable at low temperatures. Thus, the development of electrochemical capacitors has attracted great attention. In this chapter, low-temperature supercapacitors (SCs) are discussed from electrode structure design to electrolyte rationalization in the light of current knowledge about their low-temperature properties. We hope that this chapter will give the reader a wealth of knowledge about low-temperature electrochemical capacitors, from materials to systems.

6.1 Introduction

The high cost and continuous depletion of fossil fuel reserves and the global warming caused by excessive CO_2 emissions have led to the development of renewable and sustainable energy systems. Batteries and SCs are among the leading electrochemical energy storage devices. Compared to Li-ion batteries, SCs can provide high power and can be charged and discharged in seconds (10 kW kg^{-1}) at smaller energy densities (~5–10 Wh kg^{-1}). Li-ion batteries are the most widely used commercial electrochemical energy storage devices; they have the advantages of high energy density (100–265 Wh kg^{-1}) and low power density (~200–500 W kg^{-1}). As a result of the wide application of energy storage devices in industry, the military, and aerospace exploration, the requirement for energy storage devices that can operate in different environments has created new challenges in the field of energy research (figure 6.1(a)). Many countries require outstanding low-temperature energy devices for winter use in civilian, military, and high-altitude applications. With the exception of SCs, most commercial energy storage systems suffer from a significant loss of performance below −20 °C. The commercially available non-aqueous SCs are rated for a minimum operating temperature of −40 °C at the cost of power density loss. Although Li-ion batteries have high energy density, the ionic conductivity of the electrolyte is severely reduced at low temperatures. Moreover, the presence of a solid electrolyte interphase (SEI) dramatically hampers the solid-state diffusion of Li

doi:10.1088/978-0-7503-5042-6ch6

© IOP Publishing Ltd 2023

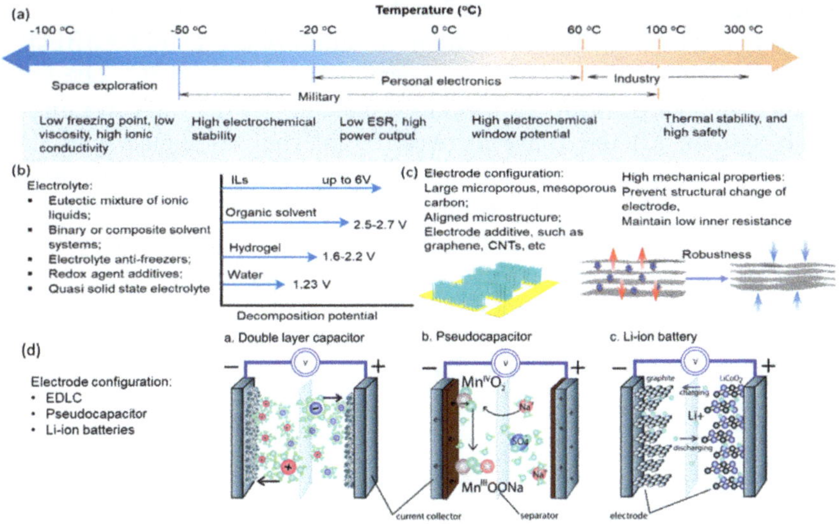

Figure 6.1. (a) The temperature ranges of energy storage devices for various applications. (b) Strategies for the optimization of the electrolyte and electrochemical window for different types of electrolyte. (c) Adjustment of the electrode structure to reach high capacitance retention at low temperatures with high mechanical robustness. (d) Schematics showing the electrochemical processes of an electric double-layer capacitor (EDLC), a pseudocapacitor, and a Li-ion battery [7], reproduced with permission © Royal Society of Chemistry.

ions. The temperature-dependent faradic reaction rates of Li-ion batteries currently limit commercial Li-ion battery products to a minimum operating temperature of −20 °C. Some of these limitations can be circumvented through the use of SCs. SCs have intrinsic advantages over batteries, such as high power density and long cyclic life. SCs are potentially promising for energy storage at low temperatures, as the entropic contribution to the Gibbs electrostatic free energy should be reduced at low temperatures, which in turn leads to a theoretical capacitance value comparable or similar to that at room temperature (RT) [1]. Moreover, the leakage current is minimized at low temperatures (<40 °C) [2]. However, the real performance of SC devices at low temperatures is usually not as good as that at RT.

SCs can be classified into two groups according to the energy storage mechanism of their electrode materials: EDLCs rely on charge accumulation at the electrode–electrolyte interface upon the application of a voltage, while pseudocapacitors are characterized by reversible faradic reactions in the electrode. EDLCs employ porous carbon as electrodes, whereas pseudocapacitor electrodes include transition-metal oxides/sulfides/hydroxides/selenides, etc. and conductive polymers, such as polypyrrole, polyaniline, etc. Currently, most commercial SC devices employ EDLCs with microporous carbon as electrodes due to their good chemical stability upon cycling. Li-ion capacitors (LICs) are hybrid SCs that combine the Li-ion intercalation mechanism of an anode/cathode with the double-layer mechanism of a cathode/anode, depending on the potential of the Li-ion intercalates.

Commercial EDLCs with activated carbon (AC) electrodes and an electrolyte composed of 1 M tetraethylammonium tetrafluoroborate ([TEA][BF_4]) salt in acetonitrile (ACN) fail to operate at temperatures below −40 °C. The key factors limiting their low-temperature performance include: (1) the high freezing point of the solvent system; (2) the rapid increase of internal resistance with decreasing temperature; (3) the sluggish ion mobility and the increasing solvent viscosity, etc. To expand the operational temperature limit of SCs, it is important to develop (1) electrodes with appropriate pore structures for the efficient generation of interfacial charge, and (2) electrolytes with low freezing points and high ionic conductivity to match the electrodes. The energy density of a SC cell is given by:

$$E = \frac{1}{2}CV^2 \tag{6.1}$$

$$P_{\text{max}} = \frac{V^2}{4R}, \tag{6.2}$$

where C is the specific capacitance (F g^{-1}), V is the electrochemical window potential (V), and R is the equivalent series resistance (ESR) in ohms of the SC device. The maximum power density relies on the window potential and the ESR of the whole device. Among the widely used electrolytes, ionic liquids (ILs) provide the largest electrochemical window potential, followed by organic solvents, hydrogel, and aqueous electrolytes (figure 6.1(b)). In order to reduce the viscosity and improve the ionic conductivity of ILs at low temperatures, ILs need to be combined with organic solvents. In addition to the design of eutectic mixtures of ILs to further reduce the freezing temperatures of electrolytes, the use of cosolvents, antifreezes, redox additive agents, etc. has been explored with the aim of designing electrolytes for low-temperature applications. Earlier studies of SCs indicated that microporous carbon with a pore size of less than than 1 nm can deliver superior capacitance in ILs [3] and aqueous and organic electrolytes [4]. However, carbon electrodes suffer from a fourfold increase in their ESR in organic electrolytes when the temperature decreases from 40 °C to −40 °C, and 18% of the capacity loss takes place from 25 °C to −35 °C [5].

At RT, ion desolvation takes place on the nanosecond scale in microporous carbon electrodes, which helps SCs to achieve high capacitance with organic electrolytes. For microporous carbon electrodes, the ion desolvation process is more sluggish at low temperatures than at RT. The design of carbon electrodes with an appropriate pore structure which can adapt to the properties of the electrolyte is of great importance.

It is noteworthy that pseudocapacitors usually have a significant capacitance decay at low temperatures, as the redox reaction rate decreases exponentially with decreasing temperature. Lithium titanate (Li_4TiO_5, LTO) has extraordinary properties such as a stable operating potential, a small volume change during Li^+ intercalation/deintercalation and a high rate performance that benefits from the fast kinetics of Li migration in the spinel LTO network. Li-ion capacitors (LICs) consist of a Li-intercalative negative electrode and an anion-adsorbing positive electrode, such as AC, graphene, etc. During the charging process, Li ions intercalate into the negative electrode, while anions are adsorbed on the interface of the positive electrode (figure 6.1(d)). At low temperatures, asymmetric SCs with LTO anodes and AC cathodes have better capacitance retention than the pseudocapacitors which employ two types of Li-ion intercalation compound in both the anode and the cathode [6].

This chapter discusses low-temperature SCs from electrode structure design to electrolyte rationalization in the light of current knowledge about their low-temperature properties.

6.2 The low-temperature behavior of SCs

Symmetric EDLC SCs use microporous carbon as an electroactive material, which is coated onto Al foil current collectors to prepare electrodes. A thin layer of carbon black or graphene-based coating is usually required to modify the Al foil to (1) prevent the corrosion of the Al foil by the electrolyte and (2) improve the adhesion between the electroactive material and the electrode. In the last decade, many reports have proposed ways to develop SCs capable of operating at subzero temperatures. However, −40 °C represents the typical rated lowest operational limit for non-water-based commercial off-the-shelf (COTS) cells. Kotz *et al* characterized the electrochemical performance of commercial capacitors at various temperatures and observed a rapid increase in the ESR with reducing temperature. When the temperature decreases from 0 °C to −40 °C, the electrochemical performance of SCs that use propylene carbonate (PC) as a solvent decreases more rapidly than the performance of SCs that use ACN as a solvent. It is noteworthy that a 10 °C increase in temperature or a potential increase of 0.1 V speeds up the capacity degradation of SCs. Measurements have determined that the lifetime of a SC is 25 years at 25 °C and 3.0 V, while the lifetime of a capacitor is 10 years at 65 °C and 2.5 V [8]. Low temperatures help in obtaining a better cyclic life, although the capacitance values are then not as high as those at room temperature [8]. Liu *et al* tested commercial SCs that employed ACN- and PC-based electrolytes. An ACN-based SC had a capacitance retention of 93% at high current and 96.8% at low current at −30 °C. A PC-based SC had a capacitance retention of 63% at high current and 71.6% at low current at −30 °C. The PC-based SC had a significant increase in resistance at the end of its charging step from 0.015 ohms at 20 °C to 0.075 ohms at −30 °C. However, the resistance of the ACN-based electrolyte over the temperature range from −30 °C to 45 °C remained at ~0.012 ohms [9]. ACN-based electrolytes have a critical challenge for operation below −40 °C due to the high melting point of ACN (−45 °C). The development of SCs with sufficiently high capacitance and power density values at temperatures below −40 °C continues to be a challenge.

It is known that EDLCs can achieve reliable low-temperature operation using organic electrolytes or ILs, as the energy storage mechanism of such electrolytes only involves non-faradic reactions. The specific surface area of AC ranges from 1000 to 2500 $m^2\ g^{-1}$, and it is the most widely used electrode material for EDLCs. Although

AC has a high surface area, some of its small micropores are inaccessible to electrolyte ions and barely contribute to the effective capacitance. At low temperatures, the electrolyte ions need to be desolvated in order to be accommodated in the micropores. The desolvation process of the ions in the electrolyte is more sluggish at low temperatures than at room temperature; therefore, carbon with large micropores or small mesopores is beneficial. Unlike EDLCs, pseudocapacitors usually have a significant capacitance decay at low temperatures because (1) the redox reaction rate decreases exponentially with decreasing temperature, and (2) conventional aqueous electrolytes have high freezing temperatures. Pasquier *et al* evaluated the rechargeability of Li-ion batteries, Li-ion SCs, and carbon-based EDLCs. Asymmetric hybrid SCs were constructed with a $Li_4Ti_5O_{12}$ (LTO) anode and a cathode selected from AC, $LiCoO_2$, $LiMn_2O_4$, etc. At a power density of 1000 W kg^{-1}, the percentage energy losses were 32.5%, 45.9%, and 53.8% for carbon/carbon, LTO/carbon, and LTO/$LiCoO_2$ SCs, respectively, while Li-ion batteries lost most of their capacity [6]. LTO-based asymmetric SCs had high power density and good low-temperature performance at −30 oC in a 2 M $LiBF_4$/ACN electrolyte [10]. As LTO has a high Li intercalation potential (~1.55 V vs Li/Li^+), Li ions can be intercalated without the risk of Li plating [6]. The downsizing of pseudocapacitive materials is another modification strategy for low-temperature applications, but it does not apply to spinel LTO, which is a 'zero-strain' material when used as a Li-ion host material. Xu *et al* chose $Ti_3C_2T_x$ as an electrode and infiltrated 40 wt.% sulfuric acid electrolyte between 2D sheets of MXene. This method allowed the electrolyte ions to reach redox-active sites unimpeded and rapidly interact with the surface oxygen groups of MXene, even at low temperatures. MXene electrodes have achieved a high capacity retention of >75% at 100 mV s^{-1} at −50 °C, indicating excellent low-temperature rate performance [11].

The electrolyte is one of the key factors that determines SC performance in terms of the ESR, power density, rate performance, cyclability, safety, etc. The ionic conductivity, solvation structure, ion size, viscosity, dielectric constant, thermal properties, and electrochemical stability of the electrolyte are the key parameters that affect the overall performance of a SC. In ILs, strong coulombic attraction takes place between ions of opposite charge. The liquid state of ILs can be maintained at temperatures above 200 °C, which allows SCs to operate at high temperatures. However, the high melting points and viscosities of IL-based electrolytes create challenges for their use at low temperatures. Eutectic mixtures of ILs have shown promising characteristics which have allowed SCs to work at temperatures tens of degrees lower than the freezing points of their individual components [12–14]. For example, Lin *et al* developed a eutectic IL-based electrolyte composed of N-methyl-N-propylpiperidinium bis(fluorosulfonyl)imide ([PIP_{13}][FSI]) and N-butyl-N-methylpyrrolidinium [FSI] ([PYR_{14}][FSI]). [PIP_{13}][FSI] and [PYR_{14}][FSI] have two different cationic species but only one type of anion. The eutectic IL had a significantly lower freezing temperature (<−80 °C) than the freezing temperatures of the individual ILs, namely 6 °C for [PIP_{13}][FSI] and −18 °C for [PYR_{14}][FSI] [13]. To decrease the temperature limit of SCs, solvents with low freezing points are usually combined with ILs or inorganic solutes to form electrolytes. Janes *et al* utilized a binary mixture of ethylene carbonate (EC) and esters, such as methyl acetate (MA), ethyl acetate (EA), and methyl formate (MF) to decrease the working temperature to −40 °C [15]. Brandon *et al* reported a spiro-(1,1′)-bipyrrolidinium tetrafluoroborate ([SBP][BF_4]) and tetraethylammonium tetrafluoroborate ([TEA][BF_4])-based electrolyte with cosolvents of acetonitrile (ACN), MF, esters, and cyclic ethers which was able to expand the low-temperature limit of a SC to −75 °C [1, 2]. However, in addition to the low melting point and high ionic conductivity which are key parameters for low-temperature electrolytes, optimum ion–solvent interactions are also important. The optimal electrolyte has ionic diameters or solvated ions whose sizes match the size of the carbon pores, thereby ensuring high areal capacitance, highly efficient pore utilization, and the saturation of active porosity. Decaux *et al* found that highly polarizing ions such as Li^+ retain their solvation shells during charging in polar solvents (alkylcarbonates). As EC is more polarizable than ACN and has a larger dipole moment, EC binds more strongly with small ions, such as Li^+. The Li^+ in EC-based solvents needs a high desolvation energy in order to diffuse into micropores; alternatively, a mesoporous electrode is required for EC-based electrolytes. On the other hand, [TEA][BF_4] or Li (trifluoromethanesulfonyl)imide (TFSI) dissolved in ACN has poorly solvated cations which can be fully desolvated in order to penetrate the small pores of activated carbon [16].

6.3 Strategies for the development of low-temperature SCs

6.3.1 Electrodes

AC is still the most widely used electrode material in EDLCs. Its specific surface area ranges from 1000 to 2500 $m^2\ g^{-1}$. AC is a carbonaceous material with high porosity, and its physicochemical properties such as its surface area, pore structure, and chemical polarity depend on the precursors and activation process used [17]. Most commercial AC is produced using fossil-fuel-based precursors (petroleum and coal). In the last decade, there has been an increasing focus on the utilization of biomass precursors which are environmentally friendly, cheaper, and renewable.

The activation process of AC plays a critical role in the pore structure of the final product. Physical and chemical methods of activation are the two mainstream methods used for the production of different types of AC. The physical activation process consists of the carbonization of carbonaceous precursors at temperatures between 400 °C and 800 °C, followed by activation using oxidizing gases, such as CO_2, air, steam, or a mixture of these gases at a temperature within the range of 600 °C–900 °C. In the chemical activation process, the precursor material is mixed with activating agents, such as KOH, $ZnCl_2$, H_3PO_4, etc. followed by heat treatment. These chemicals work as dehydrating agents and oxidants to oxidize the precursor at 300 °C–950 °C. KOH and $ZnCl_2$ are the activating chemicals most commonly used for the synthesis of SC-grade AC. Chemical activation has the advantages of producing AC with a higher surface area and a more uniform pore size distribution than that produced via the physical method. However, the post-treatment of AC to remove the residuals of the reactants makes the chemical activation method energy intensive and makes chemically derived AC expensive [17].

The Brunauer–Emmett–Teller (BET) surface area is an important parameter for EDLC electrodes. The BET specific surface area of porous carbon is measured using a probe gas, such as N_2, CO_2, or Ar, which forms a monolayer of adsorbates on the surface. Although AC has a high surface area, not all of its surface contributes to the capacitance value, as the inner pore channels of slit-shaped pores are usually inaccessible to solvated or partially solvated ions. Therefore, the ions need to be desolvated to maximize the specific capacitance of AC. This process is one of the rate-determining steps for low-temperature SCs, as the electrolyte becomes more viscous and less conductive compared with that at RT. A high surface area is highly preferable, as it can potentially improve the gravimetric specific capacitance by providing more active sites for adsorption. However, a high surface area usually leads to a high number of small micropores and creates strong polarization at high rates or low temperatures. Gualous *et al* reported a fourfold increase in the ESR of carbon electrodes in an organic electrolyte SC for a temperature change from 40 °C to −40 °C, and a capacity loss of 18% was observed from 25 °C to −35 °C [5]. Carbon with large micropores or mesopores has shown advantages over conventional microporous carbon when used in low-temperature SCs.

6.3.1.1 The effects of pore structure

Earlier studies showed that AC with a pore size smaller than 1 nm can deliver superior capacitance at RT compared to the performance of AC with 1–2 nm micropores or mesopores [18]. Chmiola *et al* developed carbide-derived carbon (CDC) electrodes and tuned their pore sizes from 0.6 to 2.25 nm by adjusting the synthetic parameters (figure 6.2). The capacitance variation of CDC electrodes with different pore diameters was evaluated in 1.5 mol l^{-1} [TEA][BF_4]/ACN [21]. When the pore size was twice the size of the solvated ions (>2 nm), a compact layer of solvated ions that was adsorbed on both (i.e. adjacent) walls of the pore contributed to the capacitance. A decrease in the pore size to 1–2 nm led to a reduction in the compact ion layers on the opposing pore walls. However, a significant increase in capacitance was observed for electrodes with pore sizes which were smaller than those of the solvated ions (<1 nm). This was the case because the distortion of the solvation shell enabled the ions to move closer to the pore walls. Raymundo-Piñero *et al* found that the capacitance of AC-based electrodes in aqueous electrolytes was optimal when the pore size was ~0.7 nm. The best performance was achieved by AC with a pore size of ~0.8 nm in a 1 M [TEA][BF_4]/AN organic electrolyte [4]. Patrice Simon's group found that the maximum capacitance obtained using carbon electrodes was achieved in solvent-free 1-ethyl-3-methylimidazolium bis(trifluoromethanesulfonyl)imide ([EMIm][TFSI]) IL when the pore size of the carbon electrode was close to the ion size (~0.7 nm) [3]. Wan Yu Tsai *et al* investigated CDC samples with two different pore sizes (0.65 and 1 nm) as electrodes and 2 M [EMIm][TFSI]/ACN as the electrolyte. They utilized an electrochemical quartz crystal microbalance (EQCM) to investigate the influence of carbon pore size on the performance of the electrode. The solvation numbers for cations were estimated to be 3–4 and 1–2 for

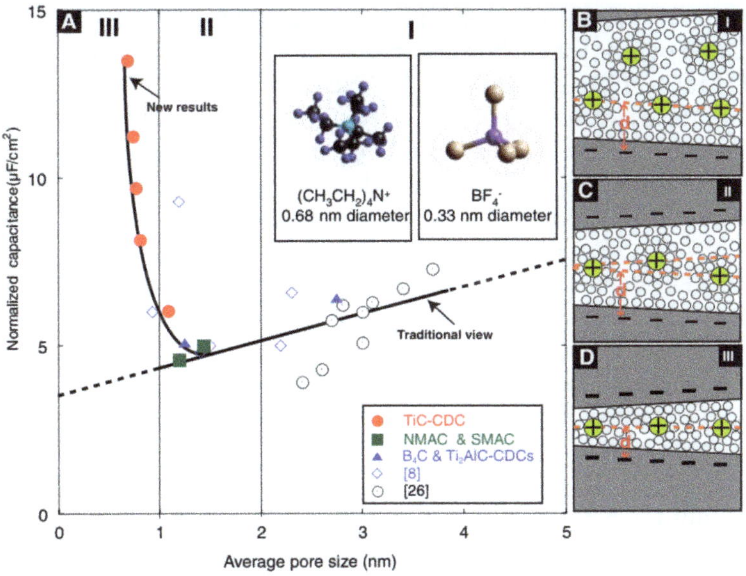

Figure 6.2. (a) The correlation between normalized capacitance changes and the average pore size of the electrodes in a 1.5 mol l^{-1} [TEA][BF_4]/ACN electrolyte. (b), (c), and (d) Schematic illustrations of solvated ions in pores with various distances between opposing walls: (b) >2 nm, (c) 1–2 nm, and (d) <1 nm [18] © American Association for the Advancement of Science.

carbon electrodes with pore diameters of 1 nm and 0.65 nm, respectively. Decreasing the pore size from 1 nm to 0.65 nm required desolvation to reduce half of the solvated molecules. However, both neat and solvated ions were less able to access the 0.65 nm pores than the 1 nm pores. The [EMIm] cations had higher mobility for adsorption on the electrode surface than the [TFSI] anions. The undesirable large [TFSI] anions adsorbed at the positive electrode by ion exchange with co-ions slowed down the ion transport and adsorption kinetics during the discharge process [19]. At ambient temperatures, the time required for the desolvation process of 1.5 M [BMIm] [PF_6]/ACN was short, namely ~1–3 ps, i.e. two magnitudes faster than that required for ion adsorption (~150–500 ps). The diffusion of ions from one pore to another involves several surface adsorption/desorption events. Each of these events can be accompanied by the loss/gain of several solvent molecules in their solvation shells. The diffusion of ions inside electrified nanoporous electrodes was an order of magnitude slower than that in bulk electrolytes due to confinement effects. When the potential was reversed, the ion diffusion rate was reduced by an additional factor of four for counterions due to the strong electrostatic attraction between the ions and the pore walls. Overall, these processes still take place at the nanosecond level, which is sufficiently rapid for fast charging/discharging processes [20].

The desolvation of ions helps carbon electrodes to achieve high capacitance at ambient temperatures. However, the desolvation process is significantly more sluggish at low temperatures compared with that at RT, which is the rate-determining step for EDLCs. Xu *et al* utilized the Arrhenius equation to construct a temperature-dependent ion desolvation model that was used to determine the desolvation rate of ions for carbon electrodes with different pore sizes in the presence of a 0.5 M [$MeEt_3N$][BF_4] in ACN/1,3-dioxolane (DIOX) electrolyte. Their results indicated that the desolvation of ion clusters failed to occur at low temperatures and that pores larger than ion clusters could not participate in charge storage [21]. It is very important to design carbon electrodes with the appropriate pore shape and structure to match the electrolyte ions. Wu *et al* recently reported that AC with large micropores (0.7–2 nm) maintained 90% of its capacitance at −40 °C compared with that at 25 °C. The capacitance retention value decreased to 40% when the majority of the pores were smaller than 0.7 nm. A gradual desolvation mechanism was proposed for the submicroporous electrode, in which the solvated ions energetically preferred to remove the solvent molecules one by one. During the gradual desolvation process, a much lower energy barrier of 0.19–0.31 eV had to be overcome at each step [22] (figure 6.3).

In addition to the sluggish desolvation of ions at the pore openings, low ion migration within pores places another limitation on the low-temperature application of microporous AC electrodes. Xu *et al* synthesized AC with a bimodal distribution of predominantly slit-shaped micropores and mesopores. Triethyl methyl ammonium ($MeEt_3N^+$) ion desolvation kinetics in an ACN/DIOX mixed solvent with carbon-based electrodes was investigated at various temperatures from RT to −100 °C. Carbon that had a pore size of ~0.9–1 nm achieved superior performance compared to AC that had a pore size of ~0.8 nm or 0.9 nm. AC that had a pore size of 1 nm

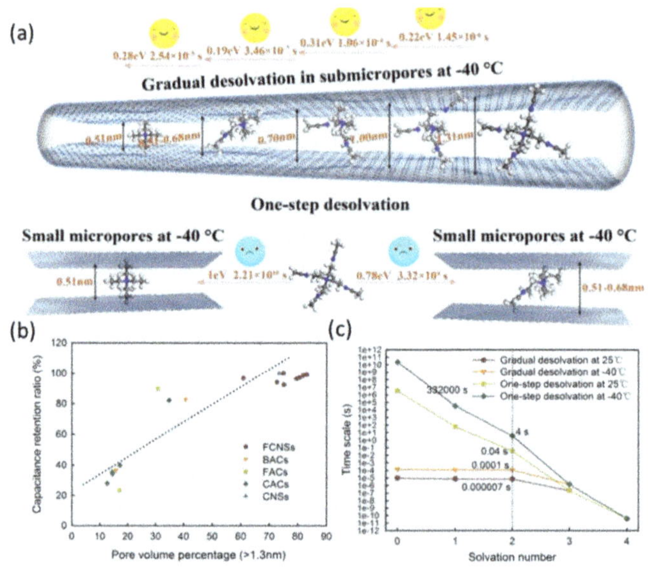

Figure 6.3. (a) A schematic diagram of the mechanisms of the gradual and one-step desolvation processes. (b) The relationship between the capacitance retention ratio and the percentage of the pore volume corresponding to a larger pore size than the size of a fully solvated ion at −40 °C. (c) Different degrees of desolvation time calculated using density functional theory (DFT) based on the gradual desolvation and one-step desolvation models at 25 °C and −40 °C [22], reproduced with permission © Springer Nature.

achieved a capacitance retention of 92.6% at 5 mV s^{-1} when the temperature was low as −100 °C. Density functional theory (DFT) simulation indicated that carbon with a pore size of 0.9 nm has a smaller desolvation time of 0.1 ms at −70 °C. However, when the pore size of carbon was smaller than 0.8 nm, only ions with an effective size of 0.6 nm or less were allowed to enter, corresponding to a solvation number of one or less. Although the desolvation process of $[MeEt_3N][BF_4]$ in ACN was a fast process at RT (less than 1 s), a desolvation time of ~3 h was required at −70 °C to overcome a minimum energy barrier of 0.73 eV [21]. Calculated results indicated that ion diffusion into micropores is temperature dependent. At low temperatures, the desolvation of ion clusters is a very sluggish process, and ions that have large solvation numbers or/and high desolvation energies barely contribute to charge storage. Wang *et al* proposed that the migration of ions from the electrolyte to the inner surface of a carbon-based SC involves three steps: in stage I, the ions migrate within the bulk electrolyte; they further migrate through the pore opening (namely the liquid–solid interface) in stage II; the ions reach the insides of the pores and migrate within the pores in stage III (figure 6.4(a)). Stage II is usually accompanied by a desolvation process if the open pore size of the AC is smaller than the solvated ions. The authors selected three types of carbon electrode material, including two types of carbide-derived carbon (CDC-400 and CDC-1000) and ordered mesoporous carbon prepared from SBA-15 (CMK-3) with average pore sizes of 0.8 nm, 1.3 nm, and 3.8 nm, respectively. The CDC-1000 was able to accommodate a fully solvated $MeEt_3N^+$ ion, while the loss of two solvent molecules was required for the same cations to be accommodated in CDC-400. Electrochemical impedance spectroscopy (EIS) of CDC-400 showed that it has a

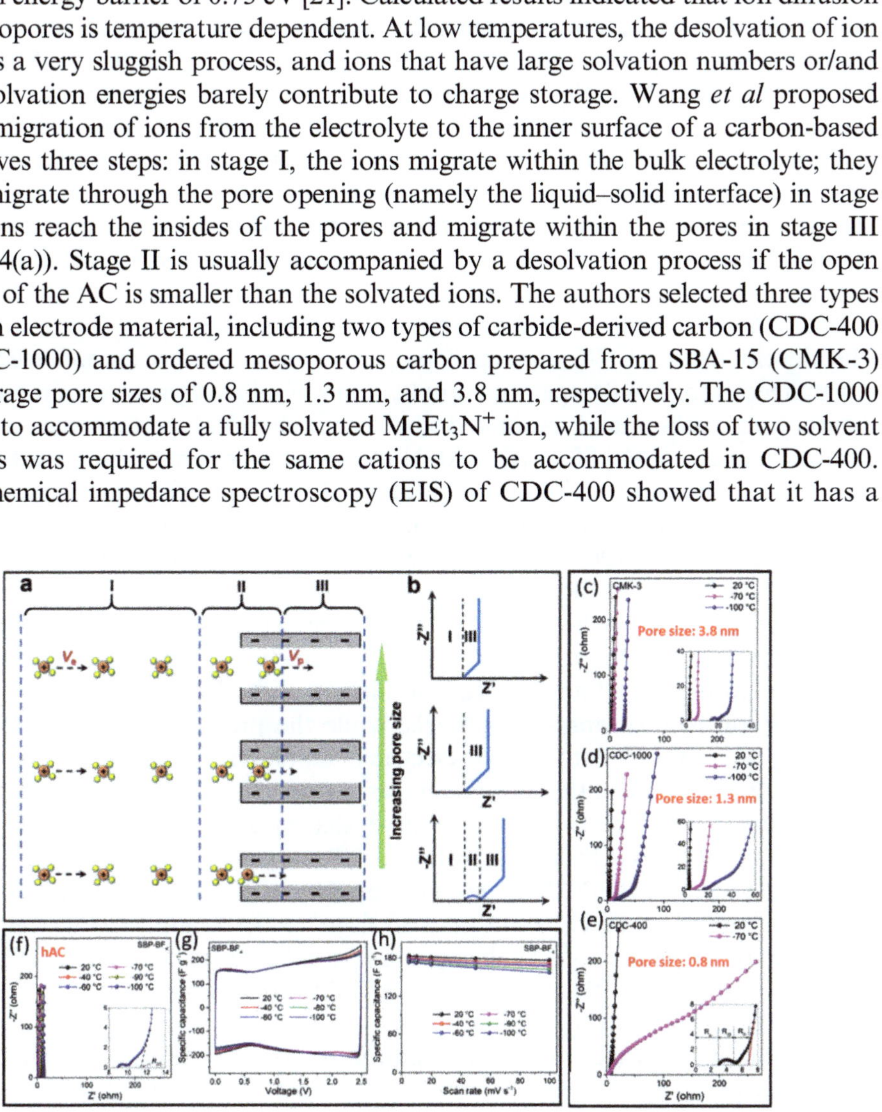

Figure 6.4. A schematic illustration of the different stages of ion migration from the electrolyte to the electrode during charging (a) and the corresponding EIS data (b). Nyquist plots of (c) CMK-3, (d) CDC-1000, and (e) CDC-400 at different temperatures. (f) Nyquist plots and (g) CV curves of hAC at a scan rate of 10 mV s^{-1} from 20 °C to −100 °C. (h) The specific capacitance vs the scan rate of hAC-based EDLCs at different temperatures [23], reproduced with permission © Royal Society of Chemistry.

high interfacial resistance, R_i, corresponding to the semicircle in the high-frequency region of figure 6.4(e). CDC-1000 exhibited a high diffusion resistance (R_d) at the intercept of the 45 °C line traced from the intermediate frequency region of the Nyquist plot (figure 6.4(d)). The CMK-3 with a pore size of 3.8 nm had both small R_i and R_d values at low temperatures (figure 6.4(c)). The authors developed high-surface-area activated carbon (hAC) with bimodal micropores (0.95 nm) and mesopores (3–5 nm) by KOH activation of carbonized soy protein. The hAC electrode had ideal capacitive behavior at various temperatures (figure 6.4(f)). Its cyclic voltammetry (CV) curves remained unchanged at 10 mV s^{-1} for temperatures ranging from 20 °C to −100 °C in 0.3 M [SBP][BF_4] in a 3:7 v/v mixture of ACN/MF electrolyte. At −100 °C, capacitances of 157 F g^{-1} at 100 mV s^{-1} and 173 F g^{-1} at 5 mV s^{-1} were reached by the hAC [23]. In order to achieve high power density, the pore opening size should not be less than that of the fully solvated ions. Electrodes with mesopores are important to maintain high rate performance, but 1–2 nm micropores with pore diameters larger than the fully solvated ions contribute significantly to the capacitance at low temperatures.

6.3.1.2 The development of advanced carbon electrodes

The development of 1–2 nm microporous or mesoporous carbon electrodes is advantageous for low-temperature applications. Tsai *et al* prepared bimodal micro/mesoporous carbon and evaluated the electrochemical performance of carbon electrodes in a 1.5 M [TEA][BF_4]/ACN electrolyte. Ordered CDCs were prepared by the chlorination of SiC precursors followed by hydrogen reduction and annealing. The resulting micro/mesoporous carbon with a surface area of 1220 m^2 g^{-1} had a high rate capability (90% capacity retention at 1 V s^{-1}), a short time constant (1 s), low ESR (0.25 Ω cm^2), and low ionic resistance (0.35 Ω cm^2), which were promising for low-temperature applications [25]. Ye *et al* prepared partially graphitized porous carbon (PGPC) by the activation and graphitization of mesocarbon microbeads (MCMBs) using NaOH and $FeCl_3$. The surface of the PGPC had numerous cracks which served as ion transport channels, while the presence of graphitized regions endowed the PGPC with better electrical conductivity than that of AC (see figures 6.5(a)–(c)). At different temperatures from 50 °C to −20 °C, PGPC-based SCs exhibited near-rectangular CV curve shapes at a scan rate of 20 mV s^{-1}; by contrast, AC-based SCs have distorted CV curves at temperatures of 0 °C and −20 °C. At −20 °C, PGPC has a specific capacity of 135 F g^{-1}. AC-based SCs have specific capacitances of 111 F g^{-1} and 86 F g^{-1} at 50 °C and −20 °C, respectively. The rate performance of AC-based SCs is inferior to that of PGPC at low temperatures due to their poor ion mobility and larger diffusion resistance (figures 6.5(f) and (g)) [24].

Pore geometry affects the performance of SCs. High-surface-area electrode materials with an open surface (onion-like carbon or nanotubes of porous graphene) can operate at low temperatures even with highly viscous ILs as electrolytes. To achieve good low-temperature performance, the pore size and shape should be tailored to facilitate ion diffusion into the surfaces of the pores.

Hierarchical carbon electrodes with interconnected pore structures are favorable for high-power and low-temperature SCs due to their higher ion diffusion rate and

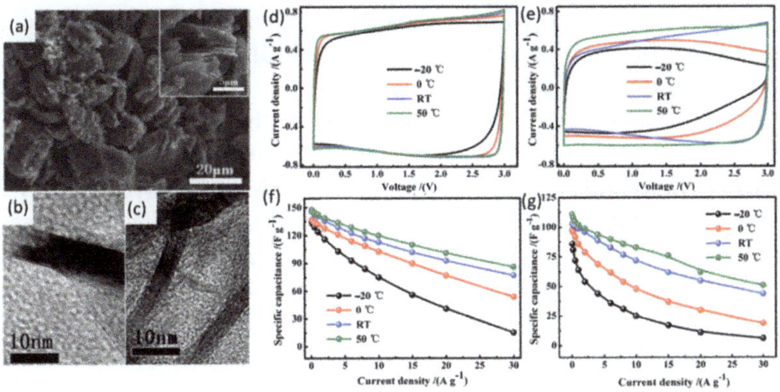

Figure 6.5. (a) SEM and (b) TEM images of PGPC. CV curves of (d) PGPC-based and (e) AC-based SCs measured at −20 °C, 0 °C, RT, and 50 °C. The rate performances of (f) PGPC-based and (g) AC-based SCs as indicated by the specific capacitance dependence on current densities from 0.1 to 30 A g^{-1} [24], reproduced with permission © Royal Society of Chemistry.

charge transfer efficiency compared with those of materials that have isolated, semi-closed, slit-shaped pores. Pan *et al* prepared nitrogen-doped porous carbon with an interconnected tubular structure ($NPTC_k$) by the polymerization of 1-vinylimidaozle by 4,4′-azobis(4-cyanovalericacid) and the use of Ti nanotubes as templates, followed by KNO_3 activation. $NPTC_k$ has a surface area of 1310 m^2 g^{-1} and pore sizes from 0.1 nm to 10 nm. The electrochemical window of $NPTC_K$-based symmetric SCs in a 6 M KOH electrolyte increased from 1.2 to 1.7 V when the temperature was reduced from −10 °C to −40 °C. This study also demonstrated that $NPTC_K$ symmetric SCs had higher specific capacitances in an aqueous electrolyte (6 M KOH) than in an organic electrolyte ([EMIm][BF_4]/ACN) at −30 °C and −40 °C [26]. Vertically aligned carbon nanotube (VA-CNT) film can be considered to be a variation of the interconnected pore approach and has been proven to deliver higher power than AC [13]. Xu *et al* coaxially coated carbon nanotubes (CNTs) with glucose and further applied chemical activation by KOH. They found that the poor capacitive performance at low temperatures was mainly caused by ion migration at the liquid–solid interface, especially for pores whose sizes were smaller than the solvated ions. When the pore size of AC was large, ions were able to freely enter the pores, resulting in a capacitance of 97 F g^{-1} at −70 °C, which was comparable with that at room temperature (108 F g^{-1}) [27].

In addition to hierarchical interconnected porous carbon and 1D CNTs, 2D graphene is potentially an excellent SC electrode material for low-temperature applications. Tsai *et al* developed activated microwave-exfoliated graphite oxide (a-MEGO) with a surface area of 1900 m^2 g^{-1} and micropores with pore sizes ranging from 1 to 4 nm [28]. At −40 °C, a-MEGO had a capacitance of 100–120 F g^{-1} at a scan rate of 1 mV s^{-1}, corresponding to ~70% of its RT capacitance. The capacitance value of a-MEGO remained greater than 100 F g^{-1} at −50 °C. However, the capacitive behavior of a-MEGO suffered from severe degradation at low temperatures, as evidenced by a large semicircle in the high-frequency region of the EIS curve and the distortion of the CV curve shape at a scan rate of 1 mV s^{-1}. The high viscosity of ionic liquids at low temperatures makes the diffusion of ions within the pores more sluggish.

MgO-templated mesoporous carbons (MPCs) with a high specific surface area of 1530 m^2 g^{-1} have been employed as electrodes in SCs. The electrolytes used were 1-butyl-3-methylimidazolium tetrafluoroborate ([BMIm][BF_4]), 1-hexyl-3-methylimidazolium tetrafluoroborate ([HMI][BF_4]), and [N,Ndiethyl-N-methyl-N-(2-methoxyethyl) ammonium bis(trifluoromethanesulfonyl) imide] ([DEME][TFSI]). The gravimetric capacitances of MPC and commercial AC (YP-17) were compared in various ionic liquids. AC exhibited a rapid capacitance decrease for each slight decrease in temperature, while the significant decrease in the capacitance of MPC started from −30 °C. MPC had a better capacitance retention capability than that of AC at low temperatures. Different IL-based electrolytes affected the specific capacitance of the electrodes and their low-temperature performance. The capacitance retention of MPC rapidly decreased at −40 °C for [BMIm][BF_4] and [DEME][TFSI], but it decreased at −30 °C for [HMI][BF_4]. At low temperatures, the ion diffusion resistance of the ILs in the pores increases, even in the pores of mesoporous carbon. It was found that the diffusion resistance of the ILs in the micropores was already very high at temperatures below 0 °C, and the migration of the electrolyte ions in the micropores was significantly limited [29].

It is worth noting that the specific gravimetric and volumetric capacitances of mesoporous carbon are not as high as those of commercial carbon electrodes at RT due to the low density of the former. Electrodes with large micropores can alleviate this problem by providing a high packing density and large pores for low-temperature charge storage. Korenblit *et al* demonstrated that microporous carbon with large micropores of 1–2 nm can operate at −70 °C with [SBP][BF_4] in 1:1 AN:MF as an electrolyte. At −70 °C, a zeolite-templated carbon (ZTC)-based microporous electrode maintained 86% of its room temperature energy density [30].

6.3.1.3 Pseudocapacitive electrodes

Since Trasatti and Buzzanca reported the typical capacitive behavior of a RuO_2 film with a rectangular CV curve shape, RuO_2 has been widely investigated as a potential pseudocapacitive material due to its unique thermal and electrical properties [31]. Liu and Pickup characterized the performance of hydrous Ru oxide in SC electrodes and found that a thinner Nafion separator and a carbon fiber paper substrate (Toray TM, TGP-H-090) could minimize the increase in resistance with decreasing temperature. RuO_2-based electrodes in a 33% H_2SO_4 electrolyte had a power density of more than 100 kW kg^{-1} at an energy density of 5 Wh kg^{-1} for a 1 cm^2 device with 10 mg RuO_2. The specific capacitance values of the RuO_2-based electrodes were 770 and 690 F g^{-1} at 40 °C and −40 °C, respectively [32]. Zheng *et al* reported that amorphous RuO_2 with 20% of conductive additives had a capacitance variation of less than 20% within the temperature range from −52 °C to 73 °C. RuO_2 in 5.26 M H_2SO_4 had an energy density of 20 Wh kg^{-1} and a power density of more than 10 kW kg^{-1} at −52 °C [33].

Spinel $Li_4Ti_5O_{12}$ is an appealing anode for Li-ion batteries which can host three Li ions per formula unit, leading to a theoretical capacity of 175 mAh g^{-1} [34]. As a 'zero-strain' material for Li-ion intercalation, LTO has shown fast charging and discharging properties when used in Li-ion capacitor electrodes. An asymmetric Li-ion SC consisting of AC and $Li_4Ti_5O_{12}$ in 2 M $LiBF_4$ in ACN showed excellent performance from RT to −30 °C. However, at −40 °C, the energy density of this Li-ion SC decreased from ~10 Wh kg^{-1} to 0.5 Wh kg^{-1} when the power density was 1000 W kg^{-1}. The energy density of this Li-ion SC was 6 Wh kg^{-1} at a power density of 400 W kg^{-1}, corresponding to a decrease of more than 40% in the energy density or capacity [10].

Most pseudocapacitive materials are semiconductors, unlike RuO_2-based precious metal oxides. In order to reduce the increase in the ESR at low temperatures, pseudocapacitive materials are usually combined with conductive additives, such as graphene or CNTs, and/or the electroactive materials are downsized to the subnano level. As in the case of VA-CNTs, the development of suitably aligned pseudocapacitive materials with ultrasmall size or ultralow thickness is another strategy for improving the ion diffusion rate and charge transfer efficiency. Freeze-dried graphene network (GN) was thermally reduced and applied as a substrate for the deposition of another thin layer of vertically aligned graphene nanosheets (VGNS) with an intersheet spacing of 50–200 nm. The MnO_2 was further deposited on the surface of vertically aligned graphene sheets. Benefiting from its unique hierarchical structure, the GN/VGNS/MnO_2 pseudocapacitor had capacitance retention ratios of 90.7% and 80.8% at 0 °C and −30 °C, respectively, compared to that at RT. At −30 °C, −10 °C, 0 °C, 10 °C, and 25 °C, the specific capacitances of the GN/VGNS/MnO_2 electrodes were 239, 263, 276, 285, and 296 F g^{-1}, respectively [35].

Xiong *et al* synthesized porous Ni–Co–Mn oxide by a hydrothermal method and investigated its temperature-dependent performance in a three-electrode configuration. Compared to its specific capacitance at RT, the specific capacitance of the electrode decreased by 18% at 4 °C, and increased by 84% at a nominal temperature of 80 °C. The temperature had a pronounced effect on the faradic reaction rate [36]. Electrochemically deposited $Ni(OH)_2$ nanosheets on a $NiCo_2O_4$ nanowire array/carbon fiber electrode had capacitances of 2100 F g^{-1} at 0 °C and 2600 F g^{-1} at 40 °C. A rapid decay in the capacitance was observed at 60 °C, indicating the occurrence of side reactions or the loss of electroactive materials [37]. Birnessite-type MnO_2 nanoflakes have been grown on carbon nanofibers to form a hierarchical coaxial structure. The MnO_2@CNF electrode thus obtained had specific capacitances of 365 F g^{-1} and 0 °C and 546 F g^{-1} at 75 °C (both at at 1 A g^{-1}). An increase in its temperature led to a decrease in its coulombic efficiency. At high temperatures, birnessite-type MnO_2 nanoflakes were found to evolve into γ-MnO_2 nanorods during a cyclic test [38]. Transition-metal oxide/hydroxide-based pseudocapacitive materials exhibited capacitance increases at high temperatures from 40 °C–60 °C at the cost of possible cyclic life reduction. The use of conventional aqueous electrolytes reduced the low-temperature limit of these pseudocapacitors to −20 °C or even higher.

Guo *et al* prepared a CNT–polypyrrole (PPy) composite film by the electrochemical deposition of PPy on a CNT forest. The CNT–PPy composite film transformed into a fiber with a diameter of 150 μm after being drawn through diamond wire-drawing dies in a wet state. The CNT–PPy fibers had an ultralow density of only 70 mg cm^3 and could withstand up to 10% tensile strain. With

H_3PO_4–polyvinyl alcohol (PVA) gel as an electrolyte, a CNT–PPy-based SC achieved a specific capacitance of 69 F g^{-1}. Its specific capacitance decreased from 49 F g^{-1} to 32 F g^{-1} as the temperature decreased from 61 °C to −27 °C. The capacitance was almost stable at 41–44 F g^{-1} from −1 °C to 40 °C [39].

Wu *et al* synthesized a hydrous Turnbull's blue analog (TBA) of $Ni[Fe(CN)_6]_{2/3}{\cdot}4H_2O$ by a precipitation method from $NiSO_4$ and $K_3Fe(CN)_6$ precursors. The NiFe–TBA delivered a reversible capacity of ~65 mAh g^{-1} with H^+ (de)insertion and had an extremely high rate performance of up to 6000 C (390 A g^{-1}) at RT in a 1.0 M H_2SO_4 electrolyte. An EQCM analysis indicated that the redox peaks of the NiFe–TBA electrode were associated with a mass reduction (due to water expulsion) and a mass increase (due to water incorporation), respectively. The water migration processes may give rise to different binding energies of protons, which are likely responsible for the redox charge storage. Wu *et al* proposed that the charges of protons transported via structural diffusion, known as the Grotthuss mechanism, were responsible for the fast proton insertion, while the breaking and formation of 'H–O' bonds of water accompanied by the building of hydrogen bonds 'H···O' with the electrodes enabled fast proton transfer. The NiFe–TBA retained half of its theoretical capacity at −40 °C in an aqueous electrolyte. X-ray absorption near-edge spectra (XANES) analysis of the NiFe–TBA electrode showed that Fe ions participated in the redox reaction, while the inert Ni sites may have stabilized the structure. The fast proton insertion kinetics gave the NiFe–TBA a high rate capacity of more than 30 mAh g^{-1} at 50 C at −40 °C, corresponding to 50% of its RT capacitance [40]. A cyclability test of the NiFe–TBA confirmed its capacity retention of ~78% after 150 cycles. However, the stability of the electrode remains one of the challenges for its application.

Redox additives help pseudocapacitors to maintain high capacitance retention ratios at low temperatures. Barua and Paul examined the consequences of temperature alteration on a multifarious nanoporous carbon (MNC) with surface-adsorbed hydroquinone (H_2Q) as the redox probe and H_2SO_4 as the electrolyte. Hydroquinone adsorbed on MNC maximized the specific capacitance of the electrode, resulting in a value of 213 F g^{-1} at −10 °C, corresponding to a capacitance retention ratio of 67% at 50 °C (319 F g^{-1}) [41].

Table 6.1 summarizes recent reports of the electrodes used in low-temperature supercapacitors. Mesoporous carbon materials have garnered significant attention due to their high capacitance retention ratio at low temperatures. Moreover, pseudocapacitive materials with customized microstructures have demonstrated their capability to operate effectively at subzero temperatures. However, developing electrodes that maintain high capacitance at both room temperature (RT) and low temperatures continues to be a challenging task.

6.3.2 Electrolytes

The design of electrolyte formulations that can achieve low melting points, high ionic conductivity, low viscosity, and an optimum solvation structure that is adapted to the pore structure of the electrodes is one of the significant challenges for low-temperature SCs. Organic electrolytes have much lower conductivities

Table 6.1. The electrochemical performances of various types of electrode at ambient and low temperatures.

Electrode					Reference
Electrode material	Surface area (m^2 g^{-1}); pore size distribution (PSD) (nm)	Electrolyte	RT capacitance, energy, and power density	Low-temperature capacitance, energy and power density	
Mesoporous carbon produced using SiO_2 template	1065; 6–12	1-ethyl-3-methylimidazolium bis(fluorosulfonyl)imide ([EMIm] [FSI])$_{0.6}$[BF_4]$_{0.1}$[TCB]$_{0.3}$	144 F g^{-1}, 34 Wh kg^{-1} @ 700 W kg^{-1}	12.1 Wh kg^{-1} @ 1000 W kg^{-1} at −30 °C; 5.2 Wh kg^{-1} @ 1000 W kg^{-1} at −50 °C	[42]
Mesoporous AC	1344; 3.6	[TEA][BF_4]/AN with DIOX, MA, MF		18–22 F g^{-1} per cell at −60 °C	[2]
Zeolite NaY-templated carbon	2025; 0.3–2	[SBP][BF_4]/AN:MF	123 F g^{-1}	123 F g^{-1} at −60 °C, 117 F g^{-1} at −70 °C	[30]
MP98B	>1500; 3.5	[EMIm][FSI]$_{0.5}$ [BF_4]$_{0.5}$	140 F g^{-1}	123 F g^{-1} at −40 °C	[43]
Activated carbon	1933, <2	Choline chloride:ethylene glycol (EG):urea = 1:2:1	74.37 Wh kg^{-1} @ 1300 W kg^{-1}	36.6 Wh kg^{-1} @ 1300 W kg^{-1} at −40 °C	[44]
Ni foam	—	1-ethyl-3-methyl imidazolium ([EMIm])$_{0.5}$1-propyl- 3-methylpyrrolidinium ([PMPyrr])$_{0.5}$ bis (trifluoromethanesulfonyl) imide ([TFSI])	100 mF cm^{-2} at 80 °C	5 mF cm^2 at −70 °C	[45]
Kuraray YP-80FP		24.8 wt% LiCl in PVA/H_2O	28 F g^{-1} at 5 mV s^{-1}	20.2 F g^{-1} @ 5 mV s^{-1} at −70 °C	[46]
Activated carbon	1933, <2	Choline chloride:EG = 1:2	~140 F g^{-1} at 1 A g^{-1}	14.13 Wh kg^{-1} at −40 °C	[47]

Kuraray YP 50F		3.5 M $Mg(ClO_4)_2$	~25 F g^{-1} at 1 A g^{-1}	~19 F g^{-1} @ 1 A g^{-1} at −40 °C	[48]
Spectracarb 2225	2169, 0.6	[SBP][BF_4] in AN:MF (1:1)	110 F g^{-1} at 1 A g^{-1}	~110 F g^{-1} at −65 °C	[1]
Onion-like carbon (microsupercapacitor)		N-methyl-N-propylpiperidinium ([PIP_{13}])$_x$N-butyl-N-methylpyrrolidinium ([PYR1_4])$_{1-x}$ [FSI]	1.1 mF cm^{-2} at 200 mV s^{-1}	0.84 mF cm^{-2} at 10 mV s^{-1}	[12]
Mesoporous carbon, microporous carbon		5 M $NaClO_4$ in 16.11% methanol and 83.9% H_2O		~16–20 F g^{-1} @ 5 mV s^{-1} at −40 °C	[49]
Activated carbon	2300,	[PIP_{13}]$_x$[PYR_{14}]$_{1-x}$[FSI] in SiO_2 ionogel	91 F g^{-1} at 5 mV s^{-1}	~34 F g^{-1} @ 5 mV s^{-1} at −40 °C	[50]
Porous carbon (PACMM 203)	1300	0.5 M [TEA][BF_4] in AN/DIOX	0.9 F at 1 mA	0.6 F at 1 mA at −70 °C	[51]
Graphene film		[PIP_{13}]$_{0.5}$[PYR_{14}]$_{0.5}$ FSI	135 F g^{-1} at 1 mV s^{-1}	100 F g^{-1} @ 1 mV s^{-1} at −30 °C	[14]
Mesoporous carbon		3.4 M LiTFSI in dimethyl sulfoxide (DMSO)/H_2O	67 F g^{-1} at 1 A g^{-1}	22 F g^{-1} @ 1 A g^{-1} at −35 °C	[52]
Reduced graphene oxide (rGO)		1.4 M Li_2SO_4 with 30% EG	~100 F g^{-1} at 1 mA cm^{-2}, 3.44 Wh kg^{-1}	~70 F g^{-1} at 1 mA cm^{-2}, 2.57 Wh kg^{-1} at −20 °C	[53]
MnO_2 nanotube//carbon		132 g ml^{-1} $(NH_4)_2SO_4$ with 30%–50% EG	~80 F g^{-1} at	40–50 F g^{-1} at −20 °C, 23–30 F g^{-1} at −30 °C	[54]
Graphene nanoplatelet-coated carbon paper	750	20–30 M [BMIm]Cl/H_2O	424 F g^{-1} at 20 mV s^{-1}	97 F g^{-1} at 20 mV s^{-1}	[55]
Polyaniline (PANI)-derived carbon	2357.7, 3.3 nm	1 M [TEA][BF_4] in ACN	100 F g^{-1} at 1 A g^{-1}	100 F g^{-1} at 1 A g^{-1} at −40 °C	[22]

(30~40 mS cm^{-1} at 25 °C) than aqueous electrolytes (550–700 mS cm^{-1} at 25 °C) [56]. However, the high freezing point of aqueous electrolytes is the key challenge for low-temperature applications.

The diversity of organic solvents enables the efficient design of suitable electrolytes specifically tailored for low-temperature supercapacitors. Table 6.2 provides a summary of the physicochemical properties of several commonly used solvents in supercapacitors. In addition to the operating temperature range, factors such as viscosity, compatibility with electrolyte salts, electrochemical stability etc. significantly impact the low-temperature performance of supercapacitors.

Commercial SCs have a typical low-temperature rating of −40 °C; [TEA][BF_4] is the most widely used electrolyte solute, while ACN, PC, EC and dimethyl carbonate (DMC) are widely used solvents. The high freezing temperature of PC (−25 °C) limits its low-temperature application as an electrolyte. ACN-based electrolytes enable SCs to operate at −40 °C to 70 °C; however, ACN is toxic and has a low flash point of 5 °C, which raises safety concerns [13]. The preferred approach for expanding the low-temperature limit of SCs is the formulation of solvent systems that include appropriate cosolvents with low melting temperatures. Such mixed solvent systems should further decrease the freezing point while retaining the high solubility of the salt, low viscosity, and favorable dielectric properties at low temperatures.

Iwama *et al* performed a floating test of 600 F commercial SCs at low temperatures. A conventional 1 M [TEA][BF_4] in ACN electrolyte was compared with ACN+MA mixed electrolytes. At temperatures from 20 °C to −40 °C, the ACN electrolyte showed higher capacitance than the capacitances of ACN and MA (25–33 vol%) binary electrolytes. However, the ACN electrolyte had no capacitance at −55 °C, whereas the 25 vol% and 33 vol% MA mixed electrolytes retained 95% and 91% of their original (20 °C) capacitances, respectively. The resistance of ACN-based cells at −55 °C was found to be about 10 000 times higher than that at 20 °C, while the MA+ACN-based SCs experienced a fortyfold increase in resistance when the temperature decreased from 20 °C to −55 °C. A floating test of the MA+ACN-based electrolyte also indicated that the electrolyte supported high-temperature operation at 60 °C [57]. The mixture of a cosolvent which has a lower melting point with a conventional electrolyte can increase the low-temperature operational range of AC-based SCs.

6.3.2.1 Eutectic ionic liquids as electrolytes

ILs composed of bulky organic cations and versatile anions have been regarded as green and safe electrolytes for SCs. ILs have negligible vapor pressure and chemical/thermal stability. The asymmetric arrangement of IL cations and anions tends to shield the intermolecular force, thereby preventing ions from aggregating into a crystalline state. Therefore, ILs have lower freezing points than inorganic solutes. However, the large ionic size of ILs leads to high viscosity at low temperatures and high resistance accompanied by reduced ionic conductivity [58]. The high solubility and affinity of ILs for different compounds enables the physicochemical properties of electrolytes to be tailored by coupling ILs with different ions, solvents, etc.

IL-based SCs are expected to perform well at high temperatures but to have restricted performance at low temperatures due to the high melting points of ILs.

Table 6.2. Commonly utilized solvents for electrochemical energy storage applications and their physicochemical features.

Solvent	Freezing point (°C)	Boiling point (°C)	Dielectric constant (ε)	Viscosity (cP)
$H_3C—C\equiv N$ ACN	−45.7	81.6	37.5	0.345 (25 °C)
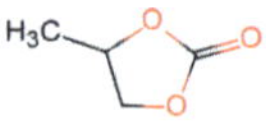 PC	−49	242	69	2.5
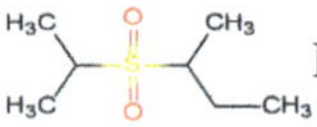 Isopropyl s-butyl sulfone [59]	−20	260	52	6.3
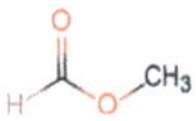 MF^2	−100	32	8.5	0.319 (29 °C)
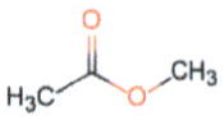 MA	−98	56.9	6.68	0.38 (20 °C)
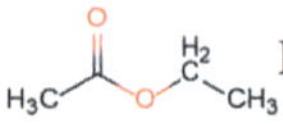 EA	−83.6	77.1	6.0	0.426 (25 °C)
 EC	36.4	243		1.93 (40 °C)
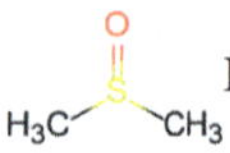 DMSO	−32	189	41.9(55 °C)	1.99 (20 °C)
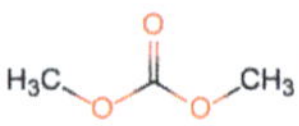 DMC	2–4	90	3.2	0.585 [60]
 1,2-Dimethoxyethane (DME)	−58	85	7.2	0.42 (20 °C)
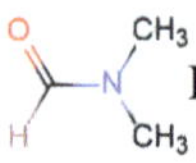 Dimethyl formamide (DMF)	−61	153	36.7	0.86 (20 °C)
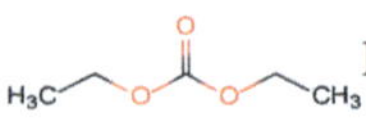 Diethyl carbonate (DEC)	−43	126	3.1	0.749
 DIOX	−95	78	7.3	0.6 (20 °C)
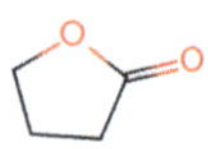 gamma-Butyrolactone (GBL)	−43	204	41.7	1.7 (25 °C)

One widely accepted strategy for overcoming this limitation is the formulation of suitable binary or ternary mixtures of ILs to form eutectic ILs, which can reduce the melting point and extend the temperature range. Eutectic ILs can be obtained by mixing two or three ILs, which may contain the same cations or anions or completely different types of ion. Eutectic ILs usually have much lower melting points than their individual components; this is mainly due to the symmetry breaking of the IL ions, which leads to the suppression of the crystallization process. Eutectic ILs have been reported to have freezing points several tens of degrees lower than those of each individual electrolyte [13].

The cations of PIP and PYR have the same number of atoms and the same molecular weights but different cationic molecular structures. The side chains of PIP and PYR consist of four carbon atoms and five carbon atoms, respectively, corresponding to five-member (piperidinium) and six-member (pyrrolidinium) heterocycles. The combination of PIP and PYR cations helps to maintain similar intermolecular interactions to those of individual PIP or PYR-based ILs but with significantly decreased symmetry; the resulting ILs maintain a more disordered state that prohibits lattice formation when the temperature decreases [13]. Lin *et al* developed eutectic ILs composed of [PYR_{14}][FSI] and [PIP_{13}][FSI]. The binary ILs achieved a freezing temperature of less than −80 °C, which was significantly lower than the freezing temperatures of the individual ILs of [PIP_{13}][FSI] (6 °C) and [PYR_{14}][FSI] (−18 °C). The performances of onion-like carbon (OLC), binder-free VA-CNTs, and commercial AC electrodes were evaluated, and high capacitance retention was observed at −50 °C. With the same electrolyte, VA-CNT electrodes had much higher capacitance retention at low temperatures than those made from spherical OLC or commercial AC (figure 6.6(c), (d). The high electrical conductivity

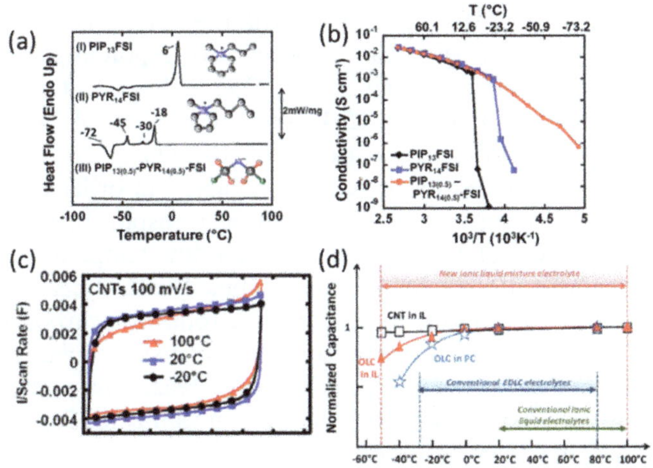

Figure 6.6. The characteristics of ILs. (a) Differential scanning calorimetry (DSC) profiles of (I) [PIP_{13}][FSI], (II) [PYR_{14}][FSI], and (III) $[PIP_{13}]_{0.5}[PYR_{14}]_{0.5}$[FSI] mixtures and their chemical structures. (b) The change of conductivity versus inverse temperature for [PIP_{13}][FSI], [PYR_{14}][FSI], and $[PIP_{13}]_{0.5}[PYR_{14}]_{0.5}$[FSI] IL mixtures [13]. Reproduced with permission © American Chemical Society.

of the VA-CNTs, their fast ion diffusion through the aligned CNT walls, and their binder-free design could be responsible for their better performance compared to that of spherical OLC [13]. A $[PIP_{13}][PYR_{14}][FSI]$ mixture had a temperature window of 150 °C (−50 °C to 100 °C), which was much higher than that of 110 °C for a conventional 1 M $[TEA][BF_4]/PC$ electrolyte (−30 °C to 80 °C) (figure 6.6(d)). The authors indicated that the electrolyte formulation and the design of the electrode structure are both important in the process of expanding the temperature range of SCs. However, the specific capacitance of the VA-CNTs was low (~30 F g^{-1} or 4 mF cm^{-2}), as the specific surface area of the VA-CNTs was not as high as that of AC.

Lin *et al* reported that the ionic conductivities of $[PYR_{14}][FSI]$ and $[PIP_{13}][FSI]$-based eutectic ILs were 28.9 mS cm^{-1} at 100 °C and 4.9 mS cm^{-1} at 20 °C. The capacitance values of graphene in eutectic ILs were 135 F g^{-1} (63 mAh g^{-1}) at 0 °C and 100 F g^{-1} (49 mAh g^{-1}) at −30 °C. CV of graphene in eutectic ILs at −40 °C exhibited a severe deviation from a rectangular shape, even at the low scan rate of 1 mV s^{-1} [14]. The high ohmic resistance and diffusion resistance caused by the high viscosity and low ionic conductivity of eutectic ILs require further research.

Yambou *et al* reported the addition of a low-viscosity tetracyanoborate ([TCB]) anion, [EMIm][TCB], to a mixture of [EMIm][FSI] and $[EMIm][BF_4]$ to form a ternary electrolyte, $[EMIm][FSI]_{0.6}[BF_4]_{0.1}[TCB]_{0.3}$. The ternary electrolyte had a glass transition at −99 °C with a relatively low viscosity of 23.6 mP s and a high conductivity of 14.2 mS cm^{-1} at 20 °C [42]. The ternary IL had a higher ionic conductivity than those of binary ILs of $[EMIm]_{0.5}[MPPyrr]_{0.5}[TFSI]$ (5.2 mS cm^{-1} at 25 °C), $[BMPyrr]_{0.5}[MPPip]_{0.5}[FSI]$ (4.9 mS cm^{-1} at 20 °C and ~0.1 mS cm^{-1} at 40 °C) [13], and $[EMIm][FSI]_{0.5}[BF_4]_{0.5}$ (12.1 mS cm^{-1} at 20 °C and 0.42 mS cm^{-1} at 40 °C) [43]. SiO_2-templated carbon with well-defined mesopores about ~9 nm in diameter was selected as the electrode material for an investigation of the low-temperature performance of ILs. At −50 °C, with the ternary IL as the electrolyte, a specific capacitance of 91 F g^{-1} was achieved, corresponding to a capacitance retention of 63% compared with that at 20 °C (144 F g^{-1}). The ionic resistance increased from 0.21 ohms at RT to 15.4 ohms at −50 °C. At −50 °C, the ionic conductivity of the electrolyte decreased by ~10 times. The viscous and less conductive electrolyte required more activation energy in order to diffuse into the mesopores. It is very important to improve both the transport properties and the ion migration kinetics of electrolytes in order to obtain low-temperature SCs.

Pyrrolidinium cations ([Pyrr]) have been combined with different mole ratios of nitrate ($[NO_3]$) and [TFSI] anions to prepare electrodes for carbon-based SCs. An electrolyte with a 0.1–0.72 mole ratio of $[Pyrr][NO_3]$ had no phase change between −60 ° C and 100 °C. The conductivity of a eutectic IL was increased by increasing the amount of $[Pyrr][NO_3]$; however, its conductivity reduced significantly to ~2 mS cm^{-1} at −10 °C. The electrochemical characterization of electrolytes of selected optimal compositions ($x[Pyrr][NO_3]$ = 0.64, 0.72, 0.80) confirmed their good capacitive behavior (with capacitances of up to 148 F g^{-1}) at potentials up to 2.0 V. Above 2.0 V, anion intercalation was observed for carbon electrodes during the first hundred cycles, which led to an increase in specific capacitance to 209 F g^{-1} [61].

$[EMIm][BF_4]$ has a small ion size (7.6 Å for EMIm and 4.8 Å for BF_4^-), low viscosity, explicit H-bond sites, and a high decomposition voltage of up to 4.6 V [62]. A eutectic IL based on a mixture of [EMIm][TFSI] and 1-propyl-3-methylpyrrolidinium [TFSI] ([PMPyrr][TFSI]) had a melting temperature of −70 °C and an electrochemical window of 3.5 V when used as an electrolyte [45]. Ni foam was applied as the electrode to measure the temperature-dependent capacitance retention from −70 °C to 80 °C. The areal specific capacitance of the Ni foam at −70 °C was 5% of that at 80 °C. The macroporous structure of Ni foam has a pore structure that is completely different from those of mesoporous or microporous (1–2 nm) carbon, which makes it difficult to compare its specific capacitance with those of other electroactive materials [45].

Microsupercapacitors are miniaturized SC devices with highly integrated electrodes that have promising applications in wearable and printable electronics. When conventional electrodes are downsized to the micrometer level, the results are a much faster diffusion process with a much smaller time constant and high power output. These properties make microsupercapacitors potentially promising for low-temperature applications. Huang *et al* reported OLC-based microsupercapacitor electrodes prepared by an electrophoretic deposition technique from a suspension containing OLC particles and 0.03 wt% of $MgCl_2$ in 95% of ethanol and 5% water. The resulting microsupercapacitor had a large voltage window of 3.7 V and a temperature range of −50 °C to 80 °C when $[PIP_{13}]_{0.5}[PYR_{14}]_{0.5}[FSI]$ was applied as an electrolyte. At −50 °C, 76% of the capacitance of the OLC-based microsupercapacitor was maintained at 10 mV s^{-1} for voltages of up to 3.7 V. However, the Nyquist plot of this microsupercapacitor at −50 °C showed a large semicircle in the high-frequency region, which may be linked to the increased viscosity and sluggish ion transfer of the electrolyte [12] (figure 6.7).

Zhou *et al* prepared eutectic mixtures of ILs from 1-butyl-4-methylpyridinium tetrafluoroborate $[BMP][BF_4]$ and 1-butyl-3-methylimidazolium tetrafluoroborate $[BMIm][BF_4]$ (figure 6.8(a)). The two electrolyte systems have the same anions but different cations. The imidazolium cation [BMIm] and the pyridinium cation [BMP] have similar structures and molecular weights. The different cations help to prevent the ordering of anions and shift the gelation and solidification temperatures of the mixture to a much lower value. These eutectic ILs enabled a SC cell to operate in a temperature range from −60 °C to 80 °C. At −50 °C, a nanoporous graphene electrode in the eutectic electrolyte had a specific capacity of 92 F g^{-1}, i.e. 85% of its RT capacitance (109 F g^{-1}). By contrast, a SC with $[BMIm][BF_4]$ as the electrolyte had a specific capacity of 14 F g^{-1} at −50 °C, i.e. only 14% of the RT capacitance (103 F g^{-1}) was retained [64].

Yambou *et al* prepared a binary mixture of [EMIm][FSI] and $[EMIm][BF_4]$ in a 1:1 molar ratio for use as an electrolyte. They compared the electrochemical performances of SC2A and MP98B electrodes, which have comparable surface area values but different pore size distributions. SC2A has a surface area of 1502 m^2 g^{-1} and pore sizes of 0.58 and 0.98 nm. MP98B has a surface area of 1556 m^2 g^{-1}, a pore size of 1.0 nm, and a mesopore size of 3.5 nm (i.e. larger pores than those of SC2A). They found that SCMP (the mixture of SC2A and MP98B) carbon, which

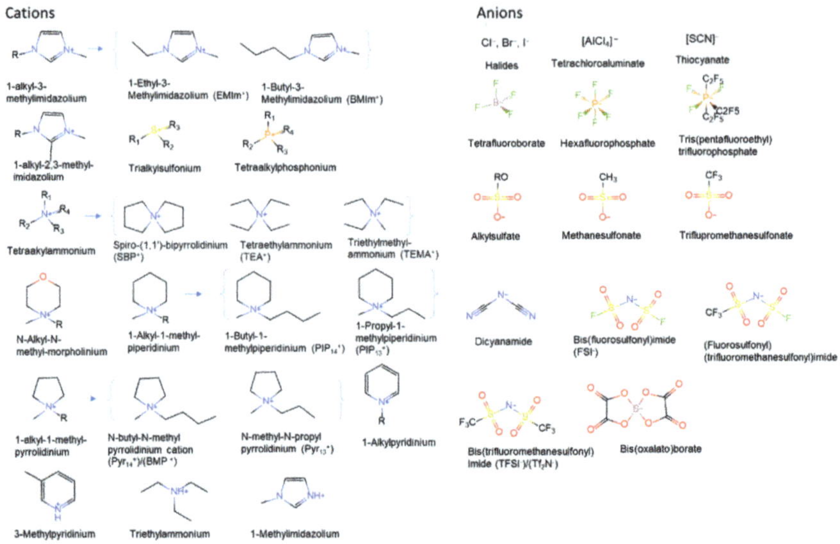

Figure 6.7. Cations and anions commonly used in ILs for energy storage (R represents an alkyl group, which may be a halogen, a carbonate, a quinone, an olefin, or piperidinyl-1-oxyl). The chemical structures were redrawn from [63], with permission © Springer Nature.

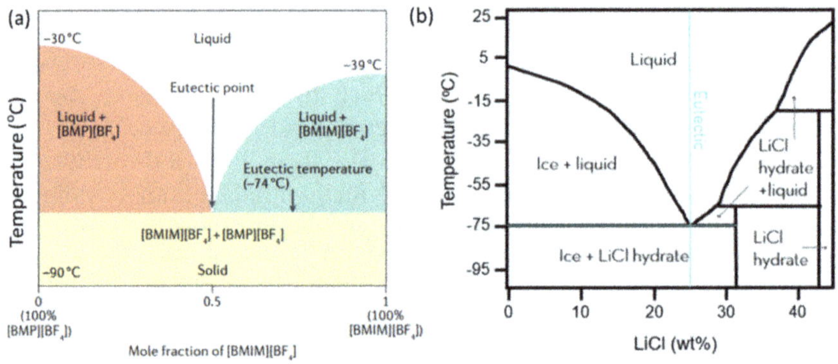

Figure 6.8. (a) A phase diagram of a eutectic mixture of 1-butyl-4-methylpyridinium tetrafluoroborate ([BMP][BF_4]) with 1-butyl-3-methylimidazolium tetrafluoroborate ([BMIm][BF_4]). The melting temperatures of [BMP][BF_4], [BMIM][BF_4], and their eutectic phase were −30 °C, −39 °C, and −74 °C, respectively [63], reproduced with permission © Springer Nature. (b) LiCl–H_2O phase diagram. At 24.8% LiCl, the solution was eutectic and had a freezing point of −74 °C [46], reproduced with permission © American Chemical Society.

contains 50 wt% of MP98B and 50% of SC2A, has a specific capacity close to that of MP98B/MP98B cells and surpassing that of SC2A-based symmetric SCs. Moreover, at low temperatures, SCMP has higher volumetric energy and gravimetric energy density than those of MP98B- or SC2A-based symmetric SCs. This indicates that the specific capacitance of SC devices can be enhanced by tuning the electrode formulation through the use of different structures and textures. An SCMP electrode in an $[EMIm][FSI]_{0.5}[BF_4]_{0.5}$ electrolyte had an energy density of 8.2 Wh kg^{-1} at −30 °C, i.e. 39% of that at RT (21 Wh kg^{-1}). Nyquist plots showed sluggish ionic diffusion in the low-frequency range and an increase in the ESR related to the reduction in electrolyte conductivity (12.1 mS cm^{-1} at 20 °C and 0.42 mS cm^{-1} at −40 °C). At a low scan rate of 1 mV s^{-1}, the CV curve of the SC was accurately maintained over a temperature range of 20 °C to −40 °C [43].

Park and Kim introduced tetramethylammonium $[BF_4]$ ($[TMA][BF_4]$) into ILs containing tetraethylammonium $[BF_4]$ ($[TEA][BF_4]$), trimethylethylammonium $[BF_4]$ ($[TMEA][BF_4]$), and $[SBP][BF_4]$. TMA has a cation size (0.28 nm) that is much smaller than that of TEA (0.34 nm). Its small ionic size increases the effective pore volume available for adsorption. Park and Kim found that a binary salt electrolyte with a small ratio of 4–5 mol% $[TMA][BF_4]$ led to a 12%–13% increase in the capacitance of AC-based SCs at 10 A g^{-1} compared to that obtained using a conventional electrolyte. The TMA ion, as the smallest quaternary ammonium ion, can enter narrow pores that are inaccessible to conventional, large electrolyte ions, and impart extra capacitance to the cell [65]. As cations in an electrolyte, TMA ions are advantageous over Li^+ due to the lower ionic conductivity of $LiBF_4$ (18 mS cm^{-1}) compared to that of $[TMA][BF_4]$ and the poor ionization of $LiBF_4$ to Li^+ in the electrolyte. The fast response of the electrolyte at 10 A g^{-1} confirmed the possible application of the [TMA] cation as an additive for low-temperature SCs. The anion size was reported to affect the capacitance of commercial carbon black, known as Cabot MONARCH 1300 pigment black (CMPB). The capacitance of a CMPB electrode was found to be related to the size of the hydrous cation ($Li^+ \rightarrow Na^+ \rightarrow K^+$) instead of the crystal radius of the electrolyte when coupled with the same Cl^- anion. In electrolytes with the same cation concentration (K^+), changing the anion from a larger dianion (SO_4^{2-}) to a smaller monoanion (Cl^-) also increased the capacitance at high potential scan rates (>50 mV s^{-1}) [66].

6.3.2.2 Cosolvent electrolytes

It may be possible to expand the working temperature range of an electrolyte by mixing a solvent that has a low freezing point with a solvent that has a high dielectric constant [67]. Organic solvents are commonly applied as working media to dilute ILs in order to reduce their viscosity and improve their ionic conductivity. The electrochemical window of the medium is usually not high as those of the ILs. Table 6.3 provides a summary of the electrolyte systems utilized in low-temperature supercapacitors. One widely adopted approach to extend the working temperature range of supercapacitors is the incorporation of ionic liquids into organic electrolytes. This combination allows for versatile selection of ionic liquids and solvents. Other electrolyte systems such as eutectic ionic liquids, gel electrolytes, 'water-in-salt' electrolytes, and aqueous electrolytes reported for low-temperature supercapacitors are included. Brandon *et al* blended a conventional $[TEA][BF_4]$/ACN electrolyte with cosolvents such as EA, MF, MA, and DIOX. The cosolvent electrolyte thus developed expanded the low-temperature range of mesoporous carbon down to −75 °C. The solvent had a pronounced effect on the ESR of the cells; however, the cell capacitance was more dependent on the concentration of the

Table 6.3. The electrolyte systems reported for low-temperature SCs.

Electrolyte	Conductivity (mS cm^{-1})	Conductivity (mS cm^{-1})	Viscosity (mPa s)	T_g (Cooling)	Reference
1 M $LiPF_6$ EC/DMC	12 at 20 °C				[70]
[EMIm][BF_4]	150				[62]
1 M [EMIm][BF_4]/(AN + MA)				<−60	[71]
[EMIm][BF_4]–GBL	250				[62]
[EMIm][TFSI]/GBL [Pyr_{13}][FSI]/GBL	~20	1–2 at −50 °C		<−80 °C	[69]
[TEA][BF_4] + ACN	60 at 20 °C			−55 °C	[72]
0.5 M [SBP][BF_4] in 1:1 AN:MF	50 at 20 °C	~5 at −60 °C			[30]
1 M [TEA][BF_4] + PC	13.0		2.6		[73]
2.3 M [PYR_{14}][BF_4] + PC	10.2		3.9		[73]
[PIP_{13}]$_{0.5}$[$Pyrr_{14}$]$_{0.5}$[FSI]	4.9 at 20 °C	0.1 at −40 °C			[13]
pyrrolidinium [Pyrr][TFSI]		0.3 at −10 °C	40 at 30 °C		[61]
[Pyrr][NO_3]	18.3		17 at 30 °C		[61]
[Pyrr][NO_3]$_{0.72}$[TFSI]$_{0.28}$		1 at −10 °C	45 at 20 °C	~−23	[61]
([PYR_{13}][FSI])$_{0.568}$([PYR_{14}][TFSI])$_{0.432}$	1	0.001 at −70 °C	40–60 at 20 °C,		[74]
[BMIm][I]/[EA][N]/water/ LiI-5/35/55/5	<50 at 20 °C	1 at −40 °C	10 at 20 °C, 1000 at −60 °C		[75]
1-propyl-3-methylpyrrolidinium [PMPyrr]$_{0.5}$[EMIm]$_{0.5}$[TFSI]	5.17 at 25 °C		44 at 25 °C		[45]
0.65 M [TEA][BF_4] in PC	10.6		2.5		[44]
0.65 M [TEA][BF_4] in EC	13.1		1.9		[44]
1.5 M [SBP][BF_4]/(PC+DMC)	16.89	~5 at −40 °C	3.14 at 20 °C, 7.5 at −10 °C		[68]

(Continued)

Table 6.3. (*Continued*)

Electrolyte	Conductivity (mS cm^{-1})	Conductivity (mS cm^{-1})	Viscosity (mPa s)	T_g (Cooling)	Reference
1.5 M [SBP][BF_4]/(PC + DMC + EC)	18.08	~5 at −40 °C	3.55 at 20 °C, 11 at −10 °C		[68]
0.5 M [SBP][BF_4] 1:1 AN:MF	27 at 20 °C	~10 at −60 °C		−73.1	[1]
[EMIm][TFSI]	8.4 at 25 °C	0.9 at −20 °C	39.21 at 20 °C, 175.6 at −10 °C	−45	[76]
[EMIm][FSI]	16.8 at 25 °C	2.9 at −20 °C	22.28 at 20 °C, 70.33 at −10 °C		[76]
[EMIm][BF_4]	14.5 at 25 °C	1.2 at −20 °C	41.20 at 20 °C, 19.31 at −10 °C	−94	[76]
[EMIm][TCB]	13.14		23.10 at 20 °C	−76	[42]
[EMIm][FSI]$_{0.6}$ [BF_4]$_{0.1}$[TCB]$_{0.3}$	14.21	0.56 at −50 °C	23.61 at 20 °C,	−99	[42]
[EMIm][FSI]$_{0.6}$[BF_4]$_{0.4}$	12.62	~2 at −15 °C, 0.42 at −40 °C	31.5 at 20 °C, 312.7 at −25 °C		[76]
[PIP_{13}]$_{0.5}$[PYR_{14}]$_{0.5}$ [FSI]	4.9	28.9 at 100 °C, 1 at −23 °C			[13]
[PIP_{13}]$_{0.5}$[PYR_{14}]$_{0.5}$ [FSI] ionogel	10 at 60 °C	0.2 mS cm^{-1} at −40 °C		<−50	[50]
[EMIm]$_{0.5}$[MPPyrr]$_{0.5}$[TFSI]	5.2 at 25 °C				
Aqueous electrolytes					
1:2 (choline chloride:EG)	8		11.5		[47]
1:2:1 (choline chloride:EG:urea)	13.4		3.39		[47]
3.5 m $Mg(ClO_4)_2$	20.06				[48]
24.8% LiCl/PVA/H_2O	180	1.83 at −70 °C			[46]
5 M $NaClO_4$ ethanol–water	102.45	12.35 at −40 °C	1.43		[49]
12.9 M LiTFSI in DMSO:H_2O	6 at 20 °C				[52]
0.7 M Li_2SO_4 in 7:3 water/methanol	36.4 at 20 °C	1.3 at −40 °C		−51.2	[77]
LiCl/PVA in EG/H_2O organohydrogel	~20 at 20 °C	2.38 at −40 °C			[78]

solute or the cosolvent used. A 1 M concentration of [TEA][BF_4] displayed a very high impedance at −45 °C, followed by concentrations of 0.75 M and 0.25 M. An electrolyte made using 0.25 M [TEA][BF_4], however, only delivered 90% of the capacitance value obtained using 0.5 M and 0.75 M electrolytes. Mesoporous carbon with a BET surface area of 1344 m^2 g^{-1} and an average pore width of 3.6 nm was used in electrodes. A 0.75 M [TEA][BF_4] 3:1 ACN-DIOX blend could be charged and discharged at temperatures as low as −75 °C while maintaining half of its RT capacitance. Electrolytes made using 3:1 ACN-MF, 3:1 ACN-MA, and 3:1 ACN-DIOX as solvents could be discharged at −50 °C to −65 °C [2], but a significant decrease was observed in their ionic conductivity.

MF has a low melting point of −100 °C. A 1:1 mixture of MF with ACN was reported to lead to a binary solvent with an onset of crystallization to a solid phase at −73 °C and a freezing point of −110 °C. When a salt is added, a further reduction in the freezing point is expected. Brandon *et al* reported the use of binder-free, activated carbon fabric as an electrode material with [SBP][BF_4] in a 1:1 mixture of ACN:MF electrolyte and measured the electrochemical performance of microporous carbon at a low temperature. The ionic conductivities of [SBP][BF_4] in ACN: MF was ~27 mS cm^{-1} at 20 °C and ~10 mS cm^{-1} at −60 °C. Spectracarb 2225 carbon with pore size ranging from 0.4 to 3 nm and a median pore width of 0.66 nm was used as electrodes; its specific capacitance was ~110 F g^{-1} at −65 °C, which was slightly higher than, or at least comparable with that at 20 °C. The charge and discharge efficiencies of SCs made using this material remained at ~100% over a temperature range from 20 °C to −65 °C [1]. Korenblit *et al* compared the electrochemical behaviors of [SBP][BF_4] and [TEA][BF_4] in ACN and ACN/MF solvents using mesoporous carbon and 1–2 nm microporous carbon as electrodes. Microporous ZTC electrodes with large micropores stored more charge when [SBP][BF_4]/(AN:MF) was used as the electrolyte than when [TEA][BF_4]/AN or [TEA][BF_4]/(AN:MF) was used as the electrolyte, as shown in figure 6.9(a). Compared with commercial mesoporous carbon with an average pore size of 3.6 nm, the ZTC electrode, which had regular pore sizes of 0.5 and 1.5 nm and a BET surface area of 2025 m^2 g^{-1}, had a more rectangular CV curve shape and ~3 times greater capacitance at −70 °C at a scan rate of 1 mV s^{-1} (figure 6.9(b)). A ZTC-based electrode with [SBP][BF_4] in a 1:1 AN:MF electrolyte was capable of achieving a capacitance retention of 86% at −70 °C compared with that at RT [30]. The concentration of salt or IL affected the low-temperature performance of this SC in different ways. A higher concentration of salt or IL increased the ionic conductivity of the electrolyte at low temperatures, but led to a steady increase in the ESR due to the precipitation of salt on the electrode surface. The authors compared the electrochemical performance of [TEA][BF_4] vs [SBP][BF_4] in 1:1 AN:MF solution (figures 6.9(c) and (d)). The [SBP][BF_4] showed significant advantages and had higher capacitance retention at −60 °C. The smaller cation size (0.42 nm) and lower desolvation energy of [SBP][BF_4] could be the main reasons for its superior performance.

Shi *et al* prepared 1.5 M [SBP][BF_4] in various ternary solvents with PC, EC, and DMC. The ternary solvent system of PC + DMC + EC had the highest conductivity

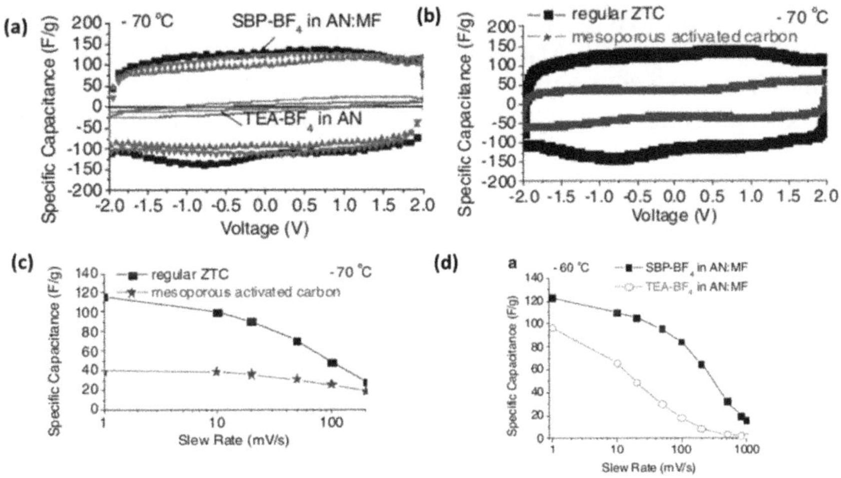

Figure 6.9. (a) CV curves of ZTCs synthesized at 700 °C in both conventional ([TEA][BF_4]/ACN; hollow symbols) and newly developed ([SBP][BF_4] in ACN:MF = 1:1; solid symbols) electrolytes at −70 °C. (b) CV of a ZTC electrode and an AC electrode with [SBP][BF_4] in ACN/MF at a scan rate of 1 mV s^{-1} recorded at −70 °C. (c) The slew rates of different carbon electrodes with [SBP][BF_4] in AN/MF as an electrolyte; the specific capacitances were calculated from the CV curves at −70 °C. (d) The rate performances of ZTC in [TEA][BF_4] and [SBP][BF_4]-based electrolytes in an ACN and MF binary solvent at −60 °C [30], reproduced with permission © John Wiley and Sons.

of 18.08 mS cm^{-1} with an electrochemical window of more than than 3.2 V. An EDLC with [SBP][BF_4] in a PC + DMC + EC electrolyte had a maximum energy density and a maximum power density of 41.5 Wh kg^{-1} and 17.0 kW kg^{-1}, respectively. The distortion of the CV curve shapes at scan rates of 50 and 100 mV s^{-1} indicates the unsatisfactory power density of this SC [68].

West *et al* reported that [TEA][BF_4] in an ACN and DIOX cosolvent had a low ESR when used as an electrolyte. The lowest ESR was achieved using a blend of 3:1 by volume of AN to DIOX. The resulting electrolyte reduced the freezing point of ACN by more than 20 °C, which was accompanied by a decrease in the RT dielectric constant of 28%. A 1:1 volume ratio of the AN:DIOX blend suppressed the freezing point of ACN by 42 °C and decreased of RT dielectric constant by 49%. A higher ratio of DIOX led to a lower freezing point but also a lower dielectric constant. The authors found that the electrolyte concentration was correlated with the ESR of the cell at low temperatures. A concentration of 0.625 M resulted in the lowest ESR among a series of [TEA][BF_4] concentrations from 0.250 M to 0.750 M. At temperatures lower than −55 °C, a greater than fourfold increase in the ESR was observed. The same trend was observed for 0.75 M [TEA][BF_4] in a 1:1 AN:DIOX binary solvent. At −65 °C, its capacitance was approximately 0.60 F, which was near the DC limit for the 0.250–0.500 M blends. However, the capacitance of the SC that used the 0.625 M blend was reduced to nearly 0.10 F, which can be ascribed to the possible precipitation of salt on the surfaces of the electrodes [51].

GBL, which has a melting point of −43.5 °C and a viscosity of 1.7 cP at 20 °C, has been reported to be a multifunction solvent. Dagousset investigated the performance of single-walled CNT (SWCNT)-based coin-cell SCs with three types of IL: [PYR_{13}][FSI], [PYR_{13}][TFSI], and [EMIm][TFSI] with GBL as the solvent over a temperature range from −50 °C to 100 °C. They found that IL/GBL mixtures based on TFSI anions had better stability than those based on FSI. SCs made using [EMIm][TFSI]/GBL and [PYR_{14}][TFSI]/GBL electrolytes had capacitance losses of 9% and 10%, respectively, after 10 000 cycles at 100 °C. A floating test was performed at 3 V for 500 h with both [EMIm][TFSI]/GBL and [PYR_{14}][TFSI]/GBL at −50 °C. A 20% reduction in capacitance was observed after 200 h for [PYR_{14}][TFSI]/GBL, while [EMIm][TFSI]/GBL had impressive stability, only losing 6.6% of its capacitance after 500 h. [EMIm][TFSI]/GBL has shown the potential to work over a broad range of temperatures (−50 °C to 100 °C) with good cyclic stability [69].

[EMIm][BF_4] is a widely used ionic liquid which transforms from a viscous liquid at 20 °C to the solid state at −50 °C. Differential scanning calorimetry (DSC) of [EMIm][BF_4] has shown that it has a supercooled region at −35 °C to 17 °C; however, [EMIm][BF_4]/GBL does not have exothermic peaks of crystallization. [EMIm][BF_4]/GBL has an ionic conductivity of ~24 S cm^{-1}, which is higher than that of neat [EMIm][BF_4] (~15 S cm^{-1}). At −50 °C, the ionic conductivity of an [EMIm][BF_4]/GBL electrolyte was 2.3 S cm^{-1}. The ionic conductivity of [EMIm][BF_4] decreased to 7.5 S cm^{-1} at 0 °C. Raman and 1H-NMR (proton nuclear magnetic resonance) spectra of neat [EMIm][BF_4] and [EMIm][BF_4]/GBL have shown that the C=O of GBL interacts strongly with the H atoms of EMI^+ via ion–dipole interactions ($^+$C–H···O=C). Analysis has indicated that neat [EMIm][BF_4] has interionic H-bonds derived from the interactions between the 2H, 4H, and 5H sites and the F–B^- anions that form H-bonds such as $^+$C–2H···F–B^-, $^+$C–4H···F–B^-, and $^+$C–5H···F–B^-. GBL significantly affected the interionic H-bonding between EMI^+ and BF_4^-, in particular targeting 2H, 4H, and 5H.

As shown in figure 6.10(b), GBL interacts with 2H, which is accompanied by the weakening or cleavage of the hydrogen bonds between 5H, 4H, and the BF_4 anion. The intense ion–dipole interactions between $EMIm^+$ and GBL can effectively weaken the hydrogen bonds between 5H and F atoms and break the hydrogen bond ($^+$C–4H···F–B^-) and interionic H-bonds in [EMIm][BF_4], thereby facilitating the separation of $EMIm^+$ and BF_4^-; the result is the formation of isolated ions or newly solvated couples (GBL–$EMIm^+$). Therefore, ion mobility and conductivity can significantly increase as the viscosity of the electrolyte decreases. The authors prepared an N-doped mesopore-dominated hierarchical carbon (NMHC) electrode using a dual-salt template method with gelatine, $Fe(NO_3)_3$, and $NaNO_3$ as precursors. NMHC has a specific surface area of 2637.4 m^2 g^{-1}, pore sizes ranging from 2 to 5 nm, and a total pore volume of 1.679 cm^3 g^{-1}. The combination of the NMHC electrode with an [EMIm][BF_4]/GBL electrolyte yielded an impressive specific energy of 61 Wh kg^{-1} at −50 °C, which was 10.7 times larger than that of a commercial YP-50-based SC [62]. The mesoporous NMHC had more rectangular CV curve shapes at 20 °C and −50 °C. The commercial microporous YP-50 carbon had lost most of its capacity at −50 °C. A Nyquist plot indicated that the mesoporous electrode had smaller R_{ct} at low temperatures. The blended electrolyte enabled the SC to operate at high rates from 0.5 to 5 A g^{-1} in low temperatures

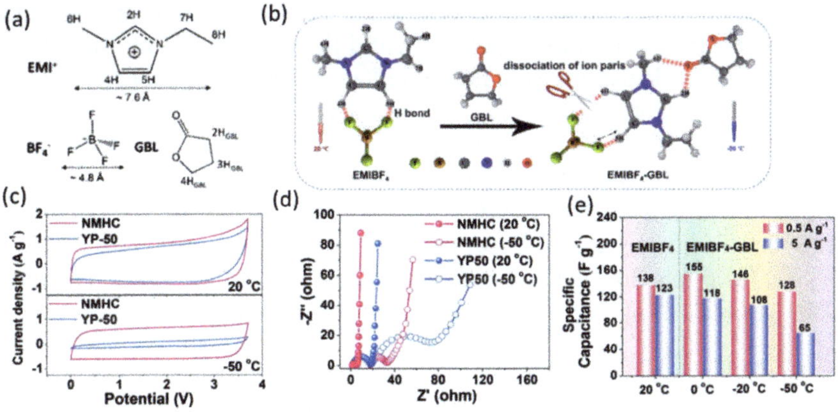

Figure 6.10. (a) The molecular structures and H sites of [EMIm][BF_4] and GBL. (b) A schematic of the H-bond breaking/weakening and dissociation of [EMIm][BF_4] ion pairs in the GBL solvent. (c) CV curves and (d) EIS plots of NMHC and YP-50 electrodes at 20 °C and −50 °C. (e) The specific capacitances of NMHC in [EMIm][BF_4] and [EMIm][BF_4]/GBL electrolytes at current densities of 0.5 and 5 A g^{-1} [62], reproduced with permission © Royal Society of Chemistry.

(figure 6.10(e)). Moreover, SCs made using the [EMIm][BF_4]/GBL electrolyte delivered a capacitance of 155 F g^{-1} at 20 °C, exceeding that of an [EMIm][BF_4] electrolyte (138 F g^{-1}) [62] (table 6.3).

6.3.2.3 Deep eutectic solvent electrolytes

A deep eutectic solvent (DES) is an ionic liquid analog that has abundant hydrogen bonds between donor molecules, such as polyols, amides, carboxylic acids, etc. and an acceptor (usually the quaternary ammonium salt) at a certain stoichiometric ratio. DES electrolytes have melting points that are significantly lower than those of individual solvents. A binary DES electrolyte composed of choline chloride (ChCl) and EG can broaden the voltage window to 2 V, which is higher than the theoretical electrolytic voltage of water molecules. At −40 °C, a DES with a ChCl, EG, and urea ratio of 1:2:1 was found to have an ionic conductivity of 6.33 mS cm^{-1}, while its conductivity was 13.41 mS cm^{-1} at temperatures of less than 115 °C [44]. Zhong *et al* developed a binary DES composed of ChCl and EG at a molar ratio of 1:2 for use as a SC electrolyte (from −40 °C to −115 °C). A specific capacitance of 362 F g^{-1} was achieved at RT, of which 102 F g^{-1} was retained at −40 °C [47]. It is noteworthy that the coulombic efficiencies of ChCl/EG-based DES were as low as 63.25%, 72.57%, and 67.47% for 1:1, 1:2, and 1:4 ratios of ChCl to EG, respectively. Moreover, Nyquist plots revealed that the SC devices with DES electrolytes had an R_{ct} of 17 Ω at RT. The unbounded hydroxyl group in the DES solution between the EG (donor) and ChCl (acceptor) formed a hydrogen bond with the oxygen-containing surface functional groups (–COOH or OH) on the surface of the AC, accelerating electrolyte ion diffusion and improving the ionic conductivity [47].

Bu *et al* reported a metal-based DES prepared using 3.5 M $Mg(ClO_4)_2{\cdot}6H_2O$, which had an ionic conductivity of 20.06 mS cm^{-1} at −40 °C. Raman spectroscopy showed that more water molecules coordinated with Mg^{2+} ions, thus breaking the hydrogen bonds among the water molecules and thereby preventing the formation of ice crystals. A SC made using this material had a window potential of 1.8 V at −40 °C and a capacitance retention of 76% at 1 A g^{-1} when the temperature decreased from 25 °C to −40 °C [48]. The CV curves of a commercial AC (Kuraray YP 50F)-based SC with a 3.5 M $Mg(ClO_4)_2$ electrolyte maintained a rectangular shape at −40 °C at a scan rate of 20 mV s^{-1}. The $Mg(ClO_4)_2$ DES electrolyte had a conductivity of 136.6 mS cm^{-1}, higher than those of 1 M $Mg(ClO_4)_2$ (104.8 mS cm^{-1}) and 1 M Na_2SO_4 (85.6 mS cm^{-1}). The DES electrolyte displayed almost no capacity decay after 6000 cycles at 4 A g^{-1} and achieved nearly 100% coulombic efficiency at −40 °C.

6.3.2.4 Aqueous electrolytes

Aqueous electrolytes are low-cost and environmentally friendly; they also enable SCs to be assembled under atmospheric conditions. At low temperatures, the hydrogen evolution reaction (HER) and oxygen evolution reaction (OER), which have a very strong (logarithmic) dependence on temperature according to the Arrhenius equation, are greatly depressed. Therefore, a higher electrochemical window potential can be achieved at low temperatures compared to that at RT. In addition to the low working voltages of aqueous capacitors, the freezing points of conventional aqueous electrolytes constitute another challenge for their application at low temperatures. Viola and Andrew demonstrated a mixture of lithium chloride (LiCl) with water as an electrolyte that could operate at temperatures as low as −80 °C. In water, 24.8 wt% LiCl has a low freezing point of −74 °C; its ionic conductivity is 180 mS cm^{-1} at RT and 1.83 mS cm^{-1} at −70 °C. SCs based on Kuraray YP-80FP as an electroactive material were able to retain a capacitance of 28 F g^{-1} at −70 °C. As the temperature changed from RT to −70 °C, the electrochemical window potential of the aqueous eutectic electrolyte expanded by 1 V and reached a maximum of 3 V at −70 °C. The aqueous eutectic electrolyte had a comparable performance to that of 0.5 M [TEA][BF_4]/ACN at −40 °C [46]. Moreover, the water molecules were segregated by surrounding hydrogen bonds between polymer chains in the aqueous electrolyte-derived polymer hydrogel network. The freezing point of hydrogel-based electrolytes can be further reduced.

Pan *et al* prepared bimodal micro/mesoporous NPTCk with a surface area of 1310 m^2 g^{-1}. The authors utilized 2.0 M [EMIm][BF_4]/ACN as a reference and compared its low-temperature performance with that of 6 M KOH. [EMIm][BF_4]/ACN had specific capacitances of 34.8, 35.5, 30.2, 30.2, and 28.4 F g^{-1} at a current density of 2 A g^{-1} at 25 °C, −10 °C, −20 °C, −30 °C, and −40 °C, respectively. $NPTC_K$ symmetric SCs in a 6 M KOH electrolyte had specific capacitance values of 33.5, 31.8, 32.4, and 31 F g^{-1} at a current density of 2 A g^{-1} at temperatures of −10 °C, −20 °C, −30 °C, and −40 °C, respectively. At lower temperatures, the capacitance of NPTCk was higher with the 6 M KOH electrolyte than with the 2.0 M [EMIm][BF_4]/ACN electrolyte. However, the capacitance retentions of $NPTC_K$ in 6 M KOH at a current density of 10 A g^{-1} were low, namely 44.7% and 53.5% after 10 000 cycles at 25 °C and −40 °C, respectively. The corrosion of the electrodes by highly concentrated

KOH and/or water hydrolysis at the high voltage may have been the reasons for the capacitance decay [26].

Aqueous electrolytes play key roles for pseudocapacitors by providing the protons and hydroxide ions for redox reactions with various pseudocapacitive materials, such as polyaniline, polypyrrole, transition-metal oxides/hydroxides/sulfides, etc. For the low-temperature applications of these electrodes, the development of a freeze-resistant aqueous electrolyte is preferred. Roberts *et al* selected EG as an antifreeze additive and developed Na_2SO_4, Li_2SO_4, K_2SO_4, and $(NH_4)_2SO_4$ solute-based electrolytes. These electrolytes had 80% of the saturation of the various solutes at 0 °C in water at corresponding concentrations of 39, 245, 59, and 132 g ml^{-1} for Na_2SO_4, Li_2SO_4, K_2SO_4, and $(NH_4)_2SO_4$, respectively. Cells based on K_2SO_4 were regarded as the least promising of those tested, and cells based on $(NH_4)_2SO_4$ had the highest specific capacitance at temperatures between 0 °C and −30 °C. The specific capacitance was observed to vary, depending on the EG concentration. The maximum capacitance was observed at 30% EG in $(NH_4)_2SO_4$ at −20 °C and −30 °C [54]. However, the viscous EG resulted in a decrease in conductivity and low specific capacitance at RT.

6.3.2.5 'Water-in-salt' electrolytes

The definition of 'water-in-salt' applies to a strong concentration of salt dissolved in water, more precisely to solutions with mass and volume ratios (m_{salt}/m_{water} and V_{salt}/V_{water}) greater than one [79]. A 'water-in-lithium' salt consisting of 31.3 M LiTFSI had an electrochemical window potential of 2.4 V. In 'water-in-lithium' electrolytes, Li^+ is surrounded by $TFSI^-$ anions together with a limited number of water molecules. Water splitting can be significantly suppressed and a maximum window potential of 3 V achieved. At a lower scan rate of 5 mV s^{-1}, 31.3 M LiTFSI provided a high energy density of 30.4 Wh kg^{-1} at 228.0 W kg^{-1}, higher than that of 0.3 M LiTFSI (18.9 Wh kg^{-1} at 170 W kg^{-1}) [80]. However, compared to less concentrated electrolytes, the power densities of SC devices operated with 'water-in-salt' electrolytes were lower because of their poorer ionic conductivity. AC-based SCs with a 31.3 M LiTFSI electrolyte had a power density of 1680 W kg^{-1} at a scan rate of 200 mV s^{-1}, which was much lower than that of 3000 W kg^{-1} for SCs with a 0.3 M LiTFSI electrolyte [80]. Tatlisu *et al* developed a 'water-in-imidazolium chloride' electrolyte based on the concept discussed above. They prepared an electrolyte composed of 30 M $[BMIm]Cl/H_2O$ with a 0.1 M 4-hydroxy-2,2,6,6-tetramethylpiperidin-1-oxyl (4hT) redox additive. An SC with graphene-coated carbon electrodes delivered a maximum capacity of 492 F g^{-1} with a window potential of 3 V at 20 mV s^{-1}. At −32 °C, the 20 M $[BMIm]Cl/H_2O$ electrolyte with 0.1 M 4hT had a specific capacitance of 97 F g^{-1} at 20 mV s^{-1}. The electrolyte resistance (R_s), defined as the point where the spectra cross the real axis, increased from 3.1 ohms at room temperature to 31 ohms at −20 °C and 104 ohms at −32 °C. The R_{ct} estimated by fitting a semicircle in the high-frequency region increased from 0.6 ohms at room temperature to 10 ohms and 13 ohms at −20 °C and −32 °C, respectively [55]. The solution resistance of the electrolyte increased more significantly than the charge transfer resistance. This indicates that the expansion of the operational temperature through the use of highly concentrated electrolytes may face a sharp increase in solution resistance due to the possible precipitation of solutes.

A LiTFSI-based 'water-in-salt' electrolyte was explored by Lu *et al*, who used a high molar ratio of LiTFSI with DMSO/H_2O as a solvent. A eutectic mixture was formed by DMSO and H_2O at a molar ratio of 1:2, and its freezing point was decreased to as low as −140 °C [52]. The ionic conductivity of LiTFSI in DMSO/H_2O decreased with an increase in the LiTFSI molar ratio. The conductivity of 3.4 M LiTFSI in DMSO/H_2O decreased from 2.97 S cm^{-1} to <1 S cm^{-1} when the temperature dropped from 20 °C to −35 °C. An electrolyte consisting of 6.6 M LiTFSI in DMSO/H_2O performed better than electrolytes with either water or DMSO as the solvent in terms of the cost-effectiveness of reaching 3.0 V [52].

6.3.2.6 Redox and antifreeze additives

The introduction of a redox additive to aqueous electrolytes is a widely accepted protocol for both pseudocapacitive and EDLC-type electrodes. Such additives enable pseudocapacitive electrodes to deliver higher capacitance at subambient temperatures. Su *et al* reported the introduction of p-phenylenediamine (PPD) into 4 M KOH to create an electrolyte for MnO_2 nanorod electrodes. At RT, the MnO_2 electrodes in 4 M KOH without PPD achieved a specific capacitance of 194 F g^{-1}, which was comparable to the value achieved using PPD. However, at −20 °C, the 4 M KOH/PPD alkaline electrolyte enabled the MnO_2 nanorods to deliver a capacitance of 156 F g^{-1}, which was significantly higher than that of 4 M KOH alone (65 F g^{-1}). The addition of PPD reduced the viscosity of the electrolyte and facilitated mass transfer [81]. The use of redox additives such as [BMIm]Cl/H_2O with 4hT for low-temperature SCs was reported by Tatlisu *et al* and was discussed earlier [55].

High concentrations of salt can decrease the freezing point by several tens of degrees Celsius; however, the precipitation of salt due to its reduced solubility at low temperatures leads to a rapid increase in the ESR and a loss of capacitance. Antifreezes are usually small-molecule solvents with a much lower freezing points than water, such as alcohols, DMSO, glycol, etc. Yin *et al* investigated the influence of alcohols such as methanol (Me), ethanol (Et), EG, and glycerol (Gl) as anti-freezing additives on a $NaClO_4$/H_2O electrolyte. The methanol/water concentrated (Me–W) $NaClO_4$, which had the highest ionic conductivity (102.5 mS cm^{-1}) and the lowest ESR at −40 °C, delivered the highest capacitance for both microporous and mesoporous carbon compared to all the other antifreeze-based electrolytes. With a 5 M $NaClO_4$/Me–W electrolyte at a scan rate of 5 mV s^{-1} at −40 °C, mesoporous and microporous carbon electrodes had specific capacitances of ~16 F g^{-1} and ~20 F g^{-1}, respectively. This study indicated that microporous carbon is capable of delivering high capacity at low rates and low temperatures compared with mesoporous carbon. The CV curves of mesoporous carbon at low temperatures have a more rectangular shape than those of microporous carbon, confirming the ability of mesoporous carbon to perform at high power [49].

Vellacheri *et al* utilized reduced graphene as an electrode and selected Li_2SO_4 as a salt with 30 wt% EG as an antifreeze and measured the performance of graphene-based symmetric SCs. At a current density of 4 mA cm^{-2}, a specific capacitance of

~43 F g^{-1} was delivered, compared with that of 80 F g^{-1} at RT. At a current density of 1 mA cm^{-2}, the specific capacitances of rGO in the Li_2SO_4-based electrolyte were 74 F g^{-1} and 99 F g^{-1} at −20 °C and 45 °C, respectively [53].

6.3.2.7 Gel electrolytes

In addition to DES-based and 'water-in-salt' electrolytes, hydrogel or ionogel-based semi-solid electrolytes have also been widely investigated use in for low-temperature SCs. The voltage windows of polymer gel devices can reach 2.2 V, enabling energy density values to be tripled compared to the values obtained at RT. Moreover, the dissociation constant of water is reduced at low temperatures due to the suppression of the HER and OER, especially in polymer hydrogel networks, where water molecules are segregated by surrounding hydrogen bonds between polymer chains [46].

Aqueous gel electrolytes are predominantly based on PVA and a poly(ethylene oxide) (PEO) matrix. Their good chemical stability enables PVA to adapt to acidic, alkaline, and neutral electrolytes, such as PVA/LiCl, PVA/KOH, PVA/H_2SO_4, etc. Conventional hydrogel electrolytes contain large numbers of water molecules, which inevitably freeze at subzero temperatures. Hydrogel electrolytes lose most of their capacity as the temperature falls below −10 °C. Increasing the concentration of salt or introducing antifreezes into hydrogel can extend the low-temperature limit. Rong *et al* developed a PVA-based freeze-resistant organohydrogel electrolyte using H_2O/EG as a dispersion medium and LiCl as the salt. The PVA in the gel matrix formed a crosslinked polymer matrix. The EG molecules formed hydrogen bonds with the PVA chains, decreasing the freezing point of the PVA-based hydrogel electrolyte. The segregation of water molecules by the EG and PVA chains led to a decrease in the freezing point of the organohydrogel to −40 °C. A PVA-based hydrogel turned into an ice-like solid at −20 °C and lost the elasticity of a polymer electrolyte. The organohydrogel electrolyte maintained excellent flexibility at a low temperature of −40 °C (figure 6.11(b)). CNT paper electrode-based SCs had a capacitance retention ratio of 70.6% at −40 °C (figure 6.11(c)) compared with that at RT and an 11.7%

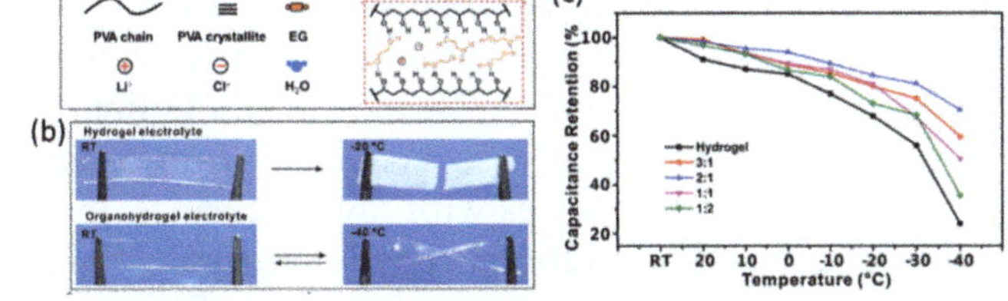

Figure 6.11. (a) A schematic illustration of the organohydrogel electrolyte applied for CNT paper electrodes. (b) Photographs of the low-temperature-tolerant performances of hydrogel and organohydrogel electrolytes: the hydrogel electrolyte entered an ice-like solid state and could easily be broken at −20 °C, but the organohydrogel electrolyte showed its ability to reversibly bend at −40 °C. (c) The capacitance retentions of different electrolyte-based SCs [78], reproduced with permission © John Wiley and Sons.

capacitance decay over 5000 continuous charge/discharge cycles at −20 °C [78]. The SCs achieved the best electrochemical performance and the highest capacitance retention at −40 °C when the electrolyte used had a mass ratio of H_2O and EG of 2:1. The ionic conductivity of the PVA/EG-based organohydrogel electrolyte was 2.38 mS cm^{-2} at −40 °C, compared with ~16–17 mS cm^{-2} at room temperature. Yu *et al* reported a hydrogen-bond-reinforced, dual crosslinked PVA, acrylic acid (AA), and H_2SO_4 hydrogel electrolyte. With 2D MoN/CNTs as electrodes, a hydrogel-electrolyte-based SC delivered a power density of 41.2 μWh cm^{-2} and a power density of 0.94 mW cm^{-2}. The freeze-resistant electrolyte enabled the SC device to be stored at −35 °C for 23 days with a capacity retention of 80% [82].

6.4 Conclusions and outlook

Temperature-tolerant SCs have been explored in the last few years for applications in extreme environments. As energy storage devices, SCs that had high power outputs and relatively high capacitance retention compared with Li-ion batteries at temperatures below −20 °C would be irreplaceable. At low temperatures, electrolyte viscosity increases, which can lead to sluggish ion diffusion kinetics in both the electrolyte and through the pores of electrodes. Moreover, the ionic conductivity of conventional SCs can be significantly reduced. The design of electrolytes with low freezing points, low viscosities, and high ionic conductivities could enable SCs to operate at low temperatures. The high power densities of SCs, however, depend on both the properties of the electrolyte and the microstructure of the electrode.

For example, the solvent used for the electrolyte has a pronounced effect on the ESR of a cell, and its specific capacitance can be affected by the concentration of the solute. Electrolytes made from high-concentration solutes help to deliver high specific capacitances at RT but face the problem of precipitation at low temperatures, which may lead to a rapid increase in resistance. The interaction between the solutes and the solvents is one of the key parameters that determines the low-temperature electrochemical performance of SCs. Highly polarizing ions, such as Li^+ in polar solvents (alkylcarbonates) retain their solvation shells during charging. The Li^+ in EC-based solvents needs high desolvation energy to diffuse into micropores. An ACN solvent with [TEA][BF_4] or LiTFSI as the solute forms poorly solvated cations, which can be fully desolvated in order to penetrate the small pores of AC. Cations such as TMA (0.28 nm), TEA (0.34 nm), and SBP (0.42 nm) have small sizes which help to increase the effective pore volume available for adsorption in the case of electrodes with micropores. In addition to the anion size, which affects the specific capacitance of SCs, chemical stability is another parameter that requires further research. Electrolytes with low viscosities, low freezing points, and high dielectric constants help to retain exceptional conductivity at low temperatures. The design of new electrolytes for low-temperature SCs remains a very active research field that includes, for example, eutectic ILs, the formulation of solvents, etc. These strategies aim to increase the ionic conductivity and decrease the freezing point and viscosity of the electrolyte in order to further enhance the interfacial properties of the electrode/electrolyte with respect to conventional electrolytes.

Mesoporous carbon has large enough pores to accommodate solvated or partially solvated ions at low temperatures. However, the specific gravimetric and volumetric capacitance of mesoporous carbon are not as high as those of microporous carbon at RT due to the low density of the former. Electrodes with 1–2 nm micropores can alleviate this problem by providing high packing density and high energy density for low-temperature applications. The combination of electrodes with different structures/textures and the development of bimodal meso/microporous electrodes are effective strategies. For example, electrodes with interconnected large micropores or small mesopores have superior electrochemical performance compared to those with small micropores (<0.7 nm). Hierarchical carbon electrodes such as VA-CNT sheets and 3D-printed mesoporous carbon film are favorable for high-power, low-temperature SCs compared with electrodes that have isolated, semi-closed, slit-shaped pores. Microsupercapacitors have the advantages of fast diffusion, low internal resistance, and high power output. These properties make microsupercapacitors potentially promising for low-temperature applications. The structural change and mechanical degradation of electrodes caused by their volume expansion and shrinkage during continuous cycling and temperature variations require more research attention.

The most promising perspectives for the development of low-temperature SCs are as follows: (1) optimize the formulation of solvents so that the solvents have physicochemical properties that match those of the electrodes; (2) take advantage of the compositional versatility of the formation processes of ILs and/or inorganic solutes, for example, by forming eutectic ILs or functionalizing side-chain structures to introduce specific interactions with the electrodes; (3) design appropriately structured electrodes with aligned orientations, open-sided pores, and/or large micropores or small mesopores; (4) select antifreezes or redox-active additives to enhance the interaction between the electrode and the electrolyte and impart extra redox charge to the electrode. In this respect, the development of temperature-tolerant electrolytes and the scaling up of the synthesis of 1–2 nm microporous and/or mesoporous carbon will lead to the future development of low-temperature energy storage devices for civilian, military, and high-altitude applications.

References

[1] Brandon E J, Smart M C and West W C 2020 Low-temperature performance of electrochemical capacitors using acetonitrile/methyl formate electrolytes and activated carbon fabric electrodes *J. Mater. Res.* **35** 113–21

[2] Brandon E J, West W C, Smart M C, Whitcanack L D and Plett G A 2007 Extending the low temperature operational limit of double-layer capacitors *J. Power Sources* **170** 225–32

[3] Largeot C, Portet C, Chmiola J, Taberna P-L, Gogotsi Y and Simon P 2008 Relation between the ion size and pore size for an electric double-layer capacitor *J. Am. Chem. Soc.* **130** 2730–1

[4] Raymundo-Piñero E, Kierzek K, Machnikowski J and Béguin F 2006 Relationship between the nanoporous texture of activated carbons and their capacitance properties in different electrolytes *Carbon* **44** 2498–507

[5] Gualous H, Bouquain D, Berthon A and Kauffmann J M 2003 Experimental study of supercapacitor serial resistance and capacitance variations with temperature *J. Power Sources* **123** 86–93

[6] Du Pasquier A, Plitz I, Menocal S and Amatucci G 2003 A comparative study of Li-ion battery, supercapacitor and nonaqueous asymmetric hybrid devices for automotive applications *J. Power Sources* **115** 171–8

[7] Jost K, Dion G and Gogotsi Y 2014 Textile energy storage in perspective *J. Mater. Chem.* A **2** 10776–87

[8] Kötz R, Hahn M and Gallay R 2006 Temperature behavior and impedance fundamentals of supercapacitors *J. Power Sources* **154** 550–5

[9] Liu P, Verbrugge M and Soukiazian S 2006 Influence of temperature and electrolyte on the performance of activated-carbon supercapacitors *J. Power Sources* **156** 712–8

[10] Plitz I, DuPasquier A, Badway F, Gural J, Pereira N, Gmitter A and Amatucci G G 2006 The design of alternative nonaqueous high power chemistries *Appl. Phys.* A **82** 615–26

[11] Xu J, Hu X, Wang X, Wang X, Ju Y, Ge S, Lu X, Ding J, Yuan N and Gogotsi Y 2020 Low-temperature pseudocapacitive energy storage in $Ti_3C_2T_x$ MXene *Energy Storage Mater.* **33** 382–9

[12] Huang P, Pech D, Lin R, McDonough J K, Brunet M, Taberna P-L, Gogotsi Y and Simon P 2013 On-chip micro-supercapacitors for operation in a wide temperature range *Electrochem. Commun.* **36** 53–6

[13] Lin R, Taberna P-L, Fantini S, Presser V, Pérez C R, Malbosc F, Rupesinghe N L, Teo K B K, Gogotsi Y and Simon P 2011 Capacitive energy storage from −50 to 100 °C using an ionic liquid electrolyte *J. Phys. Chem. Lett.* **2** 2396–401

[14] Lin Z, Taberna P-L and Simon P 2016 Graphene-based supercapacitors using eutectic ionic liquid mixture electrolyte *Electrochim. Acta* **206** 446–51

[15] Jänes A and Lust E 2005 Organic carbonate–organic ester-based non-aqueous electrolytes for electrical double layer capacitors *Electrochem. Commun.* **7** 510–4

[16] Decaux C, Ghimbeu M, Dahbi C, Anouti M, Lemordant M, Béguin D, Vix-Guterl F and Raymundo-Piñero C E 2014 Influence of electrolyte ion–solvent interactions on the performances of supercapacitors porous carbon electrodes *J. Power Sources* **263** 130–40

[17] Abioye A M and Ani F N 2015 Recent development in the production of activated carbon electrodes from agricultural waste biomass for supercapacitors: a review *Renew. Sustain. Energy Rev.* **52** 1282–93

[18] Chmiola J, Yushin G, Gogotsi Y, Portet C, Simon P and Taberna P L 2006 Anomalous increase in carbon capacitance at pore sizes less than 1 nanometer *Science* **313** 1760

[19] Tsai W-Y, Taberna P-L and Simon P 2014 Electrochemical quartz crystal microbalance (EQCM) study of ion dynamics in nanoporous carbons *J. Am. Chem. Soc.* **136** 8722–8

[20] Pean C, Daffos B, Rotenberg B, Levitz P, Haefele M, Taberna P-L, Simon P and Salanne M 2015 Confinement, desolvation, and electrosorption effects on the diffusion of ions in nanoporous carbon electrodes *J. Am. Chem. Soc.* **137** 12627–32

[21] Xu J *et al* Temperature-independent capacitance of carbon-based supercapacitor from −100 to 60 °C *Energy Storage Mater.* 2019**22** 323–9

[22] Wu J-C, Liu C, Zhou H, Liu M, Zhang D, Yu C, Zhang X, Gao H and Yang J 2020 Thermal dynamic study of the gradual desolvation in submicropores for carbon-based supercapacitor at low temperature *Ionics* **26** 4695–704

[23] Wang X, Xu J, Razal J M, Yuan N Y, Zhou X S, Wang X H, Ding J N, Qin S, Ge S H and Gogotsi Y 2019 Unimpeded migration of ions in carbon electrodes with bimodal pores at an ultralow temperature of −100 °C *J. Mater. Chem.* A **7** 16339–46

[24] Ye L, Liang Q H, Huang Z H, Lei Y, Zhan C Z, Bai Y, Li H, Kang F Y and Yang Q H 2015 A supercapacitor constructed with a partially graphitized porous carbon and its performance over a wide working temperature range *J. Mater. Chem.* A **3** 18860–6

[25] Tsai W-Y, Gao P-C, Daffos B, Taberna P-L, Perez C R, Gogotsi Y, Favier F and Simon P 2013 Ordered mesoporous silicon carbide-derived carbon for high-power supercapacitors *Electrochem. Commun.* **34** 109–12

[26] Pan H, Zhang Y, Pan Y, Lin W, Tu W and Zhang H 2020 Nitrogen-doped porous carbon with interconnected tubular structure for supercapacitors operating at sub-ambient temperatures *Chem. Eng. J.* **401** 126083

[27] Xu J, Wang X, Zhou X, Yuan N, Ge S and Ding J 2019 Activated carbon coated CNT core–shell nanocomposite for supercapacitor electrode with excellent rate performance at low temperature *Electrochim. Acta* **301** 478–86

[28] Tsai W-Y, Lin R, Murali S, Li Zhang L, McDonough J K, Ruoff R S, Taberna P-L, Gogotsi Y and Simon P 2013 Outstanding performance of activated graphene based supercapacitors in ionic liquid electrolyte from −50 to 80 °C *Nano Energy* **2** 403–11

[29] Kado Y, Imoto K, Soneda Y and Yoshizawa N 2014 Highly enhanced capacitance of MgO-templated mesoporous carbons in low temperature ionic liquids *J. Power Sources* **271** 377–81

[30] Korenblit Y, Kajdos A, West W C, Smart M C, Brandon E J, Kvit A, Jagiello J and Yushin G 2012 In situ studies of ion transport in microporous supercapacitor electrodes at ultralow temperatures *Adv. Funct. Mater.* **22** 1655–62

[31] Trasatti S and Buzzanca G 1971 Ruthenium dioxide: a new interesting electrode material. Solid state structure and electrochemical behaviour *J. Electroanal. Chem. Interfacial Electrochem.* **29** A1–5

[32] Liu X and Pickup P G 2008 Performance and low temperature behaviour of hydrous ruthenium oxide supercapacitors with improved power densities *Energy Environ. Sci.* **1** 494–500

[33] Zheng J P and Jow T R 1996 High energy and high power density electrochemical capacitors *J. Power Sources* **62** 155–9

[34] Sorensen E M, Barry S J, Jung H-K, Rondinelli J M, Vaughey J T and Poeppelmeier K R 2006 Three-dimensionally ordered macroporous $Li_4Ti_5O_{12}$: effect of wall structure on electrochemical properties *Chem. Mater.* **18** 482–9

[35] Kong J *et al* Well-aligned hierarchical graphene-based electrodes for pseudocapacitors with outstanding low-temperature stability *Chem. Electro. Chem.* 2019**6** 2788–95

[36] Xiong G P, He P G, Liu L, Chen T F and Fisher T S 2015 Synthesis of porous Ni–Co–Mn oxide nanoneedles and the temperature dependence of their pseudocapacitive behavior *Front. Energy Res.* **3** 39

[37] Zhang J, Gao H, Yang Q, Zhang X T, Zhang M Y and Xu L L 2015 Effect of temperature on pseudocapacitance performance of carbon fiber@$NiCo_2O_4$@Ni(OH)(2) core–shell nanowire array composite electrodes *Appl. Surf. Sci.* **356** 167–72

[38] Wang J G, Yang Y, Huang Z H and Kang F Y 2013 Effect of temperature on the pseudocapacitive behavior of freestanding MnO_2@carbon nanofibers composites electrodes in mild electrolyte *J. Power Sources* **224** 86–92

[39] Guo F M, Xu R Q, Cui X, Zhang L, Wang K L, Yao Y W and Wei J Q 2016 High performance of stretchable carbon nanotube-polypyrrole fiber supercapacitors under dynamic deformation and temperature variation *J. Mater. Chem.* A **4** 9311–8

[40] Wu X, Qiu S, Xu Y, Ma L, Bi X, Yuan Y, Wu T, Shahbazian-Yassar R, Lu J and Ji X 2020 Hydrous nickel–iron Turnbull's blue as a high-rate and low-temperature proton electrode *ACS Appl. Mater. Interfaces* **12** 9201–8

[41] Barua A and Paul A 2020 Unravelling the role of temperature in a redox supercapacitor composed of multifarious nanoporous carbon@hydroquinone *RSC Adv.* **10** 1799–810

[42] Yambou E P, Gorska B and Béguin F 2021 Electrical double-layer capacitors based on a ternary ionic liquid electrolyte operating at low temperature with realistic gravimetric and volumetric energy outputs *Chem. Sus. Chem.* **14** 1196–208

[43] Pameté Yambou E, Gorska B, Pavlenko V and Beguin F 2020 Fitting the porous texture of carbon electrodes to a binary ionic liquid electrolyte for the realization of low temperature EDLCs *Electrochim. Acta* **350** 136416

[44] Zhong M, Tang Q F, Qiu Z G, Wang W P, Chen X Y and Zhang Z J 2020 A novel electrolyte of ternary deep eutectic solvent for wide temperature region supercapacitor with superior performance *J. Energy Storage* **32** 101904

[45] Newell R, Faure-Vincent J, Iliev B, Schubert T and Aradilla D 2018 A new high performance ionic liquid mixture electrolyte for large temperature range supercapacitor applications (−70 °C to 80 °C) operating at 3.5 V cell voltage *Electrochim. Acta* **267** 15–9

[46] Viola W and Andrew T L 2021 An aqueous eutectic electrolyte for low-cost, safe energy storage with an operational temperature range of 150 °C, from −70 to 80 °C *J. Phys. Chem.* C **125** 246–51

[47] Zhong M, Tang Q F, Zhu Y W, Chen X Y and Zhang Z J 2020 An alternative electrolyte of deep eutectic solvent by choline chloride and ethylene glycol for wide temperature range supercapacitors *J. Power Sources* **452** 227847

[48] Bu X, Zhang Y, Sun Y, Su L, Meng J, Lu X and Yan X 2020 All-climate aqueous supercapacitor enabled by a deep eutectic solvent electrolyte based on salt hydrate *J. Energy Chem.* **49** 198–204

[49] Yin J, Qi L and Wang H 2013 Anti-freezing aqueous electrolytes for electric double-layer capacitors *Electrochim. Acta* **88** 208–16

[50] Negre L, Daffos B, Turq V, Taberna P L and Simon P 2016 Ionogel-based solid-state supercapacitor operating over a wide range of temperature *Electrochim. Acta* **206** 490–5

[51] West W C, Smart M C, Brandon E J, Whitcanack L D and Plett G A 2008 Double-layer capacitor electrolytes using 1,3-dioxolane for low temperature operation *J. Electrochem. Soc.* **155** A716

[52] Lu X, Jiménez-Riobóo R J, Leech D, Gutiérrez M C, Ferrer M L and del Monte F 2020 Aqueous-eutectic-in-salt electrolytes for high-energy-density supercapacitors with an operational temperature window of 100 °C, from −35 to +65 °C *ACS Appl. Mater. Interfaces* **12** 29181–93

[53] Vellacheri R, Al-Haddad A, Zhao H, Wang W, Wang C and Lei Y 2014 High performance supercapacitor for efficient energy storage under extreme environmental temperatures *Nano Energy* **8** 231–7

[54] Roberts A J, de Namor D and Slade A F R C T 2013 Low temperature water based electrolytes for MnO_2/carbon supercapacitors *Phys. Chem. Chem. Phys.* **15** 3518–26

[55] Tatlisu A, Huang Z and Chen R 2018 High-voltage and low-temperature aqueous supercapacitor enabled by 'water-in-imidazolium chloride' electrolytes *Chem. Sus. Chem.* **11** 3899–904

[56] Kang J, Jayaram S H, Rawlins J and Wen J 2014 Characterization of thermal behaviors of electrochemical double layer capacitors (EDLCs) with aqueous and organic electrolytes *Electrochim. Acta* **144** 200–10

[57] Iwama E, Taberna P L, Azais P, Brégeon L and Simon P 2012 Characterization of commercial supercapacitors for low temperature applications *J. Power Sources* **219** 235–9

[58] Miao L, Song Z, Zhu D, Li L, Gan L and Liu M 2021 Ionic liquids for supercapacitive energy storage: a mini-review *Energy Fuels* **35** 8443–55

[59] Chiba K, Ueda T, Yamaguchi Y, Oki Y, Shimodate F and Naoi K 2011 Electrolyte systems for high withstand voltage and durability I. Linear sulfones for electric double-layer capacitors *J. Electrochem. Soc.* **158** A872

[60] Rodríguez A, Canosa J, Domínguez A and Tojo J 2003 Viscosities of dimethyl carbonate or Diethyl carbonate with alkanes at four temperatures. New UNIFAC–VISCO parameters *J. Chem. Eng. Data* **48** 146–51

[61] Timperman L, Vigeant A and Anouti M 2015 Eutectic mixture of protic ionic liquids as an electrolyte for activated carbon-based supercapacitors *Electrochim. Acta* **155** 164–73

[62] Li J, Zhou Y, Tian J, Peng L, Deng J, Wang N, Qian W and Chu W 2020 A nitrogen-doped mesopore-dominated carbon electrode allied with anti-freezing EMIBF4–GBL electrolyte for superior low-temperature supercapacitors *J. Mater. Chem.* A **8** 10386–94

[63] Wang X, Salari M, Jiang D-e, Chapman Varela J, Anasori B, Wesolowski D J, Dai S, Grinstaff M W and Gogotsi Y 2020 Electrode material–ionic liquid coupling for electrochemical energy storage *Nat. rev. Mater.* **5** 787–808

[64] Zhou Y, Ghaffari M, Lin M R, Xu H P, Xie H Q, Koo C M and Zhang Q M 2015 High performance supercapacitor under extremely low environmental temperature *RSC Adv.* **5** 71699–703

[65] Park S and Kim K 2017 Tetramethylammonium tetrafluoroborate: the smallest quaternary ammonium tetrafluoroborate salt for use in electrochemical double layer capacitors *J. Power Sources* **338** 129–35

[66] Chae J H and Chen G Z 2014 Influences of ions and temperature on performance of carbon nano-particulates in supercapacitors with neutral aqueous electrolytes *Particuology* **15** 9–17

[67] Yao B *et al* Printing porous carbon aerogels for low temperature supercapacitors *Nano Lett.* 2021**21** 3731–7

[68] Shi Z, Yu X, Wang J, Hu H and Wu C 2015 Excellent low temperature performance electrolyte of spiro-(1,1′)-bipyrrolidinium tetrafluoroborate by tunable mixtures solvents for electric double layer capacitor *Electrochim. Acta* **174** 215–20

[69] Dagousset L, Pognon G, Nguyen G T M, Vidal F, Jus S and Aubert P-H 2017 Electrochemical characterisations and ageing of ionic liquid/γ-butyrolactone mixtures as electrolytes for supercapacitor applications over a wide temperature range *J. Power Sources* **359** 242–9

[70] Dahbi M, Ghamouss F, Tran-Van F, Lemordant D and Anouti M 2011 Comparative study of EC/DMC LiTFSI and $LiPF_6$ electrolytes for electrochemical storage *J. Power Sources* **196** 9743–50

[71] Lang J, Zhang X, Liu L, Yang B, Yang J and Yan X 2019 Highly enhanced energy density of supercapacitors at extremely low temperatures *J. Power Sources* **423** 271–9

[72] Galiński M, Lewandowski A and Stępniak I 2006 Ionic liquids as electrolytes *Electrochim. Acta* **51** 5567–80

[73] Pohlmann S and Balducci A 2013 A new conducting salt for high voltage propylene carbonate-based electrochemical double layer capacitors *Electrochim. Acta* **110** 221–7

[74] Kunze M, Jeong S, Appetecchi G B, Schönhoff M, Winter M and Passerini S 2012 Mixtures of ionic liquids for low temperature electrolytes *Electrochim. Acta* **82** 69–74

[75] Xu Y, Lin W J, Gliege M, Gunckel R, Zhao Z, Yu H, Dai L L and Dual A 2018 Ionic liquid-based low-temperature electrolyte system *J. Phys. Chem.* B **122** 12077–86

[76] Yambou E P, Gorska B and Béguin F 2020 Binary mixtures of ionic liquids based on EMIm cation and fluorinated anions: physico-chemical characterization in view of their application as low-temperature electrolytes *J. Mol. Liq.* **298** 111959

[77] Abbas Q and Béguin F 2016 High voltage AC/AC electrochemical capacitor operating at low temperature in salt aqueous electrolyte *J. Power Sources* **318** 235–41

[78] Rong Q, Lei W, Huang J and Liu M 2018 Low temperature tolerant organohydrogel electrolytes for flexible solid-state supercapacitors *Adv. Energy Mater.* **8** 1801967

[79] Suo L, Borodin O, Gao T, Olguin M, Ho J, Fan X, Luo C, Wang C and Xu K 2015 'Water-in-salt' electrolyte enables high-voltage aqueous lithium-ion chemistries *Science* **350** 938–43

[80] Lannelongue P, Bouchal R, Mourad E, Bodin C, Olarte M, le Vot S, Favier F and Fontaine O 2018 'Water-in-salt' for supercapacitors: a compromise between voltage, power density, energy density and stability *J. Electrochem. Soc.* **165** A657–63

[81] Su L, Gong L, Lü H and Xü Q 2014 Enhanced low-temperature capacitance of MnO_2 nanorods in a redox-active electrolyte *J. Power Sources* **248** 212–7

[82] Yu H, Rouelle N, Qiu A, Oh J-A, Kempaiah D M, Whittle J D, Aakyiir M, Xing W and Ma J 2020 Hydrogen bonding-reinforced hydrogel electrolyte for flexible, robust, and all-in-one supercapacitor with excellent low-temperature tolerance *ACS Appl. Mater. Interfaces* **12** 37977–85

Chapter 7

Flexible electrochemical capacitors

Wang Yang, Wu Yang, Jiaming Zeng and Xinwen Peng

The development of portable and wearable electronics has increased the demand for flexible energy storage devices. Traditional supercapacitors (SCs) have experienced remarkable progress over the last several decades, but their poor flexibility and low safety limit their further development. Benefitting from durable mechanical flexibility and stretchability, flexible supercapacitors (FSCs) with excellent electrochemical performance and safety have attracted widespread attention in the fields of portable and wearable capacitive storage. More importantly, the further research and development of FSCs requires effective synergies between electrodes and electrolytes, including improvements in the capacity performance and cyclic stability of electrodes and the mechanical strength, ionic conductivity, and electrochemical stability of electrolytes. In this chapter, we present a comprehensive overview of the research significance and the main structures of FSCs. Furthermore, the recent advances in the electrode materials and electrolytes used in FSCs are systematically reviewed.

7.1 Introduction

Nowadays, the demand for multifunctional electronic devices is increasing considerably because of their diverse applications in our daily life. In order to improve the quality of life and accommodate fast-paced work, the portability and integration of electronic devices still needs further development. Therefore, wearable electronic devices such as flexible multifunctional bracelets, smart clothing, and electronic equipment used for medical monitoring have received extensive attention recently [1]. Although flexible electronic devices have unique advantages, the related products are not mature at present, and many technical problems urgently need to be solved. Among them, flexible energy storage equipment with high energy density is the key to improving the portability of flexible electronic devices.

SCs, also known as electrochemical capacitors, store energy through the adsorption/desorption of ions or rapid oxidation–reduction reactions on the

doi:10.1088/978-0-7503-5042-6ch7

© IOP Publishing Ltd 2023

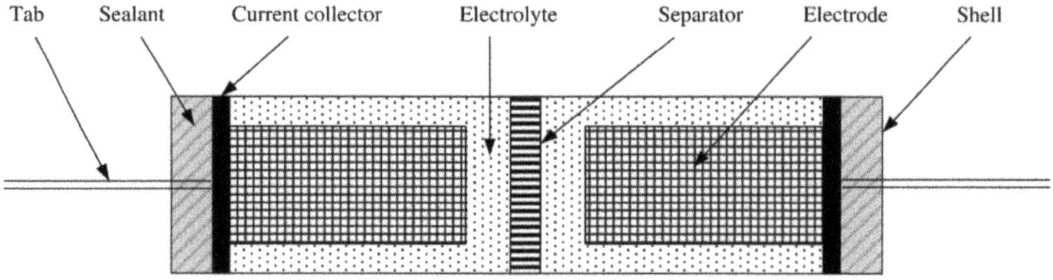

Figure 7.1. A basic structural model of an electrochemical double-layer capacitor (EDLC) [2], reproduced with permission © Elsevier.

electrode surface. SCs are usually composed of electrodes, an electrolyte, a current collector, and a separator [2], as shown in figure 7.1. Along with the development of flexible wearable electronic devices, solid-state flexible supercapacitors (SFSCs) have emerged at the historic moment at which solid electrolytes can replace the separator, preventing contact between the two electrodes. Electrode materials with excellent conductivity can not only provide abundant active sites, but also act as current collectors to simplify the device structure, reduce the device mass, and improve portability. Compared to traditional SCs, SFSCs possess the characteristics of small size, low weight, pliability, the ability to be woven, etc. which allow them to be easily integrated into electronic equipment. These unique properties make SFSCs promising alternatives for flexible electronic product energy storage devices.

This chapter mainly introduces the structure of FSCs and electrode and electrolyte materials, elucidates some advanced research results, and finally summarizes the existing problems and proposes some likely solutions.

7.2 The structures of flexible supercapacitors

The structures of SCs have a significant impact on their energy storage efficiency and flexibility. Generally, the structures of SFSCs can be divided into three types: three-dimensional (3D) laminated structures, two-dimensional (2D) interdigital structures, and one-dimensional (1D) linear structures.

7.2.1 Three-dimensional laminated structures

3D laminated structures, also known as sandwich structures, are traditional SC structures. They offer several structural advantages, as follows [3]: (1) sufficient contact between the electrode material and electrolyte in the electrochemical reaction process; (2) rapid charge transfer between the current collector and the active material; and (3) high active material loading, leading to high energy density and power density. However, the large volumes of 3D laminated SCs limit their practical application because their size and flexibility are also limited to a certain extent; they are thus incapable of twisting or knotting. Therefore, the microstructures of electrodes prepared by different strategies also have a dramatic influence on the electrochemical performance. Zang *et al* [4] reported a gas-phase method that was used to prepare flexible electrodes by depositing polypyrrole (PPy)

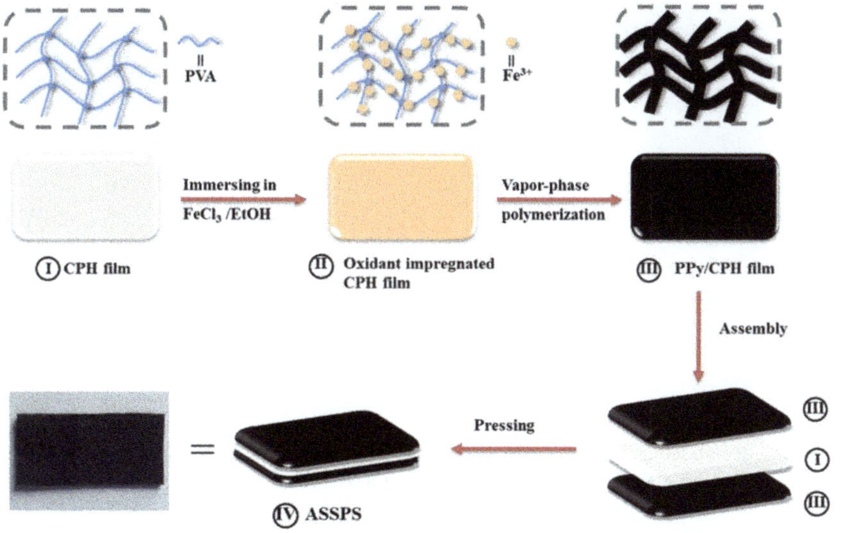

Figure 7.2. A schematic illustration of the synthetic process of the flexible all-solid-state polymer SCs [4], reproduced with permission © American Chemical Society.

on a PVA/H_2SO_4 gel substrate. SFSCs were fabricated by sandwiching a PVA/H_2SO_4 gel electrolyte between two electrodes, as illustrated in figure 7.2. As a result, the assembled SFSCs showed good electrochemical and mechanical properties: the tensile strength reached 20.83 MPa, elongation reached 377% at the breaking point, the volume specific capacitance reached 13.06 F cm^{-3}, and the capacity retention rate reached 97% after 10 000 cycles.

7.2.2 Two-dimensional interdigital structures

SCs that have interdigital structures are also called 2D coplanar SCs; they have the advantages of small size, lighter weight, and easy integration compared with 3D laminated SCs [5]. However, their lower active material loading results in low energy density and specific capacity. Their main structure contains multiple pairs of electrodes uniformly distributed on the same flexible substrate to form an interdigital structure, which is then coated with an electrolyte and finally encapsulated by a flexible material. Notably, such SCs are more suitable for integration into microelectronic devices than other types of SC owing to their unique planar configuration and their lighter, thinner, and more compact structure. Several common techniques are used for the preparation of SFSCs, including laser engraving [6], 3D printing [7], coating [8], screen printing [9], etc.

The size of an SC with an interdigital structure is much larger in the 2D direction than in the 1D direction, and its shape is closer to a planar shape. In addition, it is easier to bend and deform at a large angle. Therefore, SCs with interdigital structures are mainly applied in electronic products requiring high flexibility, such as multifunction bracelets, electronic skins, and implantable medical devices. Moreover, SCs with interdigital structures have the advantage of easy integration,

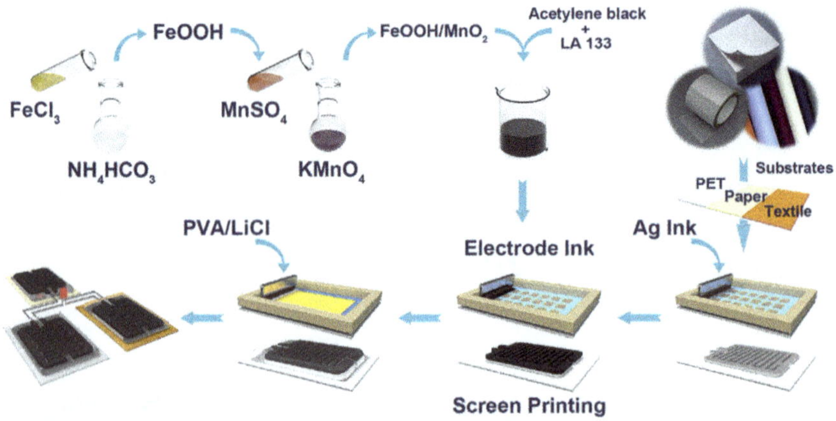

Figure 7.3. A schematic illustration of the fabrication processes used by Lu *et al* to produce SCs with interdigital structures [9], reproduced with permission © Elsevier.

which can be arranged according to the requirements of microelectronic devices, and their capacity can be effectively increased through a series–parallel structure. Lu *et al* [9] demonstrated the use of screen-printing technology to fabricate SFSCs; their preparation process for SFSCs is displayed in figure 7.3. The amorphous FeOOH/MnO_2 employed as the electrode material delivered a high specific capacity of 159.5 F g^{-1} at a current density of 20 A g^{-1} and a 95.6% capacity retention rate after 10 000 cycles. Notably, the assembled SFSCs exhibited high flexibility and were able to work as normal in stretched and bent states.

7.2.3 One-dimensional linear structures

The complex application scenarios of wearable electronic equipment make the requirements for energy storage devices more stringent; these include good capacity retention and electrochemical stability under large-scale deformation (bending, folding, knotting, etc) [10]. In this regard, linear SCs possess outstanding flexibility which is suitable for bionic electronic products and wearable electronic products. In addition, linear SCs can not only bend on a large scale, but can also be twisted at multiple angles (knotting and deformation). Although linear SCs have excellent flexibility owing to their small lateral size, the low active material loading and utilization rate result in poor cyclic performance and unsatisfactory capacity [11]. As depicted in figure 7.4, there are four common structures for linear SCs, namely parallel, twisted, coaxial, and consecutive [12]. The preparation process used for SCs with a parallel arrangement is complicated and requires an additional support substrate [13, 14]. Although the preparation process of twisted SCs is facile, their capacity and efficiency are not high [15]. Coaxial SCs have higher efficiency and a simple preparation process, but their unique skin–core structure requires higher preparation technology [16, 17]. Although consecutive SCs are safer, their efficiency is not high [18]. Different types of SC have their own advantages and disadvantages, and the appropriate structure can be employed according to the actual requirements of the application.

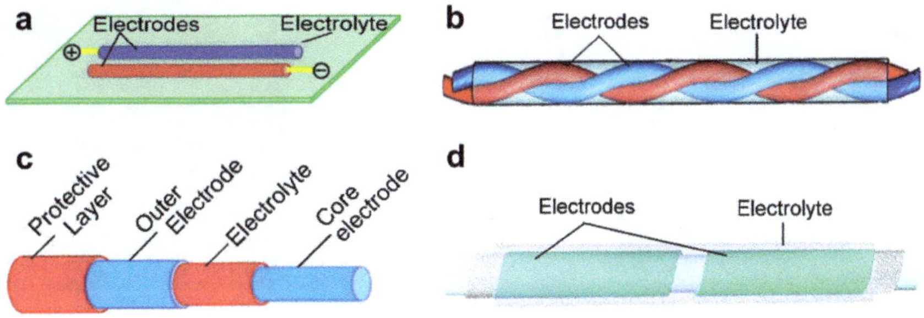

Figure 7.4. Schematic diagrams showing the different structures used for FSCs: (a) parallel, (b) twisted, (c) coaxial, and (d) consecutive [12], reproduced with permission © Elsevier.

7.3 Solid-state flexible supercapacitor materials

SFSCs are composed of flexible substrates, current collectors, electrodes, and electrolytes [19, 20]. Generally, flexible substrate materials, including polyethylene terephthalate (PET) [21], polysiloxane [22], and fabrics have good flexibility and can be bent arbitrarily [23, 24]. High conductivity is a necessary condition for current collectors, thus the commonly used current collectors are made of materials such as Cu, Ni, Ag, and carbon cloth [25]. The electrode materials and electrolyte materials have a greater impact on the performance of SFSCs.

7.3.1 Electrode materials

7.3.1.1 Carbon-based materials

Carbon-based materials, such as activated carbon (AC), carbon nanotubes (CNTs), graphene, carbon fibers (CFs), and their composite materials, have attracted increasing attention due to their large surface area, excellent electrical conductivity, high stability, and good mechanical properties [26–30]. Graphene is a very popular carbon material due to its good electrical conductivity and huge specific surface area [31]. Weng *et al* [32] reported a simple and scalable method for the fabrication of graphene–cellulose paper (GCP) membranes (figure 7.5). A graphene suspension was used to attach the graphene sheet to filter paper fibers, and the filter paper was used as the electrode material of SCs. The specific capacitance of the GCP electrode reached 120 $F{\cdot}g^{-1}$. The assembled SFSCs maintained stable capacitance performance under conditions of high flexure, and the specific capacitance reached as much as 46 $mF{\cdot}cm^{-2}$. When three of these devices were connected in series, the energy density reached 15 $\mu Wh{\cdot}cm^{-2}$. The preparation method used for the electrode material was simple and conducive to large-scale preparation. However, the graphene sheets were randomly attached to the filter paper fibers during the filtration process, which inevitably caused the agglomeration and stacking of the graphene sheets. As a result, when the fibers were used as an electrode material in SCs, the ion transport channel was blocked, which hindered the transfer of ions, thereby reducing the power performance of the SCs. Furthermore, the contact area between the electrode material and the electrolyte was also greatly reduced, thereby reducing the

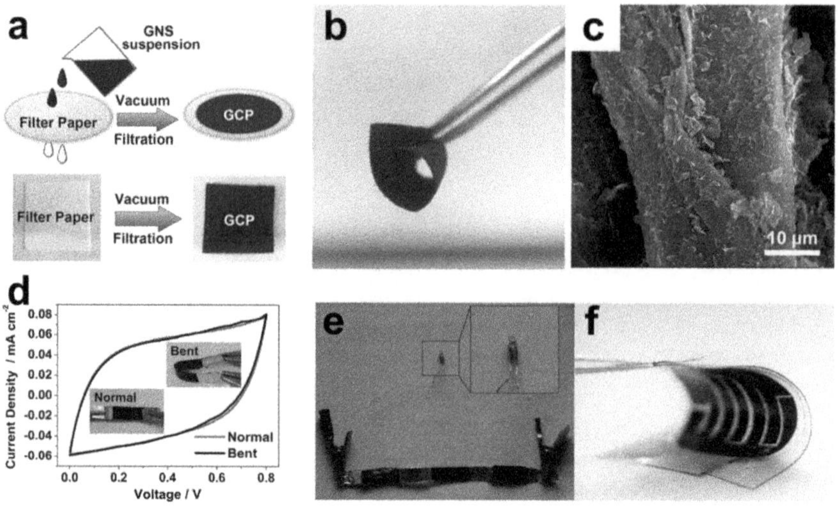

Figure 7.5. (a) A schematic illustration and (b) a photograph of the GCP membrane, (c) a scanning electron microscopy (SEM) image of the GCP membrane, (d) electrochemical stability performance testing for a flexible laminated poly-SC tested under normal and bent conditions, (e) a photograph of a red LED lit by an in-series poly-SC consisting of three units, (f) a photograph of a flexible interdigital GCP poly-SC [32], reproduced with permission © John Wiley & Sons.

active sites available for ion adsorption and lowering the specific capacitance and energy density.

Mask-assisted printing technology has also been used to prepare FSCs employing graphene nanoscrolls (GNSs) as active materials [33], which had excellent electrochemical performance and mechanical stability. Related research was reported by Jiang *et al* [34]. The high-aspect-ratio GNS played an important role in enhancing the mechanical strength of the film electrode and enhancing the stability of SCs [35]. As shown in figure 7.6, the wrinkles on the surface of the GNS gradually became flat with increasing tensile strength. Even in the 200% stretched state, no fracture occurred on the GNS film. The noodle-like microstructure of the GNS can clearly be observed on the SEM image, which shows that the length of a single GNS is greater than 100 μm and its diameter ranges from 100 to 1000 μm. In addition, a small symmetrical SC was assembled using GNS as the electrode and poly(vinyl alcohol) (PVA)/H_2SO_4 gel as the electrolyte. The assembled GNS microsupercapacitors (denoted as GNS-MSC) effectively prevented electrolyte leakage, and the absence of an additional current collector reduced the overall weight of the device. The cyclic voltammetry (CV) curves of the SC retained almost rectangular shapes, confirming good capacitive performance. The constant current charge/discharge curves at different current densities at 25–250 C resembled ideal triangles with good symmetry. Calculations indicated that the capacitances were 10.8 and 4.4 mF cm^{-2} at current densities of 0.025 and 1 mA cm^{-2}, respectively. The mechanical properties were verified by CV and galvanostatic charge/discharge (GCD) tests at different stretch rates (0%–200%). It can be seen that the CV and GCD curves did

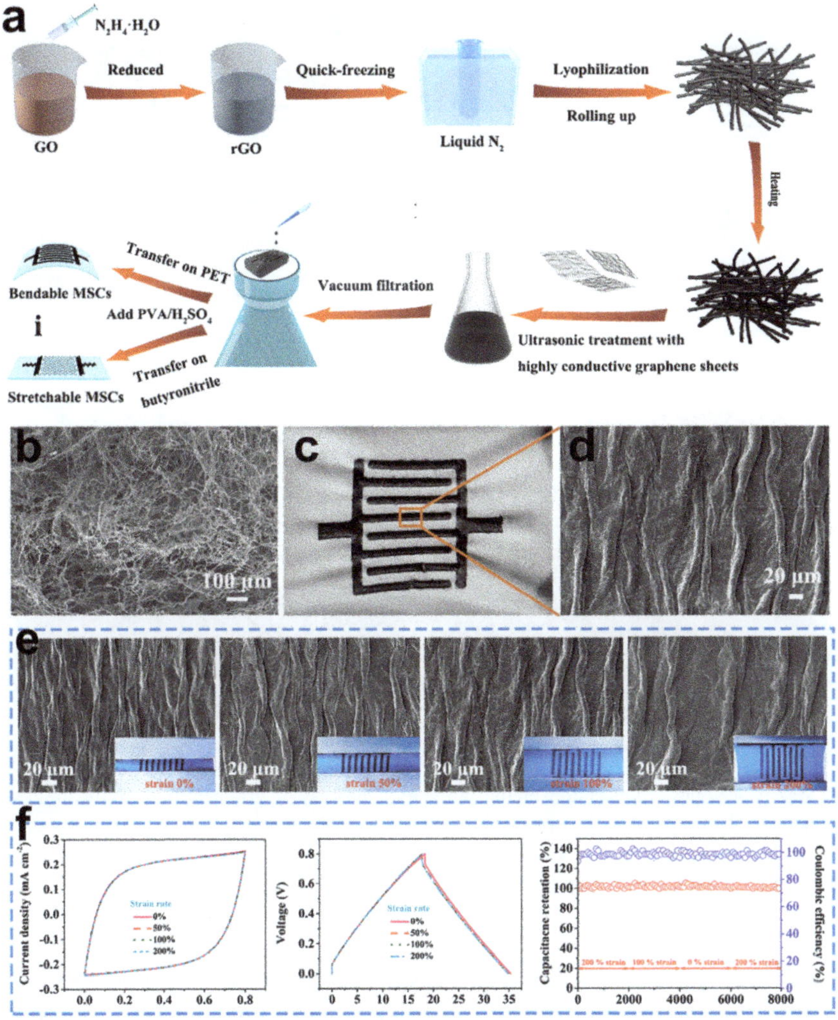

Figure 7.6. (a) A schematic diagram of the preparation processes of GNS and GNS-MSCs, (b) an SEM image of GNS, (c) a digital photograph of the stretchable film electrode, (d) an SEM image of the stretchable L-GNS film electrode, (e) SEM images of GNS-MSCs at different strain ratios, and (f) electrochemical performances of GNS-MSCs at different stretching states [34], reproduced with permission © Elsevier.

not change significantly at different stretching rates. The GNS-MSC device was cycled 2000 times at four different stretching rates (for a total of 8000 cycles). Finally, the cyclic stability of the SC was measured at a constant current density of 0.5 mA cm^{-2}, and its final capacitance retention rate was still close to 100%.

In the continuous development of graphene electrode materials, a breakthrough has been made which allows graphene to fabricated by the laser reduction of graphite oxide (GO). Kaner *et al* [36] demonstrated that a laser could be used to reduce a GO film to obtain graphene film with a specific surface area of 1520 m^2 g^{-1} (figure 7.7(a)). The film was used as a SC electrode, delivering a high areal capacity

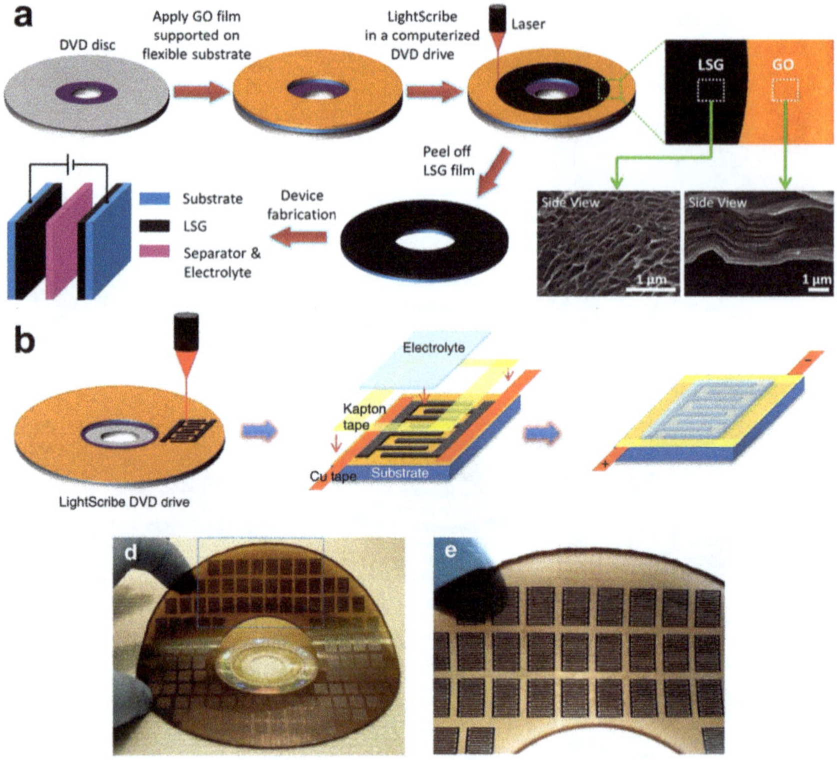

Figure 7.7. (a) A schematic illustration of the fabrication process of graphene film using the laser-scribed reduction method. Reproduced with permission from reference [36]; copyright (2012), American Association for the Advancement of Science. (b) A schematic illustration of the fabrication process and a photograph of the laser-scribed graphene microsupercapacitors (denoted as LSG-MSC) with an interdigital structure [37], reproduced with permission © Springer Nature.

of 3.67 mF cm^{-2}. Based on the above result, Kaner *et al* [37] used the same laser reduction method to prepare numerous microinterdigital SCs on a graphite oxide film under the precise control of a computer program (figure 7.7(b)). However, the entire process was time-consuming. Within 30 min, the power density of the obtained miniature SCs reached 200 W·cm^{-3}. This method is also suitable for the preparation of microelectronic devices. The use of a laser to reduce GO and obtain graphene is novel, simple, and fast, and is well suited to the mass production of devices. However, this method requires a lot of equipment, and presents major technical difficulties related to debugging the equipment and setting operational parameters.

CNT yarns were fabricated by directly drawing CNTs from superaligned CNT arrays using chemical vapor deposition (CVD) technology, as shown in figures 7.8(a)–(c) [38]. The CNT yarns thus obtained consisted of CNTs in a nearly parallel alignment with the long axis in the drawing direction. These 1D CNT yarns exhibited an improved mechanical strength of 575 MPa and an electrical conductivity of 300 S cm^{-1}. The morphology of the 1D CNTs could be also optimized by

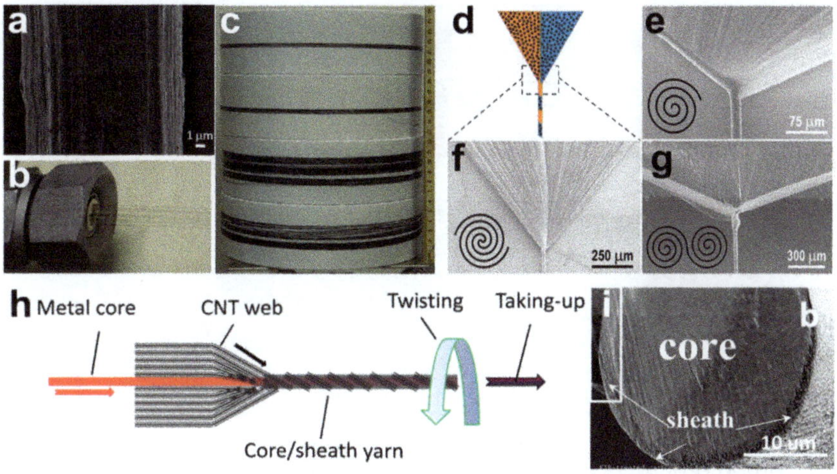

Figure 7.8. (a) An SEM image of the CNT fibers, (b) the direct spinning process used for the CNTs, (c) winding drums with collected fibers [38], © American Association for the Advancement of Science. (d) An illustration of biscrolling by twist insertion in a spinning wedge: CNT fiber with a twist-dependent extension of Archimedean spirals (e), a Fermat-type twist (f), and dual Archimedean scrolls (g) [39], © American Association for the Advancement of Science. (h) A schematic illustration of the formation of the core–sheath yarn structure, (i) a cross-section of the Pt/CNT/polyaniline (PANI) core–sheath yarn [40], reproduced with permission © American Chemical Society.

modulating the twisting process to form spiral pairs and scrolls (figures 7.8(d)–(g)) [39]. The formation of scrolls increased the strength and electronic connectivity of the 1D yarns. In order to greatly improve the strength and electronic connectivity, a metal core was introduced into the 1D fibers to form a core–shell structure, as depicted in figure 7.8(h) [40]. A 1D CNT rope with a compact structure was prepared by passing a CNT yarn through a drawing die or solvent. The 1D CNT fibers thus obtained could easily be bent, folded, twisted, or tailored and were thus suitable for application in flexible electronics.

To address the above issues, the assembly process of the CNT film was further optimized. Based on the floating-catalyst chemical vapor deposition (FCCVD) method, large-scale CNT films were prepared through restrained assembly as shown in figure 7.9(a) [41]. A CNT sock-like aerogel was first formed and then pulled out from the furnace. A densified CNT film was obtained by shrinking the CNT aerogel with ethanol. The CNT films exhibited high shape conformability and homogeneity with excellent flexibility, strength, and conductivity.

Wu's group conducted a series of studies of CNT films for use in flexible energy storage devices [42]. The efficiency of CNT preparation can be increased by adding CVD reaction tubes. Three-dimensional CNT sponges, in which CNTs were self-assembled into a 3D interconnected framework, were first synthesized using CVD (figure 7.9(b)) [43]. These sponges were very light, highly porous, hydrophobic in their pristine state and could be elastically and reversibly deformed into any shape. These intriguing properties originated from the isotropic porous CNT network, which was very flexible and able to sustain a large degree of deformation.

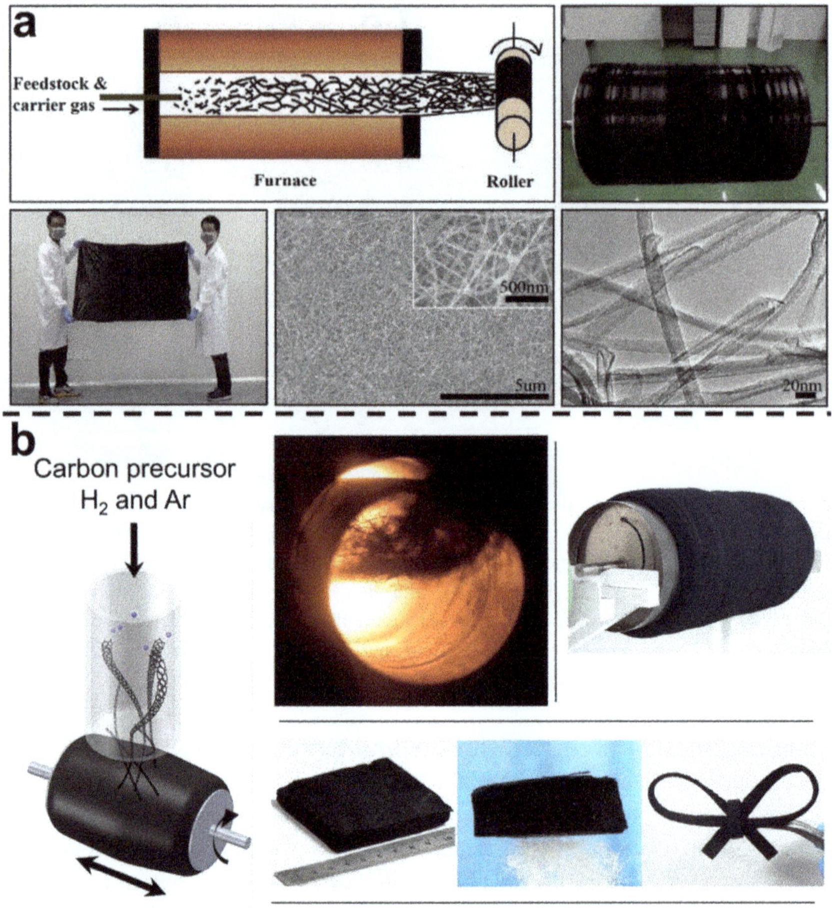

Figure 7.9. (a) A schematic of CNT film fabrication, a CNT sponge wound on a collecting roller, and the morphology of the FCCVD CNT film [41], reproduced with permission © Elsevier. (b) The fabrication process and photographs of joint-welded CNT foam [43], reproduced with permission © John Wiley & Sons.

7.3.1.2 Metal oxides

Pseudocapacitor electrode materials are usually made from metal oxides such as MnO_2, NiO, Co_2O_3, etc. Generally, metal oxides can store and release electricity by means of reversible redox reactions and insertion/extraction into/from the crystal lattices or gaps of metal oxides. The main factors affecting their electrochemical performance are their crystal form, valence, structure, and composition. In addition, metal oxide electrodes have the following shortcomings: (1) because metal oxides are ionic crystals with a strong internal ability to bind electrons, the ability of charges to move freely within them is relatively poor, resulting in low conductivity. (2) Due to the presence of numerous defects and the complex crystal structures of metal oxides, their pore distribution and specific surface area are not easy to control. (3) Since the redox reaction of the electrode is irreversible, metal oxide electrode has poor stability in long-term cycling. A previous report confirmed that the cyclic stability can be enhanced by improving the microstructure of metal oxides using, for example, a core–shell structure [44].

Ghorbani *et al* [45] designed and constructed a distinctive flexible electrode based on freestanding sandwich-type ZnO/rGO/ZnO paper via a simple low-cost sol–gel method for use in FSCs. A schematic illustration of the synthesis process is shown in figure 7.10. When freestanding GO paper was used as a new substrate for the deposition of ZnO films, the resulting ZnO/rGO/ZnO paper demonstrated great flexibility. ZnO layers were deposited on both sides of rGO paper which was prepared using a modified Hummer's method and evaporation-induced assembly. The ZnO/rGO/ZnO paper showed a dense layer with an unchanged layered structure of GO paper, which were attributed to the ZnO film deposited on the GO paper. The ZnO/rGO/ZnO flexible electrode exhibited a satisfactory capacitance of 60.63 F g^{-1} at scan rate of 5 mV s^{-1}. In addition, an SFSCs device, whose electrodes were prepared via a facile oxidation-assisted dealloying protocol used for the first time, demonstrated considerable flexibility and excellent electrochemical performance. A prototype all-solid-state asymmetric SC was assembled by Wang *et al* using the prepared Co_3O_4 flakes and γ-Fe_2O_3 nanoparticles as the positive and negative electrodes, respectively [46]. Figure 7.11 schematically reveals the fabrication process used. The flexible asymmetric supercapacitor (ASC) device possessed an extended operating voltage window of 1.7 V and a high energy density of 38.1 Wh kg^{-1}. The ASC device exhibited excellent power density and energy density and retained good charge and discharge performances under different bending conditions.

A straightforward worthwhile approach was adopted in another study for the manufacture of an interdigitated planar all-solid-state flexible microsupercapacitor [47]. The abovementioned procedure used commercial toner in a commercial laser printer with a polyethylene terephthalate sheet. The structural water concentration of hydrous ruthenium oxide was optimized at a low annealing temperature (100 °C). The painstaking design of mass-produced, micro-patterned prototype devices could be revolutionized by the use of this undemanding patterning method. The new

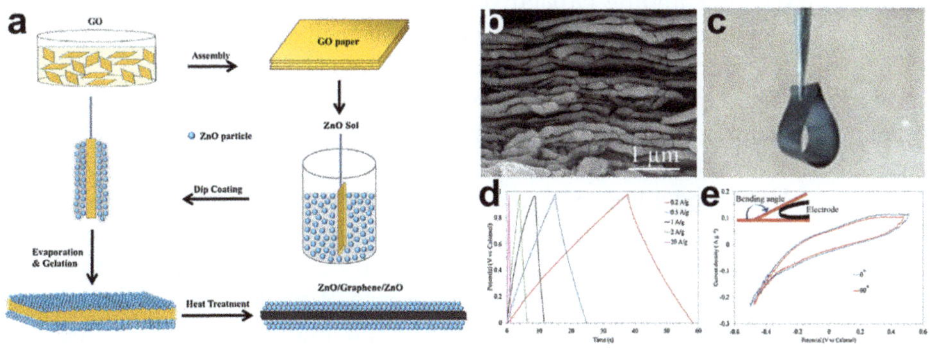

Figure 7.10. (a) A schematic illustration of the synthesis of sandwiched ZnO/rGO/ZnO paper by a sol–gel dip-coating method, (b) an SEM image of the ZnO/rGO/ZnO paper, (c) a ZnO/rGO/ZnO paper strip is bent to demonstrate its flexibility, (d) GCD curves of the ZnO/rGO/ZnO, and (e) CV curves of the ZnO/rGO/ZnO with and without bending through 90° [45], reproduced with permission © Elsevier.

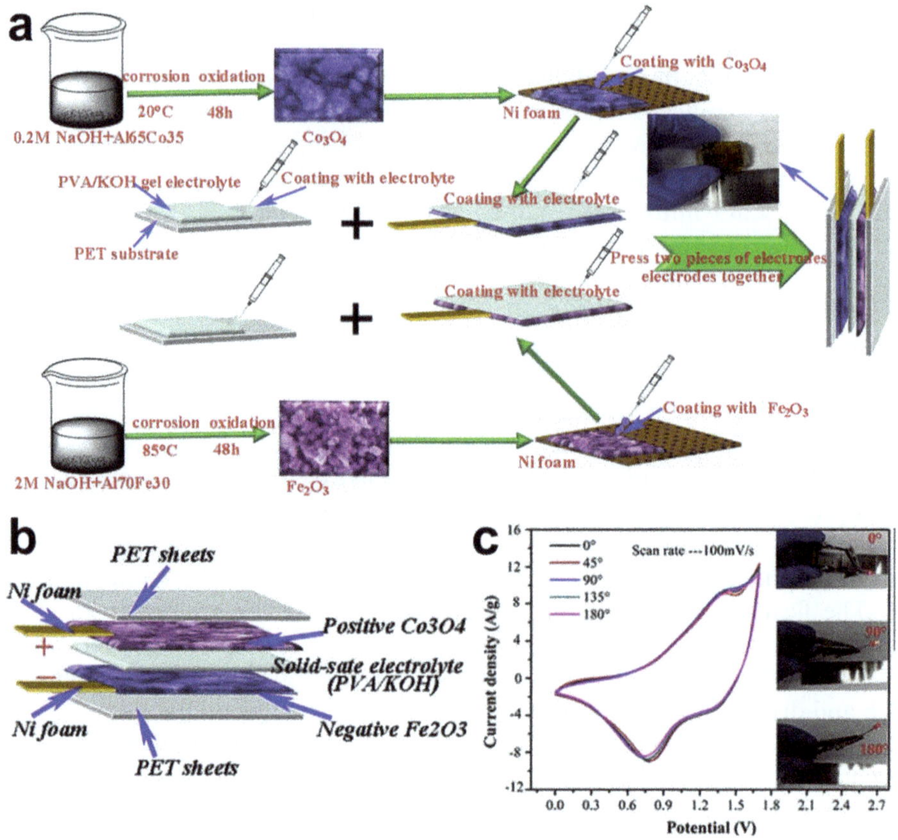

Figure 7.11. (a) A schematic illustration of the fabrication of Co_3O_4//γ-Fe_2O_3 SFSCs, (b) a schematic illustration of Co_3O_4//γ-Fe_2O_3 SFSCs, and (c) CV curves collected for the Co_3O_4//γ-Fe_2O_3 ASC under different bending conditions [46], reproduced with permission © Elsevier.

method achieved outstanding capacitive quality as well as an extraordinarily high volumetric energy density of 40.3 mWh cm^{-3} at a high volumetric power density of 1147.8 mW cm^{-3}, only reducing to 24.9 mWh cm^{-3} 24.9 mWh cm^{-3} at a high volumetric power density of 73 460 mW cm^{-3}. This result can be ascribed to the excellent characteristics of the resulting hydrous ruthenium-oxide-based μSCs (figure 7.12).

Huang *et al* [48] used a hydrothermal reaction to prepare vertically grown MnO_2 nanosheets on a titanium sheet (figures 7.13(a) and (b)) and used glucose as a carbon source to improve the conductivity of the electrode material. When the current density was 0.25 mA·cm^{-2}, the specific capacitance of the flexible electrode reached 143 mF·cm $^{-2}$. The preparation method was simple and highly operable because it did not require high equipment. Yuan *et al* [49] reported $Co(OH)_2$ nanosheet arrays uniformly grown on foam nickel by electrochemical deposition and then calcined to obtain mesoporous Co_3O_4 nanosheet arrays (figures 7.13(c)–(e)). When used in an SC, a Co_3O_4 nanosheet array electrode delivered a high specific capacitance of

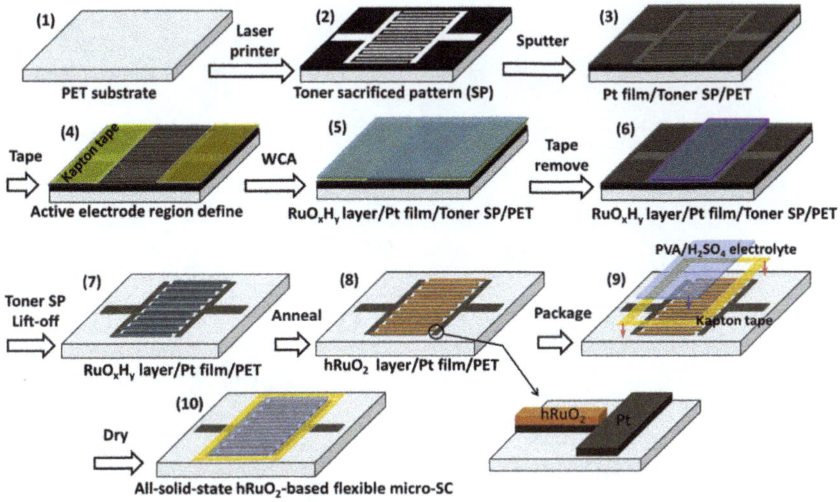

Figure 7.12. A schematic of the fabrication process used to produce hRuO$_2$-based flexible μSCs on a PET-sheet substrate [47], reproduced with permission © Elsevier.

2735 F·g^{-1}. This unique 3D electrode manifested exceptional supercapacitive performance with excellent cyclic stability. Zheng *et al* [50] prepared Ni(OH)$_2$ by a hydrothermal method and then converted it into NiO through an annealing process. The specific capacitance of the electrode material was 137.7 F·g^{-1}, and its specific capacitance dropped from 138.2 F·g^{-1} to 126.1 F·g^{-1} after 1000 cycles (the specific capacitance retention rate was 91.24%). The specific capacitance of the electrode material was mainly provided by NiO, which has poor conductivity. Therefore, the specific capacitance of the electrode material was still low. In addition, oxidation–reduction reactions continuously occurred on the surface and in the interior of the NiO during the charging and discharging processes, which consumed the electrode material, resulting in poor cyclic stability. After a certain number of cycles, the specific capacitance of the electrode material decreased significantly.

Yu *et al* [51] prepared a coaxial SC using copper wire as the electrode substrate and MnO$_2$ as the electrode material. An array of copper oxide nanowires clearly appeared on the surface as a result of high-temperature treatment, and then a layer of MnO$_2$ active material was uniformly deposited on the surface of the copper wires to obtain an internal electrode material with a large specific surface area and pseudocapacitance (figure 7.14(a)). The preparation method used for the outer electrode was exactly the same as that of the inner electrode, except that the copper wire was replaced by copper foil. The inner and outer electrodes were assembled to form a coaxial SC. The specific capacitance of this device reached 1376 F·g^{-1}, and its power and energy densities were 413 mW·cm^{-3} and 0.55 mWh·cm^{-3}, respectively. At the same time, the coaxial SC also showed better mechanical properties and electrochemical stability, which could maintain better energy storage performance after different degrees of bending and different bending times.

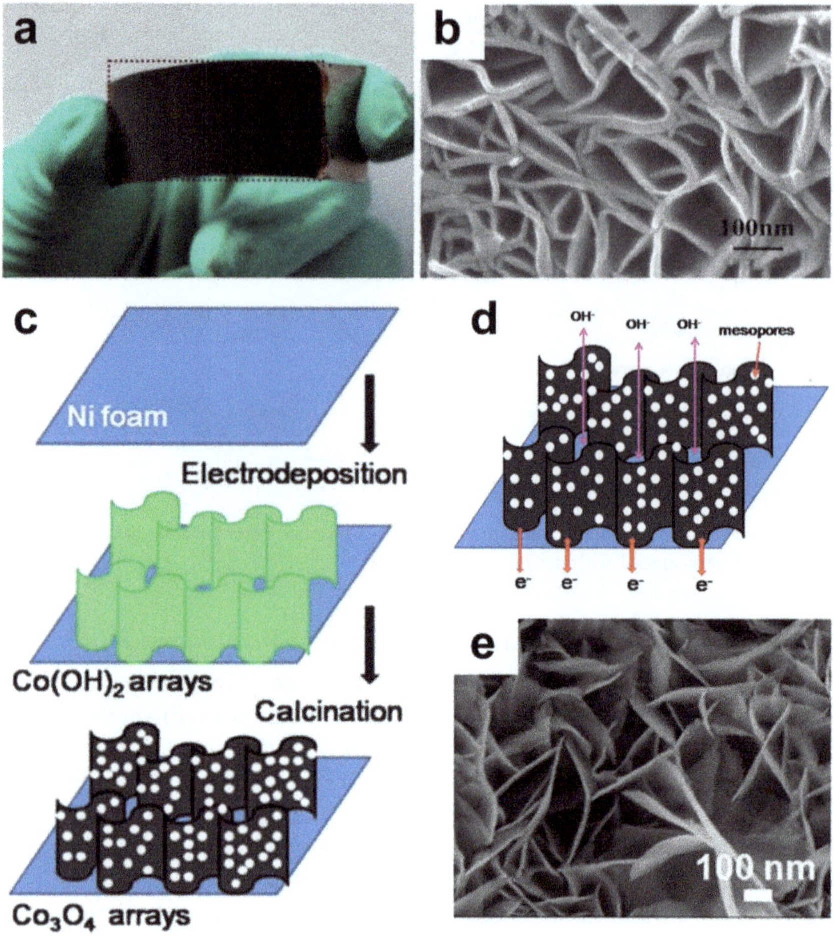

Figure 7.13. (a) An optical image of the pristine MnO_2 film supported on Ti foil, (b) an SEM image of the MnO_2 nanosheet array [48], reproduced with permission © The Royal Society of Chemistry. Schematic illustrations of (c) the general electrode design process and (d) the application advantages of Co_3O_4 ultrathin nanosheet arrays on Ni foam. (e) SEM images of the Co_3O_4 nanosheet nanoarrays supported on Ni foam [49], reproduced with permission © The Royal Society of Chemistry.

Xu *et al* [52] used gold wire as an electrode substrate with MnO_2 and carbon nanotubes as electrode materials. A novel all-solid-state, highly flexible, coaxial fiber ASC was made by scrolling a CNT/carbon paper electrode onto a flexible nanoporous gold@MnO_2 (NPG@MnO_2) wire electrode with a polymer gel. The researchers used an alloying/dealloying treatment on the gold wire to make a porous structure on the surface of the gold wire, providing more attachment sites for the active material, and then electrochemically deposited MnO_2 nanoflowers onto the obtained nanoporous gold wire, which was used as the inner electrode of an SC (figure 7.15). The NPG wire had high flexibility and satisfactory tensile strength and was used as the backbone to support the MnO_2 nanoflowers. The hierarchical nanostructures and coaxial

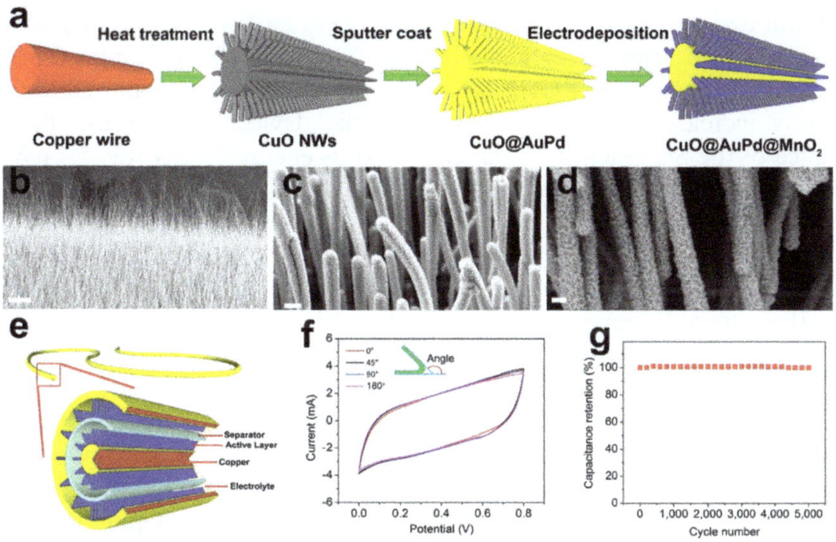

Figure 7.14. (a) A schematic illustration of the fabrication process of the CuO@AuPd@MnO_2 nanowire array. (b) SEM images of vertically grown CuO nanowires, (c) CuO@AuPd nanowires, and (d) CuO@AuPd@MnO_2 nanowires. (e) A schematic illustration of the coaxial SC based on a CuO@AuPd@MnO_2 electrode. (f) Subtle changes in the CV curves at different bending angles ranging from 0° to 180°. (g) The cyclic performance of the device over 5000 cycles [51], reproduced with permission © John Wiley & Sons.

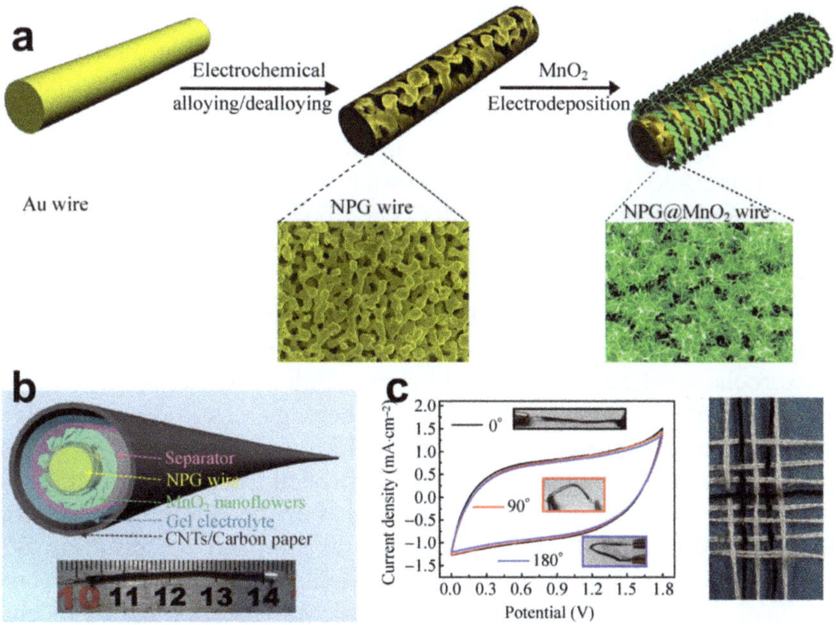

Figure 7.15. (a) A schematic illustration of the synthesis and morphology of the NPG@MnO_2 electrode, (b) a schematic illustration and a photograph of the coaxial asymmetric SC, (c) CV curves of the ASC device in different bending states [52], reproduced with permission © Springer Nature.

architectural design facilitated effective contact between the two core–sheath electrodes and the active layers with high flexibility and high performance. The biggest advantage of this type of SC is the range of its potential window, which also significantly increases the energy density of the device. As a result, the specific capacitance of the SC reached 12 mF·cm^{-2} and its energy density was 5.4 μWh·cm^{-2}. Moreover, the coaxial fiber device exhibited good cyclic durability, retaining 90% of its capacitance after 2000 cycles.

7.3.1.3 Conductive polymers

Due to their tunable conductivity and unique doping characteristics, conductive-polymer-based FSCs have been widely investigated in recent years [53]. Compared with metal oxides, conductive polymers have the advantages of easy synthesis, low density, and good flexibility. The most commonly used conductive polymers are PANI, poly(3,4-ethylenedioxythiophene) (PEDOT), polypyrrole (PPy), polythiophene, polyanthraquinone, etc. Conductive polymers can realize electrical storage and release through fast reversible redox reactions based on the doping/de-doping of n-type and p-type elements. However, the repeated insertion/extraction of electrons or ions into/from the polymer framework damages the polymer structure, resulting in serious degradation of its electrochemical performance and a significant decrease in its mechanical strength [54]. Therefore, researchers have improved the mechanical and electrochemical properties of electrode materials by doping metal oxide nanoparticles into conductive polymers and improving the preparation processes of conductive polymers [55].

PANI is a polymer material that has been studied extensively and thoroughly in conductive polymers. Chiou *et al* [56] used a chemical polymerization method to grow a neat array of polyaniline nanofibers on a flexible substrate in 2007 (figures 7.16(a) and (b)), which provided a huge specific surface area. The tip diameters of the PANI nanofibers were controlled within the range of 10–40 nm, and their average length was controlled within the range 70–360 nm. Wei *et al* [57] employed this method to prepare PANI nanofibers for use as SC electrode materials and obtained a series of research results. The abovementioned chemical polymerization method was used to uniformly grow graphene oxide nanosheets on a PANI nanofiber array (figures 7.16(c)–(e)). The composite material delivered a specific capacitance of 555 F·g^{-1}. The electrochemical results verified that the PANI-GO nanocomposite benefited from the synergistic effect of PANI and GO, exhibiting higher electrochemical capacitance and better stability. At the same time, Wei *et al* [58] demonstrated the use of PANI nanofibers as electrode materials in the preparation of microinterdigital FSCs through a combined microfabrication technology and *in situ* chemical polymerization approach (figure 7.17) [59]. The results indicated that the introduction of PANI nanofiber arrays as electrode materials and the selection of appropriate electrode substrates supplied the resulting SCs with higher energy density and considerable flexibility. The FSCs based on PANI nanowire arrays exhibited a superior volumetric capacitance of 588 F cm^{-3} and good rate capability. Subsequently, they also successfully grew PANI nanofibers in a 3D graphene film and obtained a flexible composite film electrode (figure 7.18) [60]. The specific capacitance of the composite film electrode reached 385 F·g^{-1}. The

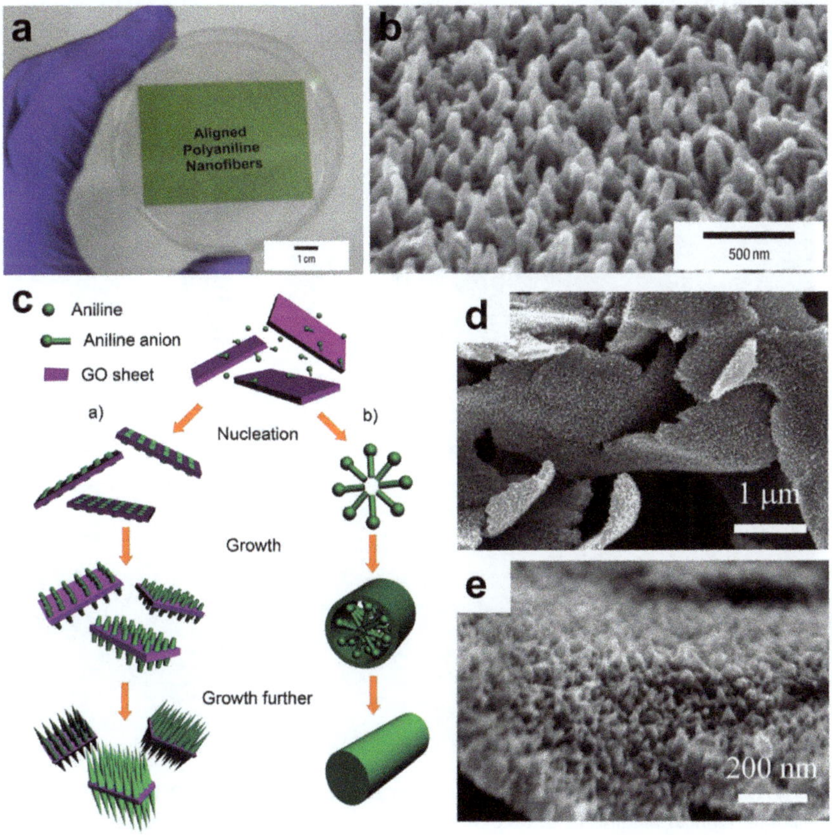

Figure 7.16. (a and b) A photograph and an SEM image of the PANI nanofiber film [56], reproduced with permission © Springer Nature. (c) A schematic illustration of the nucleation and growth mechanism of PANI nanowires. (d and e) SEM images of a PANI-GO sample [57], reproduced with permission © American Chemical Society.

assembled FSCs still had stable electrochemical performance in different bending states. In addition, PANI can also be used as an electrochromic material. Therefore, PANI-based materials are beneficial in the preparation of multifunctional electronic devices. Wei *et al* [61] used PANI nanofiber translucent film electrodes produced by chemical polymerization to assemble and form a multifunctional flexible device with energy storage, which integrated energy storage and electrochromism functions in one device. The preparation of material for multifunctional flexible devices provides an important reference.

7.3.1.4 Composite materials

EDLCs based on carbon materials have the advantages of good cyclic stability and high electrical conductivity, but their energy storage performance is relatively poor. The capacitance of pseudocapacitor materials, such as metal oxides and conductive polymers, is mainly produced by a redox reaction between the electrolyte and the active material, which possesses high charge storage capacity and excellent energy

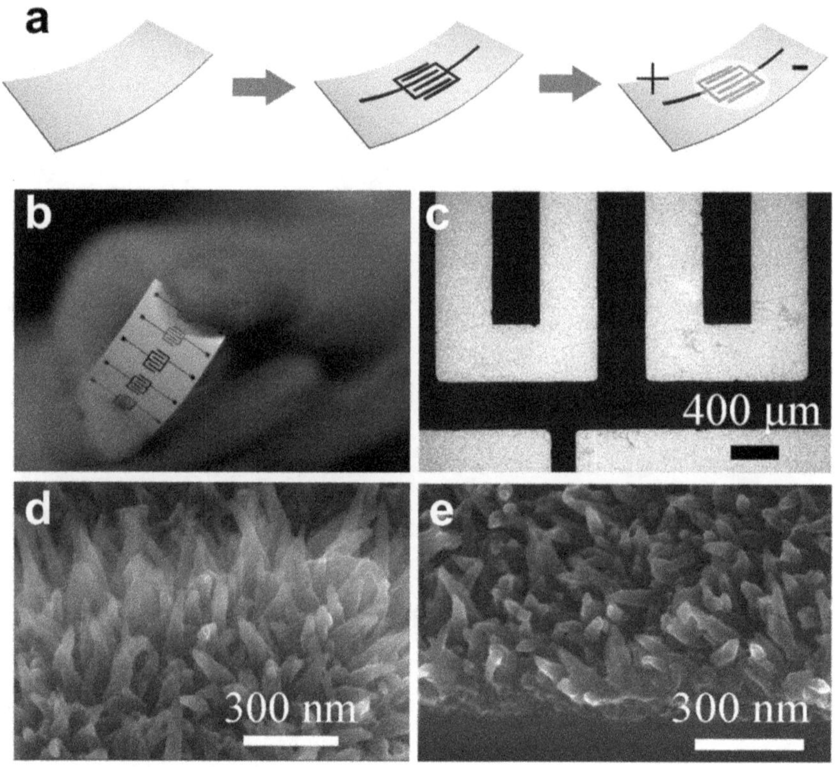

Figure 7.17. (a) A schematic image of the fabrication process used to produce PANI nanowire arrays based on MSC flexible film. Optical pictures of flexible MSC unit arrays on (b) PET film and (c) MSC 400. (d and e) SEM images of PANI nanowire arrays [58], reproduced with permission © John Wiley & Sons.

storage performance. Therefore, pseudocapacitor materials and double-layer capacitor materials are usually integrated into a single electrode at the same time, and SCs are fabricated to effectively improve the energy density and maintain cyclic stability [62–67]. Zheng *et al* [68] investigated a sandwiched carbon nanotube (CNT)/nickel cobaltate ($NiCo_2O_4$) hybrid paper electrode synthesized by a hydrothermal method using CNT paper as a raw material (figure 7.19(a)). $NiCo_2O_4$ nanosheets grew vertically on the CNTs to form a honeycomb structure with a pore size of a few microns, as illustrated by an SEM image (figure 7.19(e)). The ultrathin $NiCo_2O_4$ nanosheets were not only well interconnected with each other but also penetrated into the CNT network, thus forming a well-connected interface and electrical contact between the CNT and $NiCo_2O_4$ nanosheets. CV curves of this material obtained at different bending angles from 0 to 180° almost overlap (figure 7.19(f)), indicating that it has good flexibility, and no obvious redox peaks are visible. The CV curves also indicate that the device has good capacitive performance. The triangular GCD curves obtained under different conditions (figure 7.19(g)) also support this explanation.

Figure 7.20 shows the bending properties of a treated CNT paper layered with MnO_2 nanoparticles [10]. This arrangement served as a positive electrode and iron oxide (Fe_2O_3) functionalized on carbon fiber served as the negative material.

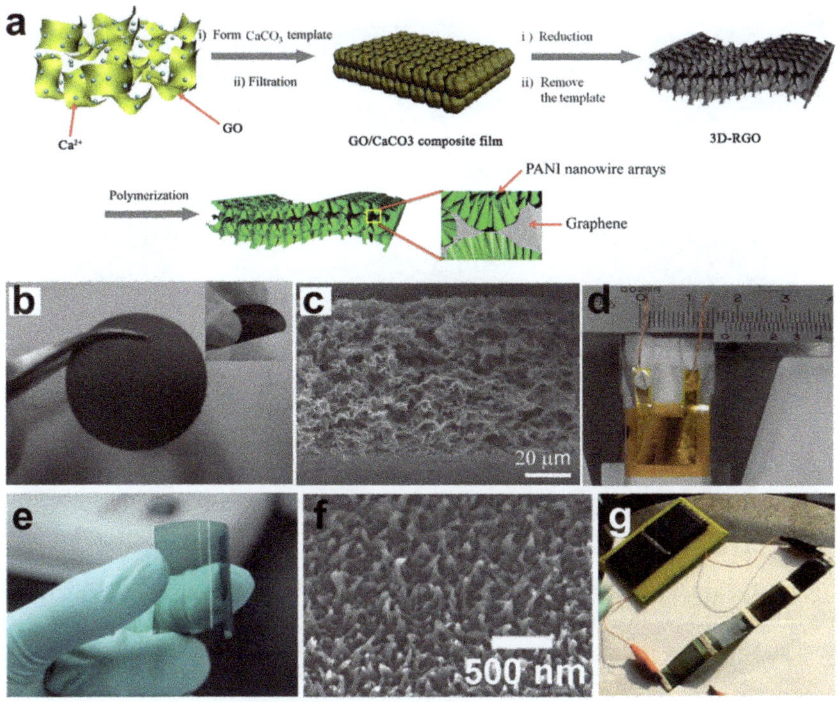

Figure 7.18. (a) The preparation of the 3D-RGO and 3D-RGO/PANI films. A digital image (b) and an SEM image (c) of 3D-RGO film. (d) A digital image of a flexible film SC [60], reproduced with permission © John Wiley & Sons. (e) A picture of a single electrode, (f) an SEM image of PANI nanowire arrays, (g) a digital image of an integrated device with solar cells [61], reproduced with permission © The Royal Society of Chemistry.

Finally, asymmetric fiber models were assembled, which were examined for a series of structural and textural benefits. The metal oxide functionalized CNTs and the carbon fiber together exhibited similar performances to those of solid-state SCs, i.e. 0.43 mWh cm^{-3} at a power density of 0.02 W cm^{-3}.

Li *et al* [69] prepared a CNT–Fe_3O_4–PANI composite by FCCVD and electrochemical deposition. The CNT–Fe_3O_4–PANI composite had excellent electrochemical properties and was used as an FSC. SEM images show that the PANI network was not only coated on the surface of the Fe_3O_4 particles but also interconnected with a single particle and acted as a bridge between adjacent particles (figure 7.21). When the electro-polymerization time was increased, the thickness of the PANI shell increased to about 30 nm. In order to study the application of this material in an energy storage device, a flexible all-solid symmetrical SC was fabricated with a CNT–Fe_3O_4–PANI electrode; the SC had excellent cyclic stability with a loss of only 3.6% of its initial specific capacitance after 10 000 cycles.

Dhilbar *et al* [62] developed a hydrothermal method to prepare a ternary nanocomposite composed of $NiCo_2O_4$/rGO/PPy (figure 7.22). The $NiCo_2O_4$/rGO/PPy composite had a sea-urchin-like microstructure with a large specific surface area and abundant active sites. $NiCo_2O_4$ was deposited on the surface of rGO and was

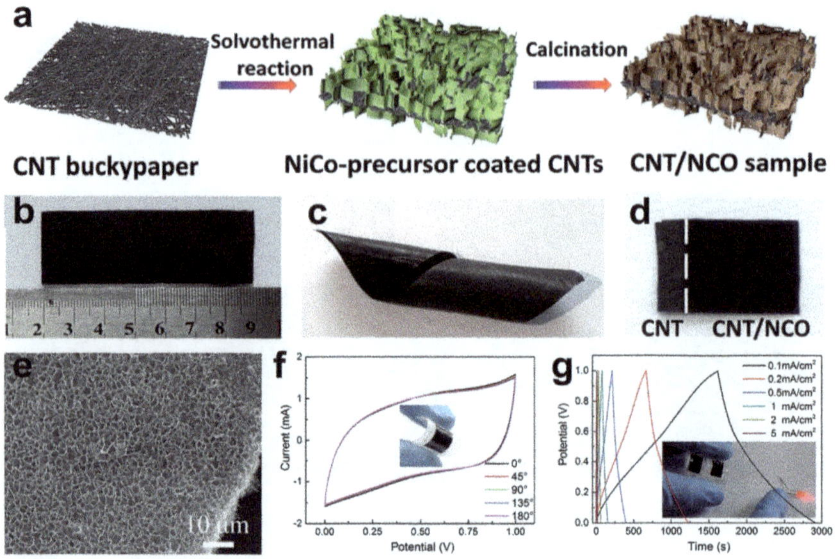

Figure 7.19. (a) The synthetic process of the CNT/$NiCo_2O_4$ hybrid paper, (b and c) digital photos of CNT buckypaper, (d) a digital photo of CNT/$NiCo_2O_4$ hybrid paper, (e) an SEM image of CNT/$NiCo_2O_4$ hybrid paper, (f and g) the electrochemical performances of all-solid-state SCs assembled using CNT/NCO electrodes [68], reproduced with permission © The Royal Society of Chemistry.

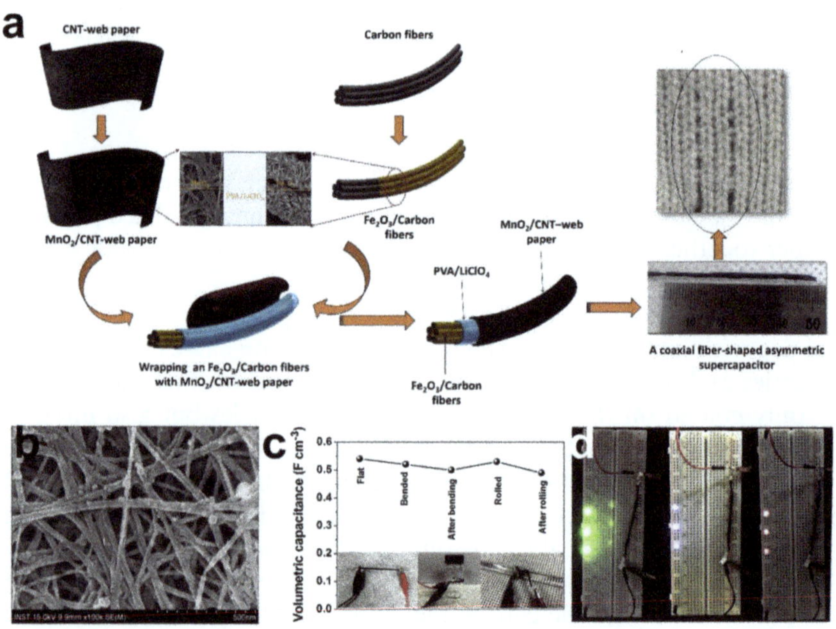

Figure 7.20. (a) A schematic illustration of the fabrication of a coaxial fiber-shaped asymmetric SC with MnO_2/CNT and Fe_2O_3/CF electrodes, (b) an SEM image of MnO_2/CNT-web paper, (c) the volumetric capacitances of the CF ASC in different bending states, (d) LEDs powered by a single CF ASC device [10], reproduced with permission © Elsevier.

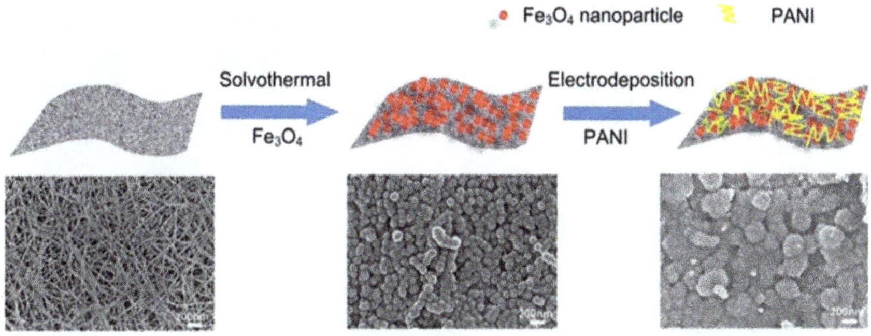

Figure 7.21. A schematic illustration of the two-step fabrication process used to produce the CNT film–Fe_3O_4–PANI electrode [69], reproduced with permission © The Royal Society of Chemistry.

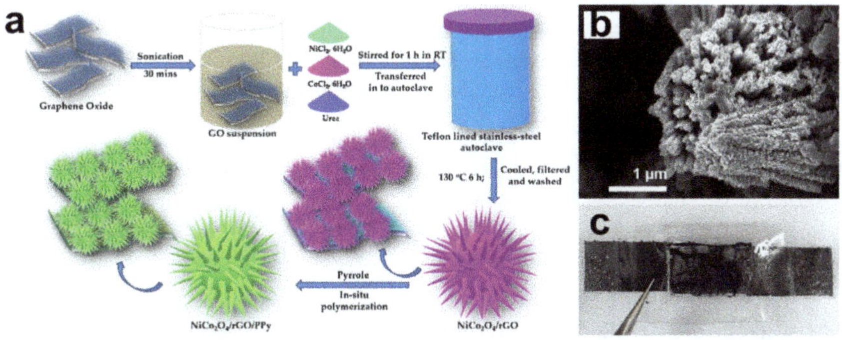

Figure 7.22. (a) A schematic representation of the synthesis of NCO/rGO/PPy, (b) an SEM image of NCO/rGO/PPy, and (c) a digital photo of the NCO/rGO/PPy nanocomposite-based flexible device [62], reproduced with permission © American Chemical Society.

evenly covered by PPy, forming a special morphology. This unique morphology further improved the electrochemical properties of nanocomposites. Subsequently, an FSC was assembled with PVA/KOH gel as the electrolyte. The device had a high energy density of 87.22 ± 0.22 Wh kg^{-1}, and its CV curves had approximately rectangular shapes at different scan rates (5–200 mV s^{-1}). Multiple redox peaks were observed, which indicated that the presence of $NiCo_2O_4$ and PPy led to the occurrence of redox behavior; in addition, the integrated area of the CV curve was larger, meaning that the specific capacitance value was higher.

Liu *et al* [63] reported a high-performance freestanding PANI/SWCNT thin-film integrated electrode obtained through a simple solution deposition method. A schematic illustration of the fabrication process of the PANI/SWCNT free-standing film is shown in figure 7.23. SEM images showed that SWCNTs were uniformly distributed in the entire thin-film electrode to form a continuous network of high conductivity, which was conducive to ion transport and realized the rapid redox reaction of SCs. As expected, this compact SC delivered a specific capacitance of 218 F g^{-1} in addition to its remarkable mechanical stability and flexibility. Moreover, the energy density of the flexible SC reached 19.45 Wh kg^{-1}

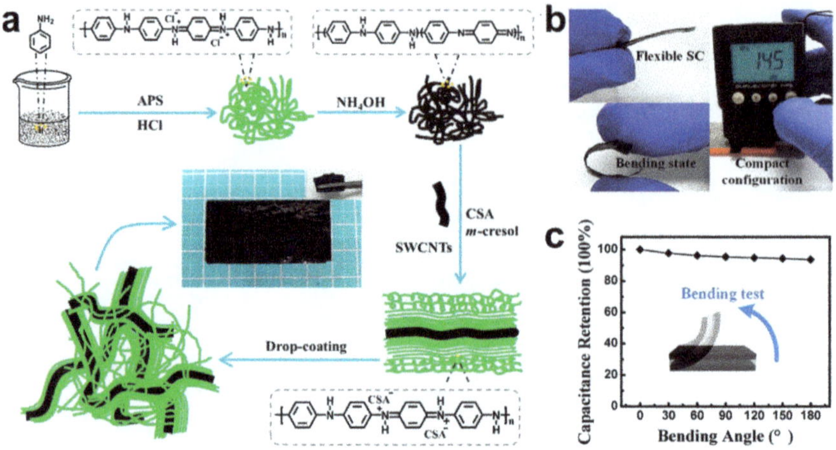

Figure 7.23. (a) A schematic illustration of the fabrication process used to produce the PANI/SWCNT freestanding film, (b) photographs of a flexible SC based on the PANI/SWCNT film electrode, (c) capacitance retentions at different bending angles [63], reproduced with permission © American Chemical Society.

at a power density of 320.5 W kg^{-1}, suggesting its promising application as an energy storage device.

7.3.2 Electrolyte materials

In addition to the electrode materials used, electrolyte materials also have important effects on the performance of FSCs, such as their operating voltage and cyclic life. The operating voltage of water-based electrolytes is generally in the range of 1–1.3 V, but organic electrolytes such as ionic liquids can reach 3.5–4.5 V [70]. The requirements for the SFSC electrolytes are as follows: (1) a wide potential window; (2) high ionic conductivity; (3) electrochemical stability and high performance; (4) good compatibility with electrode materials; (5) high thermal stability; (6) low cost; (7) environmental friendliness; and (8) good mechanical strength [71]).

Raymundo-Piñero *et al* [72] added a neutral electrolyte, Na_2SO_4, to a PVA aqueous solution and prepared a gel polymer electrolyte by the freezing and thawing method (figure 7.24). The electrochemical stability window reached 1.8 V. SCs assembled with activated carbon as the electrode exhibited good cyclic stability. The ions in the polymer matrix were able to migrate quickly under the influence of water molecules; thus, SCs based on the hydrogel electrolyte maintained the power density of the liquid electrolyte device along with high safety. Guo *et al* [73] used Fe^{3+} to crosslink acrylic acid (AA), obtaining a supramolecular hydrogel electrolyte (KCl–Fe^{3+}/PAA), the ionic and hydrogen bonds of which gave KCl–Fe^{3+}/PAA hydrogel electrolytes good self-repair ability and cyclic performance. The elongation at the breaking point reached 700%, the tensile strength was 400 kPa, and the electrical conductivity reached 0.09 S cm^{-1}. FSCs obtained by assembling the electrolyte membrane with graphene and PPy composite electrodes had excellent capacitance and electrochemical stability. At a current density of 0.5 A g^{-1}, the specific capacitance reached 87.4 F g^{-1}, and the capacity retention rate was

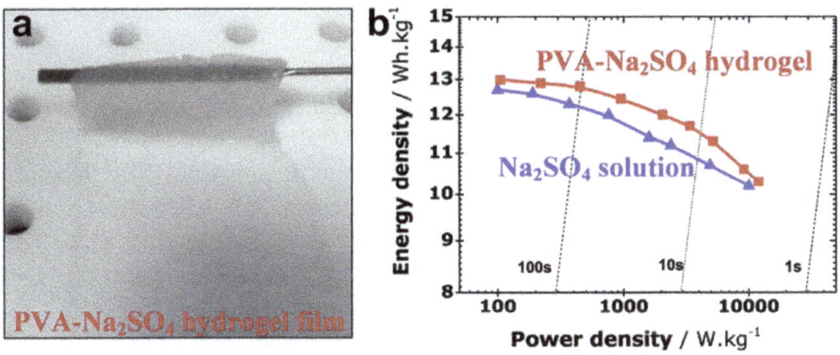

Figure 7.24. (a) A digital photo of PVA-Na_2SO_4 neutral hydrogel electrolyte, (b) Ragone plots for different electrolytes [72], reproduced with permission © Elsevier.

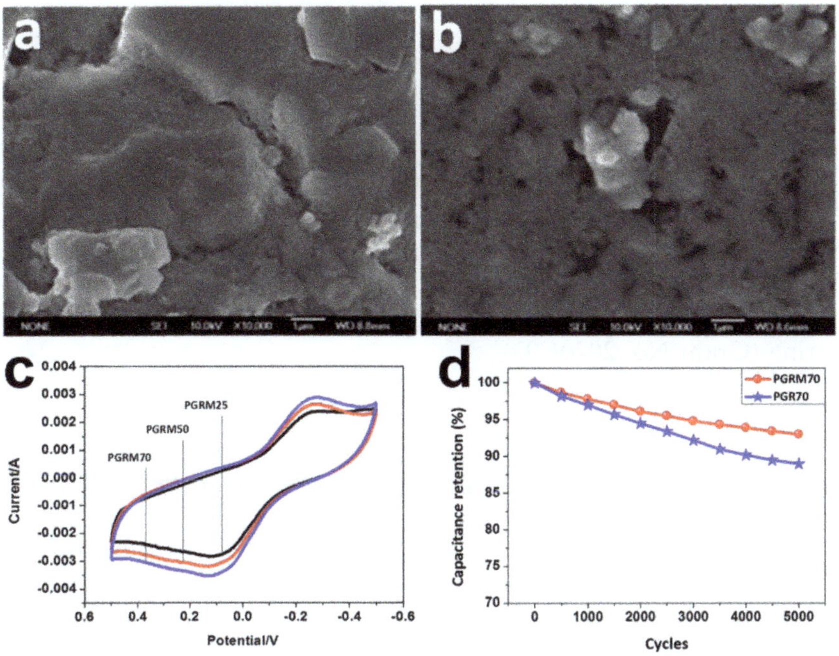

Figure 7.25. SEM images of gel electrolytes: (a) PVA-KOH-KSCN, (b) PVA-KOH-KSCN-EG. (c and d) The electrochemical performances of microsupercapacitors with gel electrolytes [74], reproduced with permission © Elsevier.

89% after 5000 cycles at 1 A g^{-1}. Zhu *et al* [74] used PVA as a matrix to prepare a gel polymer electrolyte using the action of ethylene glycol and water. The electrolyte had high thermal stability and stable electrochemical performance. The electrochemical stability window reached about 2.5 V at room temperature. An SC made by assembling the electrolyte membrane with a PPy/graphene oxide–RuO_2 composite electrode had a specific capacitance of 32.4 mF cm^{-2}, a power density of 194 mW cm^{-2}, and a capacitance retention rate of 93% after 5000 cycles (figure 7.25).

7.4 Summary

FSCs have received widespread attention in the fields of medical care, smart clothing, portable microelectronics, and even the military industry. Tremendous efforts have been devoted to improving the performance of FSCs, including changing the original morphologies of carbon materials, introducing other elements, and the production of compounds that include pseudocapacitor materials. Starting from the conventional large and portable device sizes, significant progress has been made in the novel technology of constructing FSC devices which promote wearable technologies with the help of carbon nanomaterials. Their application in future electronic devices will require SCs that are thin, flexible, lightweight, and capable of being rolled up. The tuning of doping structures, the design of all-in-one FSCs, and/or the inclusion of redox-active species can further increase the performance of conductive-polymer-based FSCs. Various types of FSC have broad application prospects in the future, but they are limited by their production processes, scale, cost, industry standards, and other issues.

Acknowledgments

We are grateful for the financial support of the China Postdoctoral Science Foundation (Grant Nos. 2020M682710, 2020M682711, 2019M652882, and 2019T120725), the Guangdong Basic and Applied Basic Research Foundation (Grant No. 2020A1515110705), the National Natural Science Foundation of China (Grant No. 31971614), the Guangzhou Science and Technology Funds (Grant No. 201904010078), and the State Key Laboratory of Pulp and Paper Engineering (Grant No. 2020C03).

References

[1] Wang F, Wu X, Yuan X, Liu Z, Zhang Y, Fu L, Zhu Y, Zhou Q, Wu Y and Huang W 2017 *Chem. Soc. Rev.* **46** 6816

[2] Zhang Y, Feng H, Wu X, Wang L, Zhang A, Xia T, Dong H, Li X and Zhang L 2009 *Int. J. Hydrog. Energy* **34** 4889

[3] Raza W, Ali F, Raza N, Luo Y, Kim K-H, Yang J, Kumar S, Mehmood A and Kwon E E 2018 *Nano Energy* **52** 441

[4] Zang L, Liu Q, Qiu J, Yang C, Wei C, Liu C and Lao L 2017 *ACS Appl. Mater. Interfaces* **9** 33941

[5] Fic K, Platek A, Piwek J and Frackowiak E 2018 *Mater. Today* **21** 437

[6] Li R Z *et al* 2016 *Energy Environ. Sci.* **9** 1458

[7] Tian X, Jin J, Yuan S, Chua C K, Tor S B and Zhou K 2017 *Adv. Energy Mater.* **7** 1700127

[8] Han W, Wu Z, Li Y and Wang Y 2019 *Chem. Eng. J.* **358** 1022

[9] Lu Q, Liu L, Yang S, Liu J, Tian Q, Yao W, Xue Q, Li M and Wu W 2017 *J. Power Sources* **361** 31

[10] Patil B, Ahn S, Yu S, Song H, Jeong Y, Kim J H and Ahn H 2018 *Carbon* **134** 366

[11] Pan Z *et al* 2018 *Adv. Energy Mater.* **8** 1702946

[12] Meng F, Li Q and Zheng L 2017 *Energy Storage Mater.* **8** 85

[13] Lokhande V C, Lokhande A C, Lokhande C D, Kim J H and Ji T 2016 *J. Alloys Compd.* **682** 381

[14] Ai Y, Lou Z, Li L, Chen S, Park H S, Wang Z M and Shen G 2016 *Adv. Mater. Technol.* **1** 1600142

[15] Li X, Li X, Cheng J, Yuan D, Ni W, Guan Q, Gao L and Wang B 2016 *Nano Energy* **21** 228

[16] Kang Q, Zhao J, Li X, Zhu G, Feng X, Ma Y, Huang W and Liu J 2017 *Nano Energy* **32** 201

[17] Cao X, Liu Y, Zhong Y, Cui L, Zhang A, Razal J M, Yang W and Liu J 2020 *J. Mater. Chem.* A **8** 1837

[18] Peng H 2015 *Fiber-Shaped Energy Harvesting and Storage Devices* (Berlin: Springer)

[19] Liu J, Huang J, Cai Q, Yang Y, Luo W, Zeng B, Xu Y, Yuan C and Dai L 2020 *ACS Appl. Mater. Interfaces* **12** 20479

[20] Cheng Y, Xiao X, Pan K and Pang H 2020 *Chem. Eng. J.* **380** 122565

[21] Li Z, Liu H, Ouyang C, Hong Wee W, Cui X, Jian Lu T, Pingguan-Murphy B, Li F and Xu F 2016 *Adv. Funct. Mater.* **26** 165

[22] Yan H, Zhong M, Lv Z and Wan P 2017 *Small* **13** 1701697

[23] Ling J, Zou H, Yang W, Chen W, Lei K and Chen S 2018 *J. Energy Storage* **20** 92

[24] Boruah B D, Mathieson A, Wen B, Jo C, Deschler F and De Volder M 2020 *Nano Lett.* **20** 5967

[25] Yu X Y and Lou X W 2018 *Adv. Energy Mater.* **8** 1701592

[26] Tian Z, Yin J, Wang X and Wang Y 2019 *J. Alloys Compd.* **777** 806

[27] Zhai S, Jin K, Zhou M, Fan Z, Zhao H, Li X, Zhao Y, Ge F and Cai Z 2020 *Colloids Surf.* A **584** 124025

[28] Tong L, Jiang C, Cai K and Wei P 2020 *J. Power Sources* **465** 228267

[29] Wang R, Yao M and Niu Z 2020 *Infomat* **2** 113

[30] Yan J, Li S, Lan B, Wu Y and Lee P S 2020 *Adv. Funct. Mater.* **30** 1902564

[31] Ma P P, Lei N, Yu B, Liu Y K and Lu Q L 2019 *Nanomaterials* **9** 1676

[32] Weng Z, Su Y, Wang D-W, Li F, Du J and Cheng H-M 2011 *Adv. Energy Mater.* **1** 917

[33] Liu X, Dang L, Nai X, Dong Y and Li W 2018 *Res. Chem. Intermed.* **44** 5697

[34] Jiang S, Zhou X, Xiao H, Chen W, Xu X and Liu Z 2021 *Chem. Eng. J.* **405** 127009

[35] Jha M K N 2018 *Appl. Surf. Sci.* **449** 461

[36] El-Kady M F, Strong V, Dubin S and Kaner R B 2012 *Science* **335** 1326

[37] El-Kady M F and Kaner R B 2013 *Nat. Commun.* **4** 1475

[38] Behabtu N *et al* 2013 *Science* **339** 182

[39] Lima M D *et al* 2011 *Science* **331** 51

[40] Zhang D, Miao M, Niu H and Wei Z 2014 *ACS Nano* **8** 4571

[41] Xu X, Zhang Y, Jiang J, Wang H, Zhao X, Li Q and Lu W 2017 *Compos. Sci. Technol.* **149** 20

[42] Wang Q H, Zhong S W, Hu J W, Liu T, Zhu X Y, Chen J, Hong Y Y and Wu Z P 2016 *J. Power Sources* **310** 70

[43] Wang H, Lu W, Di J, Li D, Zhang X, Li M, Zhang Z, Zheng L and Li Q 2017 *Adv. Funct. Mater.* **27** 1606220

[44] Zou Z, Xiao W, Zhang Y, Yu H and Zhou W 2020 *Appl. Surf. Sci.* **500** 144244

[45] Ghorbani M, Golobostanfard M R and Abdizadeh H 2017 *Appl. Surf. Sci.* **419** 277

[46] Wang R, Sui Y, Huang S, Pu Y and Cao P 2018 *Chem. Eng. J.* **331** 527

[47] Huang K-C, Lin C-H, Anuratha K S, Huang T-Y, Lin J-Y, Tseng F-G and Hsieh C-K 2019 *J. Power Sources* **417** 108

[48] Huang Y, Li Y, Hu Z, Wei G, Guo J and Liu J 2013 *J. Mater. Chem.* A **1** 9809
[49] Yuan C, Yang L, Hou L, Shen L, Zhang X and Lou X W 2012 *Energy Environ. Sci.* **5** 7883
[50] Song N, Wu Y, Wang W, Xiao D, Tan H and Zhao Y 2019 *Mater. Res. Bull.* **111** 267
[51] Yu Z and Thomas J 2014 *Adv. Mater.* **26** 4279
[52] Xu H, Hu X, Sun Y, Yang H, Liu X and Huang Y 2015 *Nano Res.* **8** 1148
[53] Gupta A, Sardana S, Dalal J, Lather S, Maan A S, Tripathi R, Punia R, Singh K and Ohlan A 2020 *ACS Appl. Energy Mater.* **3** 6434
[54] Sharma K, Pareek K, Rohan R and Kumar P 2019 *Int. J. Energy Res.* **43** 604
[55] Li Y, Meng L, Yang Y, Xu G, Hong Z, Chen Q, You J, Li G, Yang Y and Li Y 2016 *Nat. Commun.* **7** 10214
[56] Chiou N-R, Lu C, Guan J, Lee L J and Epstein A J 2007 *Nat. Nanotechnol.* **2** 354
[57] Xu J, Wang K, Zu S-Z, Han B-H and Wei Z 2010 *ACS Nano* **4** 5019
[58] Wang K, Zou W, Quan B, Yu A, Wu H, Jiang P and Wei Z 2011 *Adv. Energy Mater.* **1** 1068
[59] Singh S B, Kshetri T, Singh T I, Kim N H and Lee J H 2019 *Chem. Eng. J.* **359** 197
[60] Meng Y, Wang K, Zhang Y and Wei Z 2013 *Adv. Mater.* **25** 6985
[61] Wang K, Wu H, Meng Y, Zhang Y and Wei Z 2012 *Energy Environ. Sci.* **5** 8384
[62] Dhibar S and Malik S 2020 *ACS Appl. Mater. Interfaces* **12** 54053
[63] Liu F, Luo S, Liu D, Chen W, Huang Y, Dong L and Wang L 2017 *ACS Appl. Mater. Interfaces* **9** 33791
[64] Wang N, Han G, Song H, Chang Y, Hou W and Xiao Y 2019 *J. Electroanal. Chem.* **851** 113466
[65] Wang S, Liang Y, Zhuo W, Lei H, Javed M S, Liu B, Wang Z and Mai W 2021 *Mater. Chem. Front.* **5** 1324
[66] Chen L-M, Yu H-Y, Wang D-C, Yang T, Yao J-M and Tam K C 2019 *ACS Sustain. Chem. Eng.* **7** 11823
[67] Liu S, Yin Y, Hui K S, Hui K N, Lee S C and Jun S C 2018 *Adv. Sci.* **5** 1800733
[68] Zheng Y, Lin Z, Chen W, Liang B, Du H, Yang R, He X, Tang Z and Gui X 2017 *J. Mater. Chem.* A **5** 5886
[69] Li J, Lu W, Yan Y and Chou T-W 2017 *J. Mater. Chem.* A **5** 11271
[70] Du G and Pang H 2021 *Energy Storage Mater.* **36** 387
[71] Wang C, Xia K, Wang H, Liang X, Yin Z and Zhang Y 2019 *Adv. Mater.* **31** 1801072
[72] Batisse N and Raymundo-Piñero E 2017 *J. Power Sources* **348** 168
[73] Rong Q, Lei W, Huang J and Liu M 2018 *Adv. Energy Mater.* **8** 1801967
[74] Ping Z, XiaoJing L, Anye R and Peng G 2016 *J. Electroanal. Chem.* **782** 154

www.ingramcontent.com/pod-product-compliance
Ingram Content Group UK Ltd.
Pitfield, Milton Keynes, MK11 3LW, UK
UKHW051857100926
8556IPUK00005B/659

9 780750 350402